PADÉ AND RATIONAL APPROXIMATION

Theory and Applications

Academic Press Rapid Manuscript Reproduction

Proceedings of an International Symposium held at the University of South Florida, Tampa, Florida, December 15-17, 1976

PADÉ AND RATIONAL APPROXIMATION

Theory and Applications

Edited by

E.B. SAFF

Department of Mathematics
University of South Florida
Tampa, Florida

R.S. VARGA

Department of Mathematics
Kent State University
Kent, Ohio

Academic Press, Inc.

New York San Francisco London 1977

A Subsidiary of Harcourt Brace Jovanovich, Publishers

ACADEMIC PRESS, INC.
111 Fifth Avenue, New York, New York 10003

United Kingdom Edition published by
ACADEMIC PRESS, INC. (LONDON) LTD.
24/28 Oval Road, London NW1

Library of Congress Cataloging in Publication Data

Conference on Rational Approximation with Emphasis on Applications of Padé Approximants, Tampa, Fla., 1976.
Padé and rational approximation.

1. Padé approximant–Congresses. 2. Approximation theory–Congresses. I. Saff, E. B., Date II. Varga, Richard S. III. Title. IV. Title: Rational approximation.
QC20.7.P3C66 1976 511′.4 77-22616
ISBN 0-12-614150-9

PRINTED IN THE UNITED STATES OF AMERICA

Contents

THEORY OF RATIONAL APPROXIMATIONS

PHYSICAL APPLICATIONS

COMPUTATIONS

OPEN PROBLEMS

List of Contributors and Participants

An asterisk denotes a contributor to this volume.

WALEED AL-SALAM Department of Mathematics, University of Alberta, Edmonton, Alberta, Canada T6G 2G1

NED ANDERSON Department of Mathematics, Kent State University, Kent, Ohio 44242

R. A. ASKEY Department of Mathematics, University of Wisconsin, Madison, Wisconsin 53706

*DAVID R. AUDLEY Frank J. Seiler Research Laboratory, U.S. Air Force Academy, Colorado 80840

*GEORGE A. BAKER, JR. Service de Physique Theorique, C.E.N. Saclay B.P. No. 2–91190 Gif-sur-Yvette, France and University of California, Los Alamos Scientific Laboratory, P.O. Box 1663, Los Alamos, NM 87545

DAVID BARROW Department of Mathematics, Texas A & M University, College Station, Texas 77843

*COLIN BENNETT Department of Mathematics, McMaster University, Hamilton, Ontario, Canada L8S 4K1

*L. P. BENOFY Department of Physics, St. Louis University, St. Louis, Missouri 63103

*V. BENOKRAITIS Ballistic Modeling Division, U.S. Ballistic Research Laboratory, Aberdeen Proving Ground, Maryland, 21005

PAUL T. BOGGS DRXRO-MA-14403-M, U.S. Army Research Office, P.O. Box 12211, Research Triangle Park, North Carolina 27709

R. BOJANIC Department of Mathematics, Ohio State University, Columbus, Ohio 43210

*C. P. BOYER IIMAS, Universidad Nacional Autonoma de Mexico, Mexico 20, D.F.

*C. BREZINSKI USTL, Ver d'IEEA/Informatique, B.P. 36, 59650 Villenueve d'Ascq., France

*M. G. de BRUIN Universiteit van Amsterdam, Instituut voor Propedeutische Wiskunde, Roetersstraat 15 Amsterdam, Netherlands

PHILLIP CALLAS Department of Mathematics, Colorado School of Mines, Golden, Colorado 80401

*J. S. R. CHISHOLM Mathematical Institute, University of Kent, Canterbury, England

*C. K. CHUI Department of Mathematics, Texas A & M University, College Station, Texas 77843

CLAUDE COTE Départment de Mathématiques, Université de Montréal, Montréal, Québec, Canada H3C 3J7

STANLEY DEANS Department of Physics, University of South Florida, Tampa, Florida 33620

*ALBERT EDREI Department of Mathematics, Syracuse University, Syracuse, New York 13210

ALEXANDER S. ELDER US Ballistic Research Laboratory, ATTN: DRX-PD, Aberdeen Proving Ground, Maryland 21005

GARY FEDERICI Department of Mathematics, Syracuse University, Syracuse, New York 13210

*CARL H. FITZGERALD Department of Mathematics, University of California at San Diego, San Diego, California 92083

*J. FLEISCHER Universität Bielefeld, Fakultät für Physik, D 4800 Bielefeld, Herforder Str. 28, West Germany

*GEZA FREUD Department of Mathematics, Ohio State University, Columbus, Ohio 43210

*L. WAYNE FULLERTON C-3 Los Alamos Scientific Lab, Los Alamos, New Mexico 87544

*J. L. GAMMEL Department of Physics, St. Louis University, St. Louis, Missouri 63103

*P. M. GAUTHIER Université de Montréal, Mathématiques 207, Case Postal 6128, Montréal 101 Canada H3C 3J7

ROBERT GERVAIS Département de Mathématiques, Université de Montréal, Montréal, Québec, Canada H3C 3J7

ROBERT GILMORE Department of Physics, University of South Florida, Tampa, Florida 33620

ANDRE GIROUX Département de Mathématiques, Université de Montréal, Montréal, Québec, Canada H3C 3J7

A. W. GOODMAN Department of Mathematics, University of South Florida, Tampa, Florida 33620

RICHARD GOODMAN Department of Mathematical Sciences, University of Cincinnati, Cincinnati, Ohio 45221

*WILLIAM B. GRAGG Department of Mathematics, University of California at San Diego, P.O. Box 109, La Jolla, California 92038

*P. R. GRAVES-MORRIS Mathematical Institute, University of Kent, Canterbury, England

*P. HENRICI Eidgenössiche Technische Hochschule, Seminar für Angewandte Mathematik, Clausiusstrasse 55, CH-8006 Zürich, Switzerland

*MYRON S. HENRY Department of Mathematics, Montana State University, Bozeman, Montana 59715

JAMES L. HOWLAND Department of Mathematics, University of Ottawa, Ottawa, Ontario, Canada K1N 6N5

JOHN JONES, JR. Air Force Institute of Technology, Wright-Patterson AFB, Ohio 45433

*WILLIAM B. JONES Department of Mathematics, University of Colorado, Boulder, Colorado 80309

*J. KARLSSON Department of Mathematics, University of Umeå, Umeå, Sweden

*E. H. KAUFMAN, JR. Department of Mathematics, Central Michigan University, Mt. Pleasant, Michigan 48859

OHOE KIM Department of Mathematics, Towson State University, Baltimore, Maryland 21204

MURRAY S. KLAMKIN Department of Mathematics, University of Alberta, Edmonton, Alberta, Canada T6G 2G1

M. LACHANCE Department of Mathematics, University of South Florida, Tampa, Florida 33620

*F. M. LARKIN Department of Computing and Information Sciences, Queen's University, Kingston, Canada K7L 3N6

JOSEPH J. LIANG Department of Mathematics, University of South Florida, Tampa, Florida 33620

*G. G. LORENTZ Department of Mathematics, University of Texas at Austin, Austin, Texas 78712

*YUDELL L. LUKE Department of Mathematics, University of Missouri at Kansas City, Kansas City, Missouri 64110

*ARNE MAGNUS Department of Mathematics, Colorado State University, Ft. Collins, Colorado 80521

JOHN MAHONEY Department of Industrial & Systems Engineering, University of Florida, Gainesville, Florida 32611

*Č. MASAITIS USA Ballistic Research Lab, Aberdeen Proving Ground, Maryland 21005

ABRAHAM MELKMAN Thomas J. Watson Research Center, Post Office Box 218, Yorktown Heights, New York 10598

*E. P. MERKES Department of Mathematics, University of Cincinnati, Cincinnati, Ohio 45221

*P. MERY Centre National de la Recherche Scientific, 31, Chemin J. Aiguier, 13274 Marsielle Cedex 2, France

CHARLES A. MICCHELLI Thomas J. Watson Research Center, Post Office Box 218, Yorktown Heights, New York 10598

*W. MILLER, JR. Department of Mathematics, University of Minnesota, Minneapolis, Minnesota 55455

DIANE CLAIRE MYERS Department of Mathematics, Wesleyan College, Macon, Georgia 31201

KENT NAGLE Department of Mathematics, University of South Florida, Tampa, Florida 33620

DONALD J. NEWMAN Yeshiva University, 500 West 186th Street, New York, New York 10033

*J. NUTTALL Department of Physics, University of Western Ontario, London 72 Canada N6A 3K7

ROBERT L. PEXTON 4960 Elrod Drive, Lawrence Radiation Lab, Castro Valley, California 94546

ROGER PIERRE Départment de Mathématiques, Université de Montréal, Montréal, Québec, Canada H3C 3J7

*Q. I. RAHMAN Départment de Mathématiques, Université de Montréal, Case Postal 6128, Montréal 101 Canada H3C 3J7

ENRIQUE H. RAMIREZ AFOSR/NM, Bldg. 410, Bolling AFB, Washington, D.C. 20332

CARL H. RASMUSSEN Department of Mathematics, University of Maine at Orono, Orono, Maine 04473

*JOHN A. ROULIER Department of Mathematics, North Carolina State University, Raleigh, North Carolina 27607

*KARL RUDNICK Department of Mathematics, Texas A & M University, College Station, Texas 77843

*ARDEN RUTTAN Department of Mathematics, Kent State University, Kent, Ohio 44242

*E. B. SAFF Department of Mathematics, University of South Florida, Tampa, Florida 33620

*G. SCHMEISSER Mathematisches Institut, Universität Erlangen-Nürnberg, Bismarckstrasse 1½, 8520 Erlangen, West Germany

A. SHARMA Department of Mathematics, University of Alberta, Edmonton, Alberta, Canada T6G 2G1

*W. L. SHEPHERD Instrumental Directorate, White Sands Missile Range, New Mexico 88002

B. D. SIVAZLIAN Department of Industrial & Systems Engineering, University of Florida, Gainesville, Florida 32611

*P. W. SMITH Department of Mathematics, Texas A & M University, College Station, Texas 77843

A. D. SNIDER Department of Mathematics, University of South Florida, Tampa, Florida 33620

*L. Y. SU Department of Mathematics, Texas A & M University, College Station, Texas 77843

*G. D. TAYLOR Department of Mathematics, Colorado State University, Ft. Collins, Colorado 80523

*W. J. THRON Department of Mathematics, University of Colorado, Boulder, Colorado 80302

*J. A. TJON Institute for Theoretical Physics, University Sorbonnelaan 4, De Uithof, Utrecht, The Netherlands

*JEFFREY D. VAALER Department of Mathematics, University of Texas at Austin, Austin, Texas 78712

*CHARLES VAN LOAN Department of Mathematics, Cornell University, Ithaca, New York 14853

*H. VAN ROSSUM Universiteit van Amsterdam, Instituut voor Propedeutische Wiskunde, Roetersstraat 15, Amsterdam, Netherlands

*R. S. VARGA Department of Mathematics, Kent State University, Kent, Ohio 44242

MANJUKA VARMA Department of Mathematics, University of Florida, Gainesville, Florida 32611

*H. WALLIN Department of Mathematics, University of Umeå, Umeå, Sweden

*ROBERT C. WARD Mathematical & Statistical Research Dept., Computer Science Division, Union Carbide Corporation, Nuclear Division, Post Office Box Y, Oak Ridge, Tennessee 37830

DANIEL D. WARNER Bell Laboratories, 600 Mountain Avenue, Murray Hill, New Jersey 07974

*MARION WETZEL Department of Mathematics, Denison University, Granville, Ohio 43023

*J. A. WILSON Department of Mathematics, University of Wisconsin, Madison, Wisconsin 53706

JAN WYNN Department of Mathematics, Brigham Young University, Provo, Utah 84602

*P. WYNN School of Computer Science, McGill University, Post Office Box 6070, Station A, Montréal, Québec, Canada H3C 3G1

Preface

This volume presents the proceedings of the Conference on Rational Approximation with Emphasis on Applications of Padé Approximants, which was held December 15-17, 1976 in Tampa, Florida. More than 90 individuals attended the meeting, including participants from Canada, England, France, Germany, Israel, Netherlands, Sweden, Switzerland, as well as from the United States. The conference was organized by E. B. Saff and R. S. Varga with the assistance of J. Liang and K. Nagle.

The major goal of the conference was to bring together theoreticians as well as practitioners for the purpose of exchanging information relevant to the study of rational approximation. Accordingly, mathematicians, physicists, and government laboratory personnel participated in special sessions designed to emphasize the interplay of theory, computation, and physical applications. In addition to the invited talks, a panel discussion was held to outline future research needs. The panelists were L. Wayne Fullerton, John L. Gammel, Peter Henrici, William L. Shepherd, Richard S. Varga (chairman), and Robert C. Ward. A summary of the panel discussion is included in this volume.

The contributions to this volume include expository as well as original research papers. The reader will find many open problems mentioned, including some of particular interest to government laboratories. It is hoped that these proceedings will not only prove to be a valuable reference source for Padé and rational approximation, but will also stimulate much needed work on its theory and practical applications.

We wish to express our appreciation to the U.S. Air Force Office of Scientific Research and to the U.S. Army Research Office for having provided generous support for the conference. Accolades are due Lyn Wilson and Charolette Worthington for their dedicated secretarial work before and during the conference, Diane Gossett for her care in typing many of the manuscripts, Joseph Liang and Kent Nagle for their superb work on the arrangements committee, and A. Ruttan for his work which lead to the graph on the cover. We would also like to express our gratitude to the University of South Florida for providing the necessary facilities and a cordial atmosphere for the conference.

THEORY OF PADÉ APPROXIMANTS AND GENERALIZATIONS

PADÉ APPROXIMANTS AND ORTHOGONAL POLYNOMIALS

C. Brezinski

The general theory of orthogonal polynomials with respect to a functional defined by its moments is used to derive old and new results on continued fractions, Padé approximants, and the ε-algorithm. Orthogonal polynomials seem to be the mathematical basis on which Padé approximants and related matters are to be studied.

1 Introduction

I think that the theory of general orthogonal polynomials is the basis on which Padé approximants must be studied and the aim of this paper is to show how old and new results about Padé approximants can be derived very easily from orthogonal polynomials.

This paper is only a brief summary on the connection between these two fields and it must be considered as a preliminary report on the subject. Proofs are omitted and many developments are not mentioned. A more extensive work will be published elsewhere.

The close connection between orthogonal polynomials and Padé approximants has been known for a long time since, in fact, the theory of orthogonal polynomials arose from continued fractions. More recently several authors enlarged the scope of this connection [11,17,23].

This paper is divided into five sections. Orthogonal polynomials are defined in the second paragraph and related to Padé approximants in the third one. The fourth paragraph presents some relationships between adjacent Padé approximants as well as a new method of computing them. Section five deals with acceleration properties of Padé approximants to series of Stieltjes. The last paragraph is devoted to generalizations to the non-scalar case.

2 General Orthogonal Polynomials

Let $\{c_n\}$ be a given sequence of real numbers. We define the linear functional c acting on the space of real polynomials by $c(x^n) = c_n$ for $n = 0,1,\ldots$. A system of polynomials $\{P_n\}$ is orthogonal with respect to c (or to the sequence $c_0, c_1, \ldots$) if $c(P_n P_k) = 0$ for $n \neq k$. These polynomials are uniquely determined except for a multiplicative factor. They are given by

$$P_k(x) = \begin{vmatrix} c_0 & \cdots\cdots & c_k \\ & \cdots\cdots & \\ c_{k-1} & \cdots\cdots & c_{2k-1} \\ 1 & \cdots\cdots & x^k \end{vmatrix} .$$

P_k has the exact degree k if the Hankel determinant $H_k^{(0)}$ is different from zero, where:

$$H_0^{(n)} = 1, \qquad H_k^{(n)} = \begin{vmatrix} c_n & \cdots\cdots & c_{n+k-1} \\ & \cdots\cdots & \\ c_{n+k-1} & \cdots\cdots & c_{n+2k-2} \end{vmatrix} .$$

It will now be assumed that $H_k^{(n)} \neq 0$ for $n,k = 0,1,\ldots$.

Let us define a second kind of polynomials, named Q_k, by

$$Q_k(t) = c\left(\frac{P_k(x)-P_k(t)}{x-t}\right),$$

where c acts on the variable x and t is a parameter. Q_k has the exact degree k-1.

It is easy to prove that the polynomials P_k and Q_k satisfy most of the properties of the classical orthogonal polynomials such as the recurrence and the Christoffel-Darboux relationships. It can also be proved without any other assumptions on c that P_k and P_{k+1} have no roots in common. The same is true for P_k, Q_k as well as for Q_k, Q_{k+1}. For more details the interested reader is referred to Akhiezer [1] and Gragg [9].

3 Padé Approximants

Let us now turn to the connection between general orthogonal polynomials and Padé approximants. We define

$$f(x) = \sum_{i=0}^{\infty} c_i x^i,$$

$$\tilde{P}_k(x) = x^k P_k(x^{-1}),$$

$$\tilde{Q}_k(x) = x^{k-1} Q_k(x^{-1}).$$

Then it can be proved that

$$\tilde{Q}_k(x)/\tilde{P}_k(x) = [k - 1/k]_f(x)$$

where $[p/q]_f(x)$ denotes the Padé approximant to f whose numerator has degree p and whose denominator has degree q. The other members of the Padé table can be defined in a similar way. Let

$$f_n(x) = \sum_{i=0}^{\infty} c_{n+i} x^i,$$

then

$$[n + k/k]_f(x) = \sum_{i=0}^{n} c_i x^i + x^{n+1} [k - 1/k]_{f_{n+1}}(x).$$

Many results can be deduced from this property of Padé approximants. For example, if we write

$$v(x) = \frac{1}{P_k(t^{-1})} \frac{P_k(t^{-1}) - P_k(x)}{1 - xt} = a_0 + a_1 x + \ldots + a_{k-1} x^{k-1},$$

then

$$[k - 1/k]_f(t) = c(v) = a_0 c_0 + \ldots + a_{k-1} c_{k-1}.$$

From the orthogonality of P_k to every polynomial of degree less than k, we get

$$c_p = c(x^p(1-xt)v(x)), \quad \text{for } p = 0,\ldots,k-1.$$

Let V be the matrix of the preceding system, $a = (a_o,\ldots,a_{k-1})^T$ and $y = (c_o,\ldots,c_{k-1})^T$. Then $a = V^{-1}y$ and

$$[k-1/k]_f(t) = (y, V^{-1}y),$$

which is Nuttall's compact formula [15].

It can also be proved that

$$[k-1/k]_f(t) = \sum_{i=1}^{k} A_i^{(k)}/(1 - x_i t),$$

where $A_i^{(k)} = Q_k(x_i)/P_k'(x_i)$ for $i = 1,\ldots,k$ and where the x_i are the roots of P_k, which we assume to be distinct but not necessarily real. This formula is the Gaussian quadrature formula.

The theory of orthogonal polynomials can also be used to derive an expression for the error of Padé approximants. We obtain

$$f(t) - [k-1/k]_f(t) = \frac{1}{P_k(t^{-1})}\, c\!\left(\frac{P_k(x)}{1-xt}\right) = \frac{1}{P_k^2(t^{-1})}\, c\!\left(\frac{P_k^2(x)}{1-xt}\right).$$

These results generalize some very well known results for Stieltjes series (see, for example, [2,13]). From the preceding result we immediately get

$$f(t) - [k-1/k]_f(t) = \frac{H_{k+1}^{(0)}}{H_k^{(0)}}\, t^{2k} + O(t^{2k+1})$$

The theory of orthogonal polynomials is also related to continued fractions since the recurrence relationships of the polynomials P_k and Q_k are nothing else but the recurrence relationships between the denominators and the numerators of the successive convergents of the continued fraction associated with the series f (see, for example, [9]).

Results on the Padé table can, of course, be transformed into results on the ε-algorithm of Wynn [20]. For example if

$\{P_n\}$ are the orthogonal polynomials with respect to the sequence $\{c_n = S_{n+1} - S_n\}$ and if the ε-algorithm is applied to $\{S_n\}$ which converges to S, then

$$\varepsilon_{2k}^{(0)} = S_o + Q_k(1)/P_k(1),$$

and

$$\varepsilon_{2k}^{(0)} - S = - \frac{1}{P_k(1)} c \left(\frac{P_k(x)}{1 - x}\right).$$

If P_k is written as $P_k(x) = a_o + \ldots + a_k x^k$, then

$$\varepsilon_{2k}^{(0)} = (a_o S_o + \ldots + a_k S_k)/P_k(1).$$

Some other results on the ε-algorithm can be obtained in a similar manner.

4 Rhombus Algorithms

Let $\{P_k^{(n)}\}$ be the orthogonal polynomials with respect to the sequence $c_n, c_{n+1}, \ldots$, that is with respect to the functional $c^{(n)}$ defined by $c^{(n)}(x^i) = c(x^{n+i}) = c_{n+i}$ for $i = 0, 1, \ldots$. Then

$$[n+k/k]_f(x) = \sum_{i=0}^{n} c_i x^i + x^{n+1} \tilde{Q}_k^{(n+1)}(x)/\tilde{P}_k^{(n+1)}(x),$$

It can be proved that

$$P_k^{(n+1)}(x) = P_k^{(n)}(x) - e_k^{(n)} P_{k-1}^{(n+1)}(x)$$

$$P_{k+1}^{(n)}(x) = x P_k^{(n+1)}(x) - q_{k+1}^{(n)} P_k^{(n)}(x).$$

where the quantities $e_k^{(n)}$ and $q_k^{(n)}$ can be computed by the q-d algorithm of Rutishauser [16]. From the definition of $e_k^{(n)}$ and $q_k^{(n)}$ as the ratios of products of Hankel determinants many relationships between adjacent systems of orthogonal polynomials can be derived in a very easy way. These relationships have been found by Wynn [21]. From these, many relationships between adjacent Padé approximants can be obtained. The proofs are much easier than those habitually given because they don't assume any particular knowledge about determinants such as the Schweins development, the Sylvester identity or even the recurrence relation between Hankel determinants (see, for example, Baker [3]). Among these relationships is the cross rule of Wynn [22]:

$$(W-C)^{-1} + (E-C)^{-1} = (N-C)^{-1} + (S-C)^{-1}$$

between Padé approximants located as in the following diagram.

$$\begin{array}{ccc} & N & \\ W & C & E \\ & S & \end{array}$$

We also found an identity obtained by Cordellier [8] which is very useful to compute Padé approximants when a square block of nearly equal approximants occurs in the Padé table. Let

$$\lambda = \frac{[\tilde{P}_k^{(n+1)}(x)]^2}{\tilde{P}_k^{(n)}(x)\,\tilde{P}_k^{(n+2)}(x)} ,$$

$$\mu = -\frac{[\tilde{P}_k^{(n+1)}(x)]^2}{\tilde{P}_{k+1}^{(n)}\,\tilde{P}_{k-1}^{(n+2)}(x)} .$$

Then

$$\lambda = (W-N)(W-S)/(W-C)^2 = (E-N)(E-S)/(E-C)^2,$$

$$\mu = (N-W)(N-E)/(N-C)^2 = (S-E)(S-W)/(S-C)^2,$$

and $\lambda^{-1} + \mu^{-1} = 1, \quad \forall\, x,n,k.$

As a consequence of this identity we get

$$P_k^{(n)}(x)\, P_k^{(n+2)}(x) - P_{k+1}^{(n)}(x)\, P_{k-1}^{(n+2)}(x) = [P_k^{(n+1)}(x)]^2 .$$

Let us now turn to some recursive methods to compute sequences of Padé approximants. The first relation which can be used for this aim is the recurrence relation of the polynomials $Q_k^{(n)}$ and $P_k^{(n)}$; we thus obtain a method due to Brezinski [6] to compute approximants lying on a diagonal of the Padé table.

Let $[n+k/k]_f(x) = \tilde{N}_k^{(n+1)}(x)/\tilde{P}_k^{(n+1)}(x)$ for $k = 0,1,\ldots$ and $n = -k, -k+1,\ldots$ with

$$\tilde{P}_k^{(n)}(x) = \sum_{i=0}^{k} b_i^{(k,n)} x^i, \quad \tilde{N}_k^{(n)}(x) = \sum_{i=0}^{n+k-1} a_i^{(k,n)} x^i.$$

Then

$$(1) \quad \tilde{P}_{k+1}^{(n)}(x) = a_{n+k}^{(k,n+1)} \tilde{P}_k^{(n+2)}(x) - x\, a_{n+k+1}^{(k,n+2)} \tilde{P}_k^{(n+1)}(x),$$

$$(2) \quad \tilde{P}_k^{(n)}(x) = a_{n+k}^{(k-1,n+2)} \tilde{P}_k^{(n+1)}(x) - a_{n+k}^{(k,n+1)} \tilde{P}_{k-1}^{(n+2)}(x),$$

$$(3) \quad \tilde{P}_k^{(n+1)}(x) = \tilde{P}_k^{(n)}(x) - e_k^{(n)} x \tilde{P}_{k-1}^{(n+1)}(x),$$

$$(4) \quad \tilde{P}_{k+1}^{(n)}(x) = \tilde{P}_k^{(n+1)}(x) - q_{k+1}^{(n)} x \tilde{P}_k^{(n)}(x),$$

with $e_k^{(n)} = h_k^{(n)}/h_{k-1}^{(n+1)}$, $q_{k+1}^{(n)} = h_k^{(n+1)}/h_k^{(n)}$ and

$h_k^{(n)} = \sum_{i=0}^{k} c_{n+k+i}\, b_{k-i}^{(k,n)}$. The same relationships also hold for the polynomials $\tilde{N}_k^{(n)}$.

Starting from $[n/o]_f(x)$ the alternate use of formulae (1) and (2) forms Baker's algorithm [4] to compute Padé approximants lying on an ascending staircase of the table. If we express these relations in terms of the coefficients of the polynomials we get Longman's method [14]. Starting from $[n/o]_f(x)$ and using alternately formulae (4) and (3), we obtain Watson's method [19] for computing approximants lying on a descending staircase. It is, of course, possible to start from $[o/n]_f(x)$ and to use (3) and (4).

These two methods can be combined to compute some other sequences of Padé approximants without computing the whole q-d table. For example, the sequence [n/o], [n+1/o], [n+1/1], [n/1], [n/2], [n+1/2],... can be computed by using successively (4), (2),

(1), (3). It is also possible to follow any other path in the Padé table since formulae (1) - (4) allow us to compute the Padé approximant indicated by a star in the following figure, if the Padé approximants indicated by a small circle are known:

o		o	*		*	o	o
o	*	o		o	o		*
(4)		(1)		(2)		(3)	

Use can also be made of the relationship $[p/q]_{1/f}(x) = 1/[q/p]_f(x)$. Some additional recursive schemes can be obtained if the whole q-d table is known [21].

5 An Acceleration Result

Among the identities which can be proved between adjacent Padé approximants is:

$$[n+k/k+1]_f(x) - [n+k+1/k]_f(x) = \frac{[H_{k+1}^{(n+1)}]^2}{\tilde{P}_{k+1}^{(n)}(x)\,\tilde{P}_k^{(n+2)}(x)}\, x^{n+2k+2}.$$

Let us now assume that f is a Stieltjes series, that is

$$c_n = \int_0^{1/R} t^n \, d\alpha(t),$$

where α is bounded and nondecreasing in the finite interval $[0,1/R]$. A classical result in the theory of orthogonal polynomials is that the roots of $P_k^{(n)}$ belong to $[0,1/R]$. Thus we immediately obtain the inequalities

$$0 \le f(x) - [n+k/k+1]_f(x) \le f(x) - [n+k+1/k]_f(x), \quad \forall x \in [0,R[,$$

$$0 \le (-1)^n (f(x) - [n+k/k+1]_f(x)) \le (-1)^n (f(x) - [n+k+1/k]_f(x)),$$

$$\forall x \in]-R,0].$$

Let $f_n(x) = \sum_{i=0}^{n} c_i x^i$. Then, for such series

$$\lim_{n\to\infty} (f_{n+1}(x) - f(x))/(f_n(x) - f(x)) = x/R.$$

By using a theorem proved by Henrici [10] and the preceeding inequalities it is easy to show that the diagonals and the columns of the Padé table converge to f faster than the partial sums of the series. More precisely we get

$$\lim_{n\to\infty} ([n+k/k]_f(x) - f(x))/(f_{n+2k}(x) - f(x)) = 0, \quad k=1,2,\ldots,$$

$$\lim_{k\to\infty} ([n+k/k]_f(x) - f(x))/(f_{n+2k}(x) - f(x)) = 0, \quad n=0,1,\ldots,$$

for all $x \in]-R, R[$ where R is the radius of convergence of f.

This result can be also written in terms of the ε-algorithm for totally monotonic and totally oscillating sequences [7].

6 The Non-Scalar Case

Let us now assume that the coefficients $\{c_n\}$ belong to a topological vector space E whose dual space is E'. In [5], Brezinski has proposed a generalization of the Padé table for such series. For a sequence $\{S_n\}$ whose elements belong to E, this generalization leads to a generalization of the ε-algorithm. Let

$$\varepsilon_{2k}^{(n)} = \frac{\begin{vmatrix} S_n & \cdots & S_{n+k} \\ \langle y,\Delta S_n\rangle & \cdots & \langle y,\Delta S_{n+k}\rangle \\ \cdots & \cdots & \cdots \\ \langle y,\Delta S_{n+k-1}\rangle & \cdots & \langle y,\Delta S_{n+2k-1}\rangle \end{vmatrix}}{\begin{vmatrix} 1 & \cdots & 1 \\ \langle y,\Delta S_n\rangle & \cdots & \langle y,\Delta S_{n+k}\rangle \\ \cdots & \cdots & \cdots \\ \langle y,\Delta S_{n+k-1}\rangle & \cdots & \langle y,\Delta S_{n+2k-1}\rangle \end{vmatrix}}$$

where y is an arbitrary element of E' such that the denominator does not vanish.

The $\varepsilon_{2k}^{(n)}$ can be computed by means of the topological ε-algorithm:

$$\varepsilon_{-1}^{(n)} = 0 \in E', \qquad \varepsilon_0^{(n)} = S_n \in E \qquad n = 0,1,\ldots,$$

$$\varepsilon_{2k+1}^{(n)} = \varepsilon_{2k-1}^{(n+1)} + y/\langle y, \varepsilon_{2k}^{(n+1)} - \varepsilon_{2k}^{(n)}\rangle, \quad n,k=0,1,\ldots,$$

$$\varepsilon_{2k+2}^{(n)} = \varepsilon_{2k}^{(n+1)} + (\varepsilon_{2k}^{(n+1)} - \varepsilon_{2k}^{(n)})/\langle\varepsilon_{2k+1}^{(n+1)} - \varepsilon_{2k+1}^{(n)},$$

$$\varepsilon_{2k}^{(n+1)} - \varepsilon_{2k}^{(n)}\rangle, \quad n,k=0,1,\ldots .$$

The theory of the topological ε-algorithm can also be explained by that of orthogonal polynomials. Let $P_k(x) = a_o +\ldots+ a_k x^k$ be the orthogonal polynomial of degree k with respect to the sequence $\{c_n = \langle y,\Delta S_n\rangle\}$. Then

$$\varepsilon_{2k}^{(0)} = (a_o S_o +\ldots+ a_k S_k)/P_k(1).$$

It can also be proved that P_k is the characteristic polynomial of the operator A_k obtained by the application of the moment method to $\{S_n\}$ [18].

Let us now turn to the solution of a system of linear equations $Bx = b$ where B is a symmetric positive definite matrix. Let $\{x_k\}$ be the sequence of vectors obtained by applying the conjugate gradient method to the system with $x_o = 0$ [12]. On the other hand, we consider the sequence $S_o = 0$, $S_{n+1} = AS_n + b$ with $B = I - A$. If the topological ε-algorithm is applied to $\{S_n\}$ with $y = b$ then we get:

$$\varepsilon_{2k}^{(0)} = x_k = v_k, \quad \text{for } k = 0,\ldots,p,$$

$$\varepsilon_{2p}^{(0)} = x_p = v_p = B^{-1}b,$$

where p is the dimension of the system and where $\{v_k\}$ is the sequence produced by the moment method. If the matrix B is non-symmetric the vectors $\{\varepsilon_{2k}^{(0)}\}$ are identical to those produced by the biconjugate gradient method.

References

1. Akhiezer, N.I., The classical moment problem, Oliver and Boyd, Ltd., Edinburgh and London, 1965.

2. Allen, G.D., C.K. Chui, W.R. Madych, F.J. Narcowich and P.W. Smith, Padé approximation of Stieltjes series, J. Approximation Theory, 14 (1975), 302-316.

3. Baker, G.A. Jr., Essentials of Padé approximants, Academic Press, Inc., New York, 1975.

4. Baker, G.A., Jr., The Padé approximant method and some related generalizations, in The Padé approximant in theoretical physics, G.A. Baker Jr., and J.L. Gammel, eds., Academic Press, Inc., New York, 1970.

5. Brezinski, C., Généralisations de la transformation de Shanks, de la table de Padé et de l' ε-algorithme, Calcolo, 12 (1975), 317-360.

6. Brezinski, C., Computation of Padé approximants and continued fractions, J. Comput. Appl. Math., 2 (1976), 113-123.

7. Brezinski, C., Convergence acceleration of some sequences by the ε-algorithm, to appear.

8. Cordellier, F., to appear.

9. Gragg, W.B., Matrix interpretations and applications of the continued fraction algorithm, Rocky Mountains J. Math., 4 (1974), 213-225.

10. Henrici, P., Elements of numerical analysis, Wiley and Sons, New York, 1964.

11. Henrici, P., The quotient-difference algorithm, NBS applied mathematics series, 49 (1958), 23-46.

12. Hestenes, M.R. and E. Stiefel, Method of conjugate gradients for solving linear systems, J. Res. NBS, 49 (1952), 409-436.

13. Karlsson, J. and B. von Sydow, The convergence of Padé approximants to series of Stieltjes, Ark. för Mat., 14 (1976), 44-53.

14. Longman, I.M., Computation of the Padé table, Intern. J. Comp. Math., 3B (1971), 53-64.

15. Nuttall, J., Convergence of Padé approximants for the Bethe-Salpeter amplitude, Phys. Rev., 157 (1967), 1312-1316.

16. Rutishäuser, H., Der quotienten-differenzen algorithmus, Mittlg. Inst. f. Angew. Math., ETH no7, Birkhäuser Verlag, Basel, 1956.

17. Van Rossum, H., A theory of orthogonal polynomials based on the Padé table, thesis, University of Utrecht, Van Gorcum, Assen, 1953.

18. Vorobyev, Yu. V., Method of moments in applied mathematics, Gordon and Breach, New York, 1965.

19. Watson, P.J.S., Algorithms for differentiation and integration, in Padé approximants and their applications, P.R. Graves-Morris, ed., Academic Press, Inc., New-York, 1973.

20. Wynn, P., On a device for computing the $e_m(S_n)$ transformation, MTAC, 10 (1956), 91-96.

21. Wynn, P., The rational approximation of functions which are formally defined by a power series expansion, Math. Comp., 14 (1960), 147-186.

22. Wynn, P., Upon systems of recursions which obtain among the quotients of the Padé table, Numer. Math., 8 (1966), 264-269.

23. Wynn, P., A general system of orthogonal polynomials, Quart. J. Math. Oxford (2), 18 (1967), 81-96.

C. Brezinski
Uer d'IEEA - informatique
Université de Lille I
BP 36
59650 - Villeneuve d'Ascq
France

CONVERGENCE ALONG STEPLINES IN A GENERALIZED PADÉ TABLE

M.G. de Bruin

In this paper convergence is proved for sequences of Padé approximant n-tuples taken from steplines in a simultaneous Padé table for an n-tuple of formal power series. The method of proof generalizes the connection between approximants on a stepline in the ordinary Padé table and certain continued fractions.

1 The Padé-n-table

Consider an n-tuple of formal power series ($n \geq 2$) in an indeterminate x with complex coefficients:

$$(1.1) \quad f^{(i)}(x) = \sum_{\nu=0}^{\infty} c_\nu^{(i)} x^\nu, \; c_0^{(i)} \neq 0 \; (i = 1,2,\dots,n).$$

Let $\rho_0,\rho_1,\dots,\rho_n$ be n+1 non-negative integers (i.e. they belong to $\mathbb{N}_0 = \mathbb{N} \cup \{0\}$) and consider the problem of finding an n-tuple of rational functions $\{P^{(i)}(x)/P^{(o)}(x)\}_{i=1}^n$ with $P^{(o)} \not\equiv 0$ and

$$(1.2) \begin{cases} \text{degree } P^{(i)} \leq \sigma - \rho_i \; (i = 0,1,\dots,n; \; \sigma = \rho_0+\rho_1+\dots+\rho_n). \\ P^{(o)}(x)f^{(i)}(x) - P^{(i)}(x) = O(x^{\sigma+1}) \text{ as } x\to 0 \; (i = 1,2,\dots,n). \end{cases}$$

It is obvious that this problem always has a solution with $P^{(o)} \not\equiv 0$ because equating coefficients of the powers $\sigma-\rho_i+1$, $\sigma-\rho_i+2,\dots,\sigma$ ($i = 1,2,\dots,n$) of x in the second line of (1.2) leads to $\sigma-\rho_0$ homogeneous linear equations in $\sigma-\rho_0+1$ unknowns (the coefficients of $P^{(o)}$). However, unlike the case of the ordinary Padé table ($n = 1$), the n-tuple of rational functions does not need to be unique anymore, even when the rational functions are written with $\gcd(P^{(o)},P^{(1)},\dots,P^{(n)}) = 1$.

Let us now assume that for each point $(\rho_o,\rho_1,\ldots,\rho_n) \in \mathbb{N}_o^{n+1}$ with $\rho_o \geq \rho_i$ $(i = 1,2,\ldots,n)$, the problem (1.2) has a unique solution $\{P^{(1)}(x)/P^{(o)}(x),\ldots,P^{(n)}(x)/P^{(o)}(x)\}$ when the n-tuple of rational functions is written subject to the condition

(1.3) $\gcd(P^{(o)},P^{(1)},\ldots,P^{(n)}) = 1, \; P^{(o)}(0) = 1.$

The unique n-tuple is then placed at the point $(\rho_o,\rho_1,\ldots,\rho_n)$ and the arising configuration is called the Padé-n-table.

The sequence of points $(k,0,\ldots,0)$, $(k+1,0,\ldots,0)$, $(k+1,1,0,\ldots,0),\ldots,$ $(k+1,1,\ldots,1)$, $(k+2,1,1,\ldots,1),\ldots$ is called the stepline of order k $(k \in \mathbb{N}_o)$. There is a connection between the approximant n-tuples on a stepline and a certain generalized continued fraction (section 2) and it is not very surprising that this continued fraction can be used to prove convergence results (section 3).

Finally we define the notion normal for a point from the Padé-n-table, by requiring that the n-tuple of rational functions connected with the point in question, does not belong to any other point in the table; this is equivalent to equality signs in the first line of (1.2) for all i and an equality sign in the second line for at least one value of i.

Of course it is possible to extend the definition of the Padé-n-table in such a way that it contains all points of $\mathbb{N}_o^{n+1}$ by omitting the condition on ρ_o. It can then be seen as the set of the so called "German polynomials" for the functions $1,f^{(1)},\ldots$ $\ldots,f^{(n)}$ (compare [5]; for further references see also [2], [3]).

2 N - fractions

In this section the notion of a generalized continued fraction is given. The generalization in question can be seen as a form of the Jacobi - Perron algorithm (see [6]; for further references [1] and [8]).

Let expressions $b_o^{(i)}$, $a_\nu^{(i)}$, b_ν $(i = 1,2,\ldots,n;\ \nu \in \mathbb{N})$ be

given (Later on they will depend on the indeterminate x).
An <u>n-fraction</u> then is a <u>sequence of approximant-n-tuples</u>, $(\{A_\nu^{(i)}/A_\nu^{(o)}\}_{i=1}^{n})_{\nu=o}^{\infty}$, with recurrence relation and initial values:

(2.1) $A_\nu^{(i)} = b_\nu A_{\nu-1}^{(i)} + a_\nu^{(n)} A_{\nu-2}^{(i)} + \ldots + a_\nu^{(1)} A_{\nu-n-1}^{(i)}$ $(i=0,1,\ldots,n)$

(2.2) $A_{-j}^{(i)}=\delta_{i+j,n+1}$ $(0\leq i\leq n,\ 1\leq j\leq n)$; $A_o^{(i)}=b_o^{(i)}$ $(1\leq i\leq n)$, $A_o^{(o)}=1$.

In the sequel an n-fraction will be denoted by

$$(2.3)\quad (\{A_\nu^{(i)}/A_\nu^{(o)}\}_{i=1}^{n})_{\nu=0}^{\infty} = \begin{bmatrix} & a_1^{(1)} \ldots a_\nu^{(1)} \ldots \\ b_o^{(1)} & \cdot \qquad \cdot \\ \cdot & \cdot \qquad \cdot \\ \cdot & \cdot \qquad \cdot \\ \cdot & \\ \cdot & a_1^{(n)} \ldots a_\nu^{(n)} \ldots \\ b_o^{(n)} & b_1 \quad \ldots b_\nu \quad \ldots \end{bmatrix}.$$

It is also possible to define the n-fraction completely analogous to the definition of an ordinary continued fraction as given in the famous textbook by O. Perron [7] (compare [2]). However, a far more elegant way is to use a generalization of the method of successive linear fractional transformations (which is extremely useful in deriving convergence results) as described by W.J. Thron [9] for the ordinary continued fraction.

Consider the transformations $s_\nu^{(i)}: \mathbb{C}^n \to \mathbb{C}$ $(i=1,2,\ldots,n;\nu\in\mathbb{N})$

$$(2.4)\quad s_\nu^{(1)}(x_1,\ldots,x_n)=\frac{a_\nu^{(1)}}{b_\nu+x_n},\quad s_\nu^{(i)}(x_1,\ldots,x_n)=\frac{a_\nu^{(i)}+x_{i-1}}{b_\nu+x_n}\quad (2\leq i\leq n).$$

From this sequence of transformations, we construct another one in the following inductive way:

$$(2.5)\quad S_\nu^{(i)}(x_1,\ldots,x_n) = S_{\nu-1}^{(i)}(s_\nu^{(1)},\ldots,s_\nu^{(n)})\quad (1\leq i\leq n;\ \nu\in\mathbb{N}\setminus\{1\}),$$

with

(2.6) $S_1^{(i)}(x_1,\dots,x_n) = s_1^{(i)}(x_1,\dots,x_n)$ $(i = 1,2,\dots,n)$.

THEOREM 2.1. The quantities defined in (2.5) and (2.6) satisfy

$$(2.7)\quad S_\nu^{(i)}(x_1,\dots,x_n) = \frac{A_\nu^{(i)} + x_n A_{\nu-1}^{(i)} + x_{n-1}A_{\nu-2}^{(i)} + \dots + x_1 A_{\nu-n}^{(i)}}{A_\nu^{(o)} + x_n A_{\nu-1}^{(o)} + x_{n-1}A_{\nu-2}^{(o)} + \dots + x_1 A_{\nu-n}^{(o)}}$$

for $1 \le i \le n$, $\nu \in \mathbb{N}$. The A's follow from (2.3) with $b_o^{(i)}=0$ $(1 \le i \le n)$.

Proof. Follows easily by induction on ν.

COROLLARY 2.2. The sequence of approximant n-tuples of (2.3) is the sequence $(\{b_o^{(i)} + S_\nu^{(i)}(0,\dots,0)\}_{i=1}^{n})_{\nu=o}^{\infty}$ ($S_\nu^{(i)}$ from (2.7)).

The existence of certain subsets of $\mathbb{C}^n$ which are invariant under S_ν: $(x_1,\dots,x_n) \to (S_\nu^{(1)},\dots,S_\nu^{(n)})$ and at the same time contain the origin as an interior point, can now be proved. Consider n real numbers $\beta_1,\beta_2,\dots,\beta_n$ with $0<\beta_n<1$, $0<\beta_{j-1}<\beta_j(1-\beta_n)$ $(j = 2,3,\dots,n)$, which will be chosen to suit our purposes.

THEOREM 2.3 Let $a_\nu^{(i)}$ $(1 \le i \le n;\ \nu \in \mathbb{N})$ satisfy the inequalities

(2.8) $|a_\nu^{(1)}| \le \beta_1(1-\beta_n)$, $|a_\nu^{(i)}| \le \beta_i(1-\beta_n)-\beta_{i-1}$ $(2 \le i \le n)$ for $\nu \in \mathbb{N}$.

Consider $S_\nu(x_1,\dots,x_n) = (S_\nu^{(1)},\dots,S_\nu^{(n)})$ with $b_\nu = 1$ $(\nu \in \mathbb{N})$ and let $V = \{(x_1,\dots,x_n) \mid |x_i| \le \beta_i\ (i = 1,2,\dots,n)\}$. Then

(2.9) $S_\nu(V) \subset V$ $(\nu \in \mathbb{N})$.

Proof. Follows from (2.4) and (2.5).

Finally we give a theorem concerning the connection between approximants on a stepline in the Padé-n-table and an n-fraction.

THEOREM 2.4 Let the stepline of order k in the Padé-n-table for the formal power series (1.1) be normal; the point $(k,0,\dots,0)$ is given the number 0, the other points are numbered consecutively. Then the approximant n-tuples from the stepline are the approximants of an n-fraction of the following form:

$$\begin{bmatrix}
 & \alpha_1^{(1)}x^{k+1} & \alpha_2^{(1)}x^{k+2} & \cdots & \alpha_n^{(1)}x^{k+n} & \alpha_{n+1}^{(1)}x^{n} & \cdots & \alpha_\nu^{(1)}x^{n} & \cdots \\
\Sigma_{j=1}^{k}\, c_j^{(1)}x^j & \vdots & \vdots & & \alpha_n^{(2)}x^{n-1} & \alpha_{n+1}^{(2)}x^{n-1} & \cdots & \alpha_\nu^{(2)}x^{n-1} & \cdots \\
\vdots & & & & \vdots & \vdots & & \vdots & \\
 & \cdot & \alpha_2^{(n-1)}x^{k+2} & \cdots & \alpha_n^{(n-1)}x^{2} & \alpha_{n+1}^{(n-1)}x^{2} & \cdots & \alpha_\nu^{(n)}x^{2} & \cdots \\
\cdot & \alpha_1^{(n)}x^{k+1} & \alpha_2^{(n)}x & \cdots & \alpha_n^{(n)}x & \alpha_{n+1}^{(n)}x & \cdots & \alpha_\nu^{(n)}x & \cdots \\
\Sigma_{j=1}^{k}\, c_j^{(n)}x^j & 1 & 1 & \cdots & 1 & 1 & \cdots & 1 & \cdots
\end{bmatrix}$$

with $\alpha_\nu^{(1)} \neq 0$ $(\nu \in \mathbb{N})$. The α's are constants which can be expressed in terms of the coefficients of the power series on the right hand side of the second line of (1.2), written down for the points of the stepline in question.

Proof. The proof follows from the uniqueness of the solution of problem (1.2) and the normality of the stepline (see [2]).

3 Convergence

In this section the main convergence theorem is given and some examples for n = 2. The theorem can of course be applied to all Padé-n-tables, known to be normal, for which the denominators of the rational functions in the table are explicitly known, or for which the n-fraction from Theorem 2.4. is known.

Let ρ be the unique real number defined by

$$(3.1) \quad \rho^n + 2^{-1}\rho^{n-1} + 2^{-2}\rho^{n-2} + \ldots + 2^{-n+1}\rho - 2^{-n-1} = 0, \quad \frac{1}{6} < \rho \leq \frac{1}{4}.$$

($\rho=\rho(n)$ satisfies: $\rho(1)=\frac{1}{4}$, $\rho(n+1)<\rho(n)$ for all n, $\rho(n)\to\frac{1}{6}$ for $n\to\infty$)

THEOREM 3.1. Let the n-tuple of formal power series (1.1) have a normal stepline of order k and assume that the power series are the series expansions of n functions, holomorphic at $x = 0$ and meromorphic on $\mathcal{D} \subset \mathbb{C}$. Denote the Padé approximant n-tuples on the stepline by $\{P_\nu^{(i)}(x)/P_\nu^{(o)}(x)\}_{i=1}^{n}$ $(\nu \in \mathbb{N}_o)$ and the coefficients

<u>of the n-fraction from Theorem 2.4. by</u> $\alpha_\nu^{(i)}$ $(i=1,2,\ldots,n;\ \nu\in\mathbb{N})$. <u>Finally let</u> $\alpha_i = \limsup_{\nu\to\infty}|\alpha_\nu^{(i)}| < \infty\ (1\le i\le n)$; <u>omit</u> α_i <u>if it is zero</u>. <u>Then</u> $\lim_{\nu\to\infty} P_\nu^{(i)}(x)/P_\nu^{(o)}(x) = f^{(i)}(x)\ (1\le i\le n)$, <u>uniformly in x on each compact subset of</u> $\{x\in\mathbb{C}\mid |x|<\rho.\min(\alpha_i^{-1/(n+1-i)})\}\cap\mathcal{D}$ <u>that contains no poles of the functions</u> $f^{(1)}, f^{(2)},\ldots, f^{(n)}$.

<u>Proof</u>. For a sketch of the proof see section 4.

EXAMPLE 1. Consider $e^{\alpha x}$, $e^{\beta x}$ $(\alpha,\beta\in\mathbb{C}\setminus\{0\},\ \alpha\neq\beta)$. Then $\mathcal{D}=\mathbb{C}$ and $\alpha_1=\alpha_2=0$ (see [2]); thus we get convergence uniformly in x on each compact subset of $\mathbb{C}$. This result is contained in a paper by A.J. Goddijn [4] (a young mathematician who devoted himself to music after graduation). He proves convergence for arbitrary sequences for which $\sigma\to\infty$, without conditions on the ρ's.

EXAMPLE 2. Consider $(1-x)^\alpha$, $(1-x)^\beta$ $(\alpha,\beta\in\mathbb{C}\setminus\mathbb{Z},\ \alpha-\beta\notin\mathbb{Z})$. Then $\mathcal{D}=\{x\in\mathbb{C}\mid |x|<1\}$ and $\alpha_1=1/27$, $\alpha_2=1/3$ (see [2]); thus we get convergence uniformly in x on each compact subset of $\{x\in\mathbb{C}\mid |x|<3(\sqrt{3}-1)/4\}$. Because $3(\sqrt{3}-1)/4=0.549\ldots$, we see that the convergence theorem is not yet best possible.

Although the second example shows that the convergence theorem needs to be improved upon, it is not very likely that much better bounds can be attained by using the method of proof outlined in section 4. Much work is still to be done on the subject of analytic behaviour of generalized continued fractions.

Finally it might be interesting to notice that Theorem 3.1. for n = 1 reduces to the well known result due to E.B. van Vleck and A. Pringsheim (see [7]; Satz 3.27, page 148).

4 Sketch of Proof

In this section only a very short sketch will be given, further details will appear in a forthcoming publication.

A. Using the method of [7](Satz 3.27,page 148), it appears to be sufficient to consider an n-fraction (2.3) with $b_0^{(i)} = 0$ $(1\leq i\leq n)$, $a_\nu^{(i)} = \alpha_\nu^{(i)} x^{n+1-i}$ $(1\leq i\leq n; \nu \in \mathbb{N})$, $b_\nu = 1$ $(\nu \in \mathbb{N})$ with $|\alpha_\nu^{(i)}| \leq \alpha_i$ for $\nu \in \mathbb{N}\setminus\{1\}$ and $1\leq i\leq n$. The poles enter when the initial and end part of the n-fraction are being put together again.

B. Using a generalization of the method of [7], (Satz 2.25, page 64), it is possible to prove the existence of a (maybe very small) circle around the origin on which the n-fraction converges uniformly to n analytic functions. Furthermore, the denominators of the n-fraction are zero-free on this circle.

C. In (2.8) we choose $\beta_n = 2^{-1}$, $\beta_{n-i} = 2^{-i-1} - 2^{-i+1}\rho - 2^{-i+2}\rho^2 - \ldots - \rho^i$ $(1\leq i\leq n-1)$ after which the conditions transform into the following

$$|x^n| \leq \tfrac{1}{2}\beta_1\alpha_1^{-1}, \quad |x^{n+1-i}| \leq (\tfrac{1}{2}\beta_i - \beta_{i-1})\alpha_i^{-1} \quad (2 \leq i \leq n).$$

Because of the choice of ρ, β_1, $\beta_2, \ldots, \beta_n$ we have $\rho = (\tfrac{1}{2}\beta_1)^{1/n} = (\tfrac{1}{2}\beta_i - \beta_{i-1})^{1/(n+1-i)}$ $(2\leq i\leq n)$. Thus, for the values of x given in Theorem 3.1, all approximants of the n-fraction belong to V (of Theorem 2.3.) and therefore remain bounded (Corollary 2.2.). Because the approximants are rational functions, they must be analytic on the set of x-values in question.

D. Apply the Stieltjes - Vitali theorem in the following form: let $\{f_\nu(x)\}_{\nu=1}^{\infty}$ be a sequence of uniformly bounded analytic functions on a domain G for which $\lim f_\nu(x)$ for $\nu \to \infty$ exists for $x \in H$, H a subset of G with infinitely many points and having at least one accumulation point in G; then the sequence converges uniformly in x on each compact subset of G.

References

1 Bernstein, L., The Jacobi - Perron algorithm, its theory and applications, Springer Verlag, Berlin-Heidelberg-New York, Lect. Notes in Math. 207, 1972.

2 de Bruin, M.G., Generalized C-fractions and a multidimensional Padé table (Thesis), Amsterdam, 1974.

3 de Bruin, M.G., Three new examples of generalized Padé tables which are partly normal, Univ. of Amsterdam, Department of Mathematics, Report 76-11, 1976.

4 Goddijn, A.J., Enkele convergentie - eigenschappen van twee meerdimensionale Padé-tafels, Univ. of Amsterdam (dutch; handwritten manuscript), 1972.

5 Mahler,K., Perfect Systems, Compositio Math. 19(1968), 95-166.

6 Perron,O., Grundlagen für eine Theorie des Jacobischen Kettenbruchalgorithmus, Math.Ann. 64 (1907), 1-77.

7 Perron, O., Die Lehre von den Kettenbrüchen II, BG Teubner Verlag, Stuttgart, 1957.

8 Schweiger, F., The metrical theory of the Jacobi - Perron algorithm, Springer Verlag, Berlin-Heidelberg-New York, Lect. Notes in Math. 334, 1973.

Marcel G. de Bruin
Universiteit van Amsterdam
Instituut voor Propedeutische Wiskunde
Roetersstraat 15 Amsterdam, Netherlands

N-VARIABLE RATIONAL APPROXIMANTS

J.S.R. Chisholm

"Desirable properties" of a two-variable generalization of Padé approximants are laid down. The "Chisholm approximants" are defined and are shown to obey nearly all of these properties; the alternative ways of completing a unique definition are discussed, and the "prong structure" of the defining equations is elucidated. Several generalizations and variants of Chisholm approximants are described: N-variable diagonal (Chisholm and McEwan), 2-variable simple off-diagonal (Graves-Morris, Hughes Jones and Makinson), N-variable simple and general off-diagonal (Hughes Jones), and rotationally covariant 2-variable approximants (Chisholm and Roberts). All of the 2-variable approximants are capable of representing singularities of functions of two variables, and of analytically continuing beyond the polycylinder of convergence of the double series.

1 Introduction

The subject of this paper is the generalization of Padé approximation to power series in any number of variables. I think that it is important to put this work in perspective, by mentioning its motivation. First, the general motivation: Padé approximants have provided a means of solving a wide variety of problems. These include: several areas of numerical work, such as numerical evaluation of standard functions and numerical integration; certain difficult problems in theoretical physics, in particular the low-lying boson resonances, accurate work in quantum potential theory, and the calculation of critical indices; some problems in fluid mechanics - though many hard problems remain unsolved; and, of course, electrical engineers have a very close association with Padé approximants through continued fractions. In solving this variety of problems, several generalizations of Padé approximants have been used: Baker's D log method and the Gammel-Gaunt-Gutman-Joyce (G^3J) generalization, the Shafer quadratic approximant, the Gammel-Baker and Common approximants, and the generalizations to two-point and interpolating approximants, to which Saff, Jones and Thron have contributed.

Given the successes of the Padé method in defining easily computable rational approximants from power series in one variable, it is natural to ask whether the Padé scheme can be generalized to define N-variable rational approximants, given a power series in N variables (N=2,3,...), since one can attempt to solve a vast range of problems by using series in two or possibly more variables. I discussed this problem with John Gammel in 1966, and thought about it, without success, several times in the following years. Then, during three days in 1972, I solved the problem of defining the two-variable diagonal approximants, and a group of us in the University of Kent generalized this result to N variables and off-diagonal in a surprisingly short time. I have wondered why I did not solve the problem on previous occasions. The answer seems to be, as with many other mathematical problems, that formulation of the problem was the first major step in solving it: I had to ask quite specifically, "what properties should a generalization of Padé approximants have?", and enumerate them one by one. The whole problem consists of choosing arrays of points on the integer lattice in the plane (or in N-space); there is an enormous variety of possible choice, and until specific constraints are placed on the arrays to restrict the choice severely (providing a unique solution if possible), the situation is very confused. The basis of the original paper [1] was therefore the specification of "desirable properties" which an N-variable generalization of Padé approximants should satisfy. This initial work was restricted to 2-variable diagonal approximants: using two variables is easier than three or more, and diagonal approximants are the most restricted class, requiring the satisfaction of one property (homographic covariance) which off-diagonal approximants could not be expected to satisfy. In other words, diagonal approximants are the most closely specified set of approximants. This first paper defined what have been called "Chisholm Approximants", which satisfy all of the properties laid down, with one fairly minor exception, relative scale covariance. The configurations of points on the integer lattice in the plane were uniquely specified, and a unique set of approximants defined; small modifications of the original approximants were proposed later in order to satisfy additional properties such as relative scale covariance. The original paper [1] ended by setting out a program of future work, including the search for N-variable and off-diagonal generalizations.

2 Statement of the Problem

The "desirable properties" of 2-variable diagonal approximants which I set out to satisfy are:

(i) Defining Property: The approximants are rational fractions whose power series contain a finite number of terms agreeing with the given power series. The ratios of coefficients in an approximant are uniquely determined.

(ii) Symmetry Property: The approximants are symmetrical between the two variables. We shall see that there are at least three meanings of "symmetry".

(iii) Projection Property: If one variable is equated to zero, a 2-variable approximant reduces to the corresponding Padé approximant of the reduced series.

(iv) Reciprocal Covariance: An approximant to the reciprocal of the given series is the reciprocal of an approximant to the series.

(v) Homographic Covariance: If z_1, z_2 are the variables in the power series, the diagonal approximants are covariant under the group of transformations

$$(1) \quad z_r = \frac{A_r w_r}{1+B_r w_r} \qquad (r=1,2)$$

of the variables, where A_r ($\neq 0$) and B_r are complex numbers. This property is best expressed as a combination of the two properties.

(v)' Restricted Homographic Covariance: The diagonal approximants are covariant under the group

$$(2) \quad z_r = \frac{Aw_r}{1+B_r w_r} \qquad (r=1,2)$$

where $A \neq 0$.

(vi) Relative Scale Covariance: The approximants are covariant under the group

$$(3) \quad z_r = A_r w_r \qquad (A_1 A_2 = 1;\ r = 1,2)\ .$$

3 Two-Variable Diagonal Approximants

In order to define the 2-variable Chisholm approximants, we use the notation

$$(4a) \quad \underset{\sim}{z} = (z_1, z_2), \quad \underset{\sim}{\gamma} = (\gamma_1, \gamma_2),$$

$$(4b) \quad \underset{\sim}{m} = (m, m) \quad , \quad \underset{\sim}{\infty} = (\infty, \infty) \quad ,$$

and

$$(5) \quad z_1^{\gamma_1} z_2^{\gamma_2} = \underset{\sim}{z}^{\underset{\sim}{\gamma}}\ .$$

The given power series is then

$$(6) \quad f(\underset{\sim}{z}) = \sum_{0}^{\underset{\sim}{\infty}} c_{\underset{\sim}{\gamma}} \underset{\sim}{z}^{\underset{\sim}{\gamma}}\ ;$$

in practice, only a finite number of coefficients of the series may be known, as for Padé approximants. A diagonal 2-variable approximant is of the form

$$(7)\quad f_{\underset{\sim}{m}/\underset{\sim}{m}}(\underset{\sim}{z}) = \frac{\sum_{\underset{\sim}{\alpha}=\underset{\sim}{0}}^{\underset{\sim}{m}} a_{\underset{\sim}{\alpha}}\, z^{\underset{\sim}{\alpha}}}{\sum_{\underset{\sim}{\beta}=\underset{\sim}{0}}^{\underset{\sim}{m}} b_{\underset{\sim}{\beta}}\, z^{\underset{\sim}{\beta}}}\ ;$$

in this expression, the first choice of range of suffixes is made: the values of $\{\underset{\sim}{\alpha}\}$ and $\{\underset{\sim}{\beta}\}$ are allowed to range over the lattice square S_1 in Fig. 1. This choice is ultimately justified because the approximants can be made to satisfy all of the properties (i)-(v). To determine (7) uniquely, we need to specify

$$(8)\quad 2(m+1)^2 - 1$$

ratios of the coefficients $\{a_{\alpha}\}$, $\{b_{\beta}\}$. Note that this is half the number of points in the square with opposite corners at $\underset{\sim}{0}$ and $(2m+1, 2m+1)$, less one.

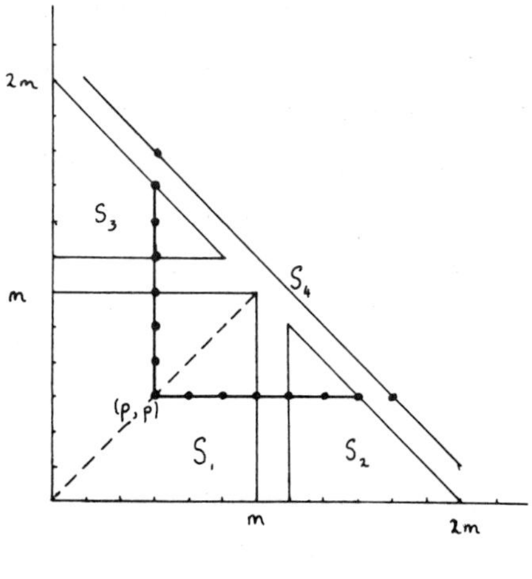

Fig.1 Parameter regions : 2-variable diagonal

Now consider some of the "desirable properties". The defining property tells us that a certain set of coefficients in the expression

$$(9) \quad E(\underset{\sim}{z}) = \left[\sum_{\underset{\sim}{0}}^{\underset{\sim}{m}} b_{\underset{\sim}{\beta}} \underset{\sim}{z}^{\underset{\sim}{\beta}}\right] \cdot \left[\sum_{\underset{\sim}{0}}^{\underset{\sim}{\infty}} c_{\underset{\sim}{\gamma}} \underset{\sim}{z}^{\underset{\sim}{\gamma}}\right] - \sum_{\underset{\sim}{0}}^{\underset{\sim}{m}} a_{\underset{\sim}{\alpha}} \underset{\sim}{z}^{\underset{\sim}{\alpha}}$$

should be equated to zero. The symmetry property suggests that the corresponding integer lattice points $\underset{\sim}{\lambda}$ should form a symmetrical pattern in the plane; this is the first meaning of "symmetry".

Putting $z_2 = 0$ in (6) and (7) gives the series

$$(10) \quad \sum_{\gamma_1=0}^{\infty} c_{\gamma_1,0} \; z_1^{\gamma_1} ,$$

and the rational fraction with coefficients $\{a_{\alpha_1,0};\alpha_1=0,1,\ldots,m\}$ and $\{b_{\beta_1,0};\beta_1=0,1,\ldots m\}$, of the form of a Padé approximant in z_1. In order that this rational fraction is the [m/m] Padé approximant to (10), the set of lattice points $\underset{\sim}{\lambda}$ must include the points $(\lambda_1,0)$ with $\lambda_1=0,1,\ldots,2m$ but <u>not</u> the point $(2m+1,0)$. The equations corresponding to these points ensure that (7) reduces to the (unique) Padé approximant of (10). Likewise, the projection property for $z_1=0$ requires the inclusion of the points with $\lambda_1=0$ and $\lambda_2=0,1,\ldots,2m$, but not $\lambda_2=2m+1$.

Since all the points $\underset{\sim}{\lambda}$ in S_1 must be included in order to define every $a_{\underset{\sim}{\alpha}}$, these three properties suggest including all the points $\underset{\sim}{\lambda}$ in the symmetrical triangular region $S_1 \oplus S_3$. On the next diagonal

$$(11) \quad \lambda_1 + \lambda_2 = 2m+1 ,$$

the end-points $(2m+1,0)$ and $(0,2m+1)$ are excluded; so we need to choose exactly half the $2m$ points on the line S_4 $(\lambda_1>0,\lambda_2>0)$ to bring the total number of equations to (8). To preserve symmetry, I chose to equate to zero the <u>sum</u> of coefficients in (9) corresponding to symmetrical pairs of points on S_4, giving "symmetrized equations". The system of equations then has symmetrical appearance, giving a second meaning to "symmetry". It is fairly clear that this choice of equal weights for the points in a pair ensures that symmetrical series have symmetrical approximants. However, Dick Hughes Jones and Peter Graves-Morris pointed out that this choice of weights was not invariant under relative scale transformations (3). So for non-symmetrical series, the choice was essentially arbitrary. For this reason, I shall introduce arbitrary weights into the symmetrized equations.

In writing down the system of equations, it is convenient to define

(12) $b_{\underset{\sim}{\beta}} = 0 \qquad (\beta_1 > m \text{ or } \beta_2 > m)$.

The equations corresponding to points $\underset{\sim}{\lambda} \subset S_1$ in Fig. 1 are

(13) $e_{\underset{\sim}{\lambda}} \equiv \sum_{\underset{\sim}{\beta}=\underset{\sim}{0}}^{\underset{\sim}{\lambda}} b_{\underset{\sim}{\beta}} c_{\underset{\sim}{\lambda}-\underset{\sim}{\beta}} - a_{\underset{\sim}{\lambda}} = 0$,

and the equations corresponding to $\underset{\sim}{\lambda} \subset S_3$ are

(14) $e_{\underset{\sim}{\lambda}} \equiv \sum_{\underset{\sim}{\beta}=\underset{\sim}{0}}^{\underset{\sim}{\lambda}} b_{\underset{\sim}{\beta}} c_{\underset{\sim}{\lambda}-\underset{\sim}{\beta}} = 0$.

For $\underset{\sim}{\lambda} \subset S_3$, the range of $\underset{\sim}{\beta}$ for given $\underset{\sim}{\lambda}$ is over the shaded rectangle in Fig. 2. The symmetrical equations for pairs of points on S_4 are

(15) $\mu_{\lambda;1}\, e_{\lambda,2m+1-\lambda} + \mu_{\lambda;2}\, e_{2m+1-\lambda,\lambda} = 0, \quad (\lambda=m+1,\ldots,2m)$

with $e_{\underset{\sim}{\lambda}}$ given by (13) and (14). For symmetrical series,

(16) $\mu_{\lambda;1} = \mu_{\lambda;2} = 1$

for all λ . If one chooses, one can break the symmetry of (15) by taking $\mu_{\lambda;1} = 0$ or $\mu_{\lambda;2} = 0$ (all λ), defining "one-sided approximants".

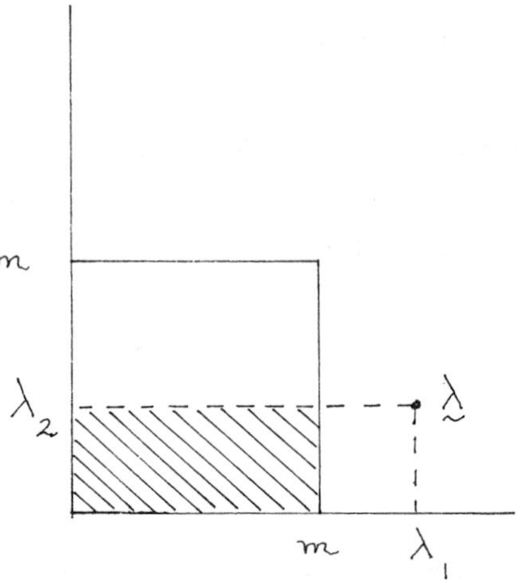

Fig.2. Range of $\underset{\sim}{\beta}$ for given $\underset{\sim}{\lambda}$

Now consider the fourth property, reciprocal covariance. Provided $c_{\underset{\sim}{0}} \neq 0$, a unique series $r(\underset{\sim}{z})$, formally reciprocal to $c(\underset{\sim}{z})$, exists; it satisfies the identity

$$f(\underset{\sim}{z})\, r(\underset{\sim}{z}) = \left[\sum_{\underset{\sim}{0}}^{\underset{\sim}{\infty}} c_{\underset{\sim}{\gamma}}\, \underset{\sim}{z}^{\underset{\sim}{\gamma}}\right] \left[\sum_{\underset{\sim}{0}}^{\underset{\sim}{\infty}} c_{\underset{\sim}{\gamma}}\, \underset{\sim}{z}^{\underset{\sim}{\gamma}}\right] \equiv 1 .$$

Multiplying $E(\underset{\sim}{z})$ formally by $-r(\underset{\sim}{z})$ therefore gives

$$(17) \quad E'(\underset{\sim}{z}) = \left[\sum_{\underset{\sim}{0}}^{\underset{\sim}{m}} a_{\underset{\sim}{\alpha}}\, \underset{\sim}{z}^{\underset{\sim}{\alpha}}\right] \left[\sum_{\underset{\sim}{0}}^{\underset{\sim}{\infty}} r_{\underset{\sim}{\beta}}\, \underset{\sim}{z}^{\underset{\sim}{\beta}}\right] - \sum_{\underset{\sim}{0}}^{\underset{\sim}{m}} b_{\underset{\sim}{\beta}}\, \underset{\sim}{z}^{\underset{\sim}{\beta}} ,$$

with coefficients

$$(18) \quad e'_{\underset{\sim}{\lambda}} = \sum_{\underset{\sim}{\alpha}=\underset{\sim}{0}}^{\underset{\sim}{\lambda}} e_{\underset{\sim}{\alpha}}\, r_{\underset{\sim}{\lambda}-\underset{\sim}{\alpha}} .$$

The coefficient $e'_{\underset{\sim}{\lambda}}$ depends only on $\{e_{\underset{\sim}{\alpha}}\}$, with $\underset{\sim}{\alpha}$ lying within the rectangle S_λ with corners at $\underset{\sim}{0}$, $\underset{\sim}{\lambda}$; this is "Rectangle Rule A". Then equations (13)-(15) imply that

$$e'_{\underset{\sim}{\lambda}} = 0 \qquad (\underset{\sim}{\lambda} \subset S_1 , S_3)$$

and that for $\underset{\sim}{\lambda} \subset S_4$

$$\mu_{\lambda;1}\ e'_{\lambda,2m+1-\lambda} + \mu_{\lambda;2}\ e'_{2m+1-\lambda,\lambda}$$

$$= -r'_{\underset{\sim}{0}} [\mu_{\lambda;1}\ e_{\lambda,2m+1-\lambda} + \mu_{\lambda;2}\ e_{2m+1-\lambda,\lambda}]$$

$$= 0,$$

establishing covariance of the system of equations. The last result depends on "Rectangle Rule B", that symmetrical equations are covariant if all other equations in $S_{\underset{\sim}{\lambda}}$ are <u>not</u> symmetrized. This rule restricts points of symmetrized pairs to lines making an angle $\frac{3}{4}\pi$ with the axes. Together with the defining property, projection property and symmetry property, the rectangle rule ensures that the configuration of Fig. 1 is unique, although the weight ratios $\mu_{\lambda;1} : \mu_{\lambda;2}$ are not determined.

It can be shown that <u>restricted</u> homographic covariance is also ensured by the rectangle rule. So each of the properties (i)-(v) is satisfied, except that relative scale covariance is not guaranteed. The question arises: can the weights be chosen so that relative scale covariance is satisfied? We shall return to this question later.

Following circulation of this first paper, some further properties of Chisholm approximants were pointed out. Alan Common and Peter Graves-Morris showed that [5]:

(vii) formally unitary series have unitary diagonal approximants. (This follows from reciprocal covariance.)

(viii) a series of the form $f(z_1)\ g(z_2)$ has $[\underset{\sim}{m}/\underset{\sim}{m}]$ approximant equal to the product $f_{m/m}(z_1)\ g_{m/m}(z_2)$ of diagonal Padé approximants. This depends upon the projection property. John Gammel [7] also noted that

(ix) the $[\underset{\sim}{m}/\underset{\sim}{m}]$ approximant to a series of form $f(z_1) + g(z_2)$ is the sum $f_{m/m}(z_1) + g_{m/m}(z_2)$ of the Padé approximants.

After the first paper had been produced, a group of us in the University of Kent set to work on various extensions which I had suggested. Dick Hughes Jones, Gordon Makinson and Peter Graves-Morris produced a series of four papers; the first of these [12], written by Hughes Jones and Makinson, analyzed the algebraic structure of the equations I had proposed. We assume that $b_{\underset{\sim}{0}} \neq 0$, and normalize by taking

(19) $b_{\underset{\sim}{0}} = 1.$

First note that, by the projection property, the sets of coefficients $\{b_{\beta_1,0};\ \beta_1 = 0,1,\ldots,m\}$ and $\{b_{0,\beta_2};\ \beta_2 = 0,1,\ldots,m\}$ are determined by the two sets of "Padé equations" of type (14), with

$$\underset{\sim}{\lambda} = (m+1,0),\ldots,(2m,0) \text{ and } \underset{\sim}{\lambda} = (0,m+1),\ldots,(0,2m).$$

These sets of coefficients are denoted by the broken lines on the axes in Fig. 3, and the $\underset{\sim}{\lambda}$-values by the solid lines on the axes. These solid and broken lines make up the "prong" labelled "0". The fact that the number of equations equals the number of unknowns corresponds to the fact that the broken and solid parts of the prong are of equal length.

Now consider the equations corresponding to points $\underset{\sim}{\lambda}$ on the solid part of prong "1" in Fig. 3. Remembering Fig. 2 and the definition (12), the corresponding equations (14) involve the (2m-1) new variables $b_{1,1}, b_{2,1},\ldots,b_{m,1}, b_{1,2},\ldots,b_{1,m}$, as well as the variables $\{b_{0,\beta_2}\}$, $\{b_{\beta_1,0}\}$ already determined. If the symmetrical equation (15) corresponding to points (1,2m) and (2m,1) is included, there are exactly (2m-1) linear equations to determine the (2m-1) new unknowns. Likewise, the solid lines of prong "2" give (2m-3) equations for (2m-3) new unknowns,

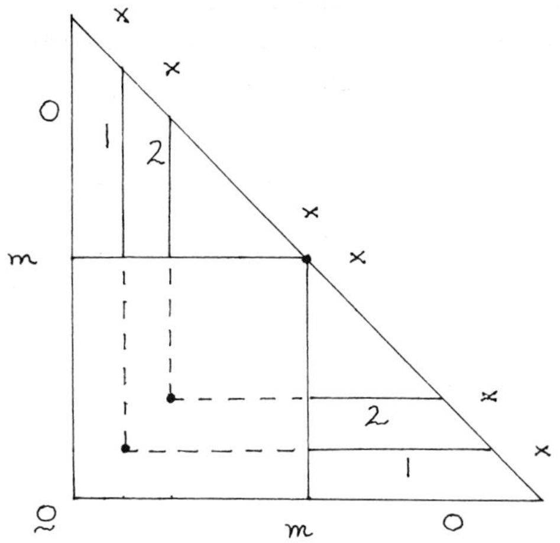

Fig. 3. Prong Structure of Chisholm Equations

corresponding to the broken lines of the prong. This continues, and finally the unknown $\underset{\sim}{b}_m$ is determined by the last symmetrical equation corresponding to the points (m+1,m) and (m,m+1). The prong structure shows that the matrix M of equations (14) and (15) is of lower triangular block form

$$(20) \quad M = \begin{bmatrix} G_0 & & & & 0 \\ \cdots & D_1 & & & \\ \cdots\cdots\cdots & & D_2 & & \\ \cdots\cdots\cdots\cdots\cdots & & & & \\ \cdots\cdots\cdots\cdots\cdots\cdots & & & & D_m \end{bmatrix} ;$$

here G_0 is a diagonal block matrix with the two "Padé matrices" as the blocks, and the equal lengths of the two parts of each prong ensures that $D_1, D_2, \ldots, D_m$ are square matrices. The matrices D_r are in fact closely connected with Padé matrices, and recur when m increases to (m+1). The very simple structure of M ensures that the solution of equations (14) and (15) requires little more computation than calculating two sets of Padé approximants up to order m. The fact that G_0 and $D_1, \ldots, D_m$ are diagonal matrices enabled Hughes Jones and Makinson to write down simple existence and uniqueness theorems for Chisholm approximants.

4 Generalized Chisholm Approximants

In the original paper, I made the suggestion of looking for certain natural generalizations of Chisholm approximants. These were:

(a) N-variable approximants defined from series in N variables,

(b) simple off-diagonal approximants (S.O.D.'s), with equal maximum powers for different variables, but different numerator and denominator orders m and n.

(c) general off-diagonal approximants (G.O.D.'s), with arbitrary maximum powers in each variable in both numerator and denominator. Later, David Roberts and I defined a variant of 2-variable diagonal and S.O.D.'s,

(d) "rotationally covariant" approximants.

While Hughes Jones and Makinson were elucidating the algebraic structure already described, John McEwan and I were studying [2] the problem of defining 3-variable and N-variable diagonal approximants satisfying generalizations of the properties (i)-(v). We chose the range of numerator and denominator suffixes in the 3-variable approximants to be over the cube S_1 in Fig. 4; the expression $E(\underset{\sim}{z})$ is again defined by (9), but with 3-component suffixes and summation limits. The coefficients of $\underset{\sim}{z}^{\underset{\sim}{\lambda}}$ in $E(\underset{\sim}{z})$ are equated to zero over S_1 (giving $a_{\underset{\sim}{\alpha}}$) and the three regions S_3; the remaining equations for $\{b_{\underset{\sim}{\beta}}\}$ are doubly symmetrized equations from pairs of points on the clipped triangular surfaces S_4, and triply symmetrized ones corresponding to three symmetrical points on the lines where these three surfaces meet. The volume formed by joining the point (m,m,m) to the points on the axes distant 2m from the origin is twice the volume of the cube, and this relationship carries over to analogous hypervolumes in N-space. As in 2-space, the configuration of Fig. 4 is unique. The projection property ensures that the points on each coordinate plane must have the configuration and symmetry of Fig. 1. The rectangle rule prevents the planes S_4 from leaning out at an angle greater than $\frac{1}{2}\pi$; by taking them perpendicular to the coordinate planes, we just manage to satisfy all of the properties (i)-(v), again with the exception of relative scale covariances. As before, the weighting factors in symmetrized equations are only determined for symmetrical series.

The equations corresponding to Fig. 4 turn out to be of a very simple algebraic form, and it is not difficult to see how to generalize them to the N variable case. The projection property for N-variable Chisholm approximants states: if (N-k) variables are equated to zero, an N-variable approximant reduces to the corresponding k-variable approximant of the reduced series. Since N can be arbitrarily large, we have defined an infinite-dimensional space of approximants, with a natural projection property; the relation between diagonal Padé approximants and this space of approximants is analogous to the relation between the complex plane and Hilbert space.

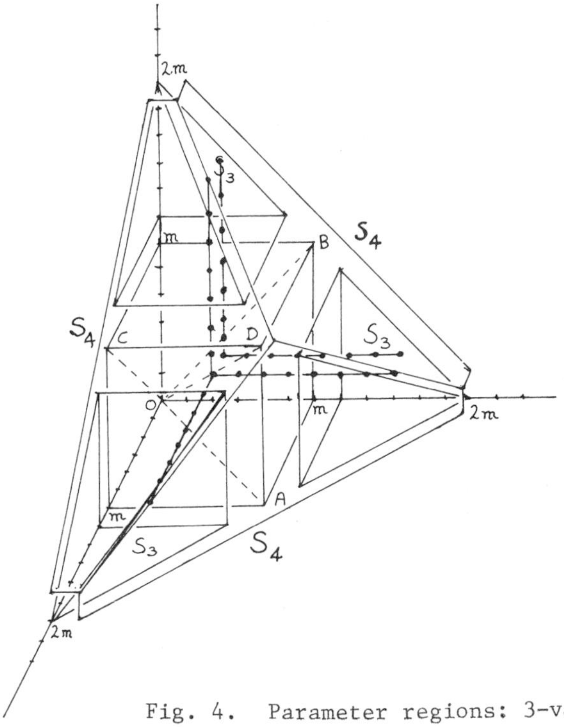

Fig. 4. Parameter regions: 3-variable diagonal

Following on the study [12] of the algebraic structure of the equations for the Chisholm approximants, Graves-Morris, Hughes Jones and Makinson wrote two papers [8,9], defining 2-variable S.O.D.'s, writing computer programs and studying diagonal and S.O.D. approximants to the β-function. The lattice configuration defining the equations when $m > n$ is shown in Fig. 5. The two squares represent the range of suffixes of numerator and denominator coefficients. The number of points in the triangles S_3 and on half of S_4 is one less than the number in S_1, as required. The form of S_3 and S_4 is dictated by the projection property, the prong structure and the rectangle rule: the numbers of equations (points in S_3 and half S_4) equals the number of new denominator unknowns (points in S_1), and the symmetrized equations must correspond to the end-points on S_4. The computation of diagonal and S.O.D. approximants to the β-function of two real variables showed that both types approximated the pole and zero lines nearest the origin, and approximated the function outside the polycylinder of convergence of the series.

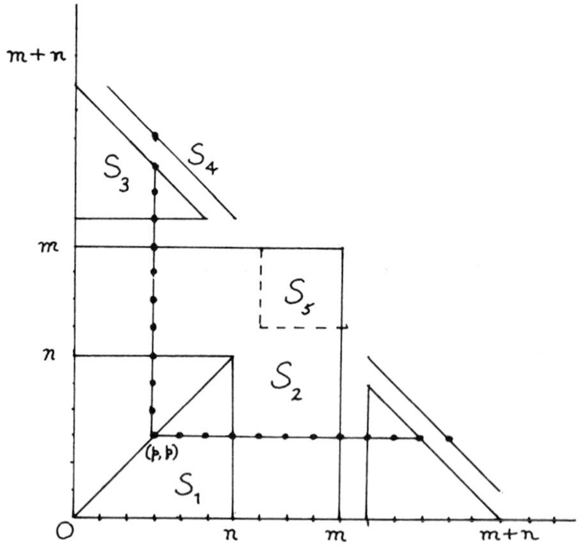

Fig. 5 Parameter regions: 2-variable S.O.D.

The definition of N-variable S.O.D.'s and the further generalization to G.O.D.'s was carried out in an extremely elegant paper by Hughes Jones [13]. The principles used to define the arrays of points are again those based on the Chisholm equations: the projection property, the prong structure (ensuring a good algebraic structure), and the rectangle rule (ensuring reciprocal covariance). The complicated structure of the arrays in N-space are described through a very neat vector notation. If the numerator and denominator orders in Hughes Jones' approximants are all equal, we obtain the Chisholm-McEwan approximants, which of course satisfy the extra property of reciprocal covariance. Since all the generalized Chisholm approximants satisfy the projection property, the infinite-dimensional space of approximants extends to include the N-variable S.O.D.'s and G.O.D.'s. In Fig. 6,7,8, typical lattice and prong structures are shown for 2-variable G.O.D.'s and for 3-variable S.O.D.'s and G.O.D.'s; the figures are based upon Hughes Jones' original diagrams.

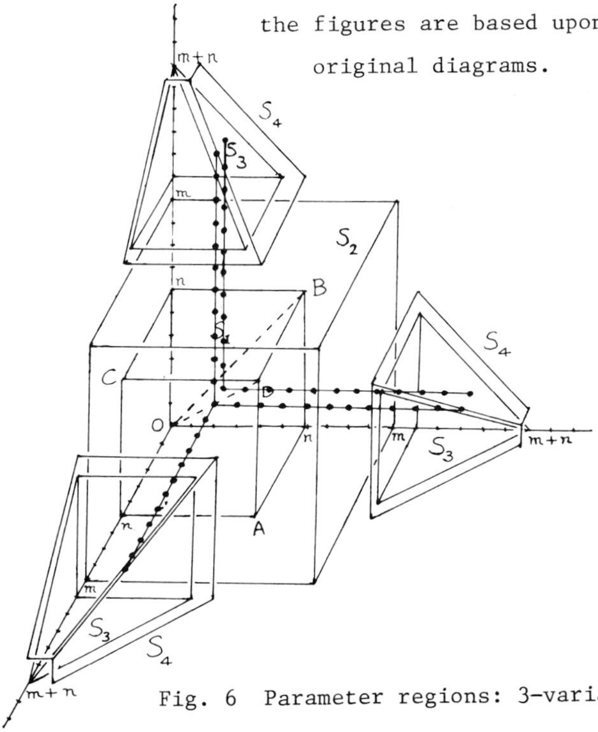

Fig. 6 Parameter regions: 3-variable S.O.D.

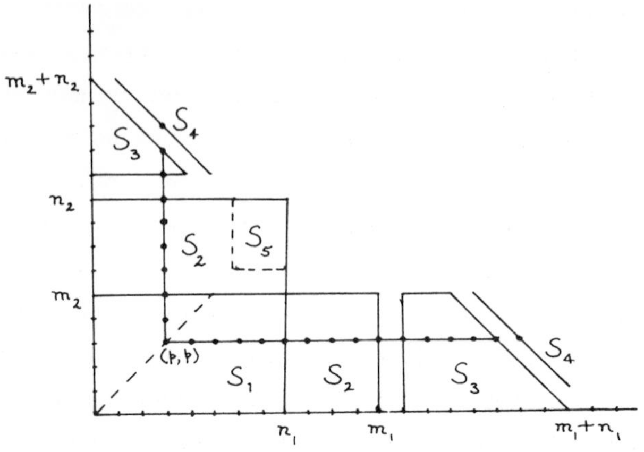

Fig. 7 Parameter regions: 2-variable G.O.D.

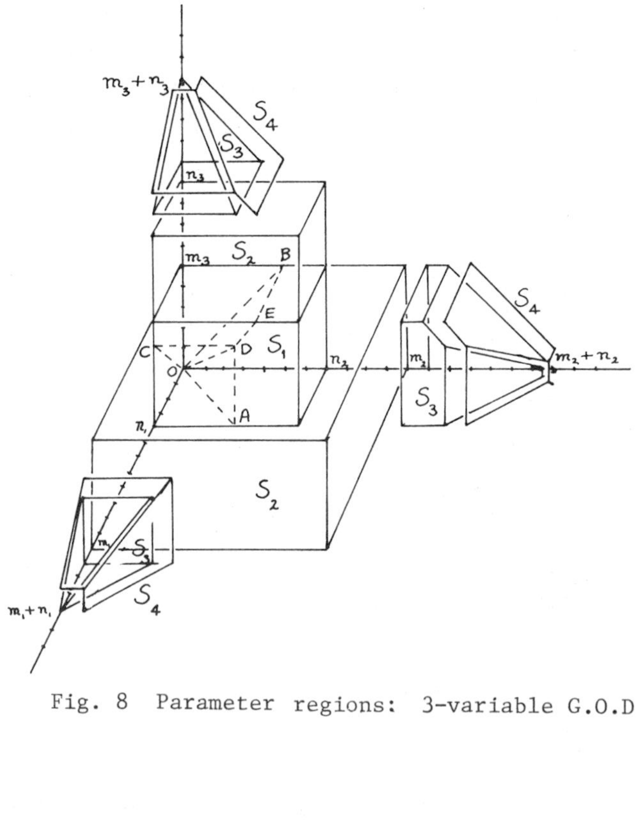

Fig. 8 Parameter regions: 3-variable G.O.D.

In diagonal and S.O.D. approximants, the weighting factors in symmetrized equations are unity for symmetrical series. But for series with no symmetry, the weighting factors in all of the above approximants are undetermined by properties (i)-(iv) and (v)′. Two suggestions have been made for determining the weighting factors, both giving symmetrical diagonal and S.O.D. approximants for symmetrical series. Graves-Morris suggested [10] that the determinant of the system of linear equations should be maximized subject to a normalizing condition; for the 2-variable equation (15), this condition is

$$(21) \quad |\mu_{\lambda;1}|^2 + |\mu_{\lambda;2}|^2 = 1.$$

This postulate determines all the weighting factors; the choice of weights is called "PETCH", and ensures maximum numerical advantage. The covariance group of diagonal approximants is the restricted homographic group, given by (2) with r=1,2,...,N, and is characterized by six real parameters.

The alternative suggestion [3] by Chisholm and Hughes Jones for determining the weights is to impose relative scale covariance on the approximants; this condition, known as "SCINCH", also determines the weights uniquely, and ensures the condition (v) of full homographic covariance (an 8-parameter group for two variables). The completion of the original list of properties seems to be very satisfactory, but this choice has a disadvantage: for real series, the weights are exactly the inverses of those giving maximum numerical advantage. So by ensuring relative scale covariance, one risks numerical problems in solving the equations.

The approximants described above are not covariant under linear transformations of the variables $\{z_r;\ r=1,2,\ldots,N\}$. For double series, Chisholm and Roberts [4] have shown that it is possible to define approximants which are covariant under transformations which are rotations in the (z_1,z_2) plane, for real

z_1, z_2. The homographic covariance group of diagonal Chisholm-Roberts approximants is given by (1) with $|A_1| = |A_2|$, intermediate between (1) and (2). The relative scale transformation group (3) is therefore restricted to changes of relative phase. The complete covariance group is characterized by eight real parameters, as is the covariance group (1) of Chisholm approximants with the scale covariant choice of weights.

Rotationally covariant approximants are defined by introducing the real matrices

$$(22) \quad I = \begin{bmatrix} 1 & 0 \\ 0 & 1 \end{bmatrix}, \quad J = \begin{bmatrix} 0 & -1 \\ 1 & 0 \end{bmatrix},$$

isomorphic to (1,i), and JI-numbers

$$(23) \quad z = z_1 I + z_2 J, \quad \bar{z} = z_1 I - z_2 J.$$

The original series (6), multiplied by I, is expressed as a series in $z, \bar{z}$; then diagonal and S.O.D. approximants are defined as before. One advantage of Chisholm-Roberts approximants is that the two choices of weights PETCH and SCINCH are now the same; so ensuring relative phase invariance gives maximum numerical advantage. So far, only 2-variable rotationally covariant approximants have been defined. To define analogous N-variable approximants would require the use of a real representation of the N-dimensional rotation group.

A number of examples have been studied using the various 2-variable approximants [11,14,15]. All of the approximants are capable of representing singularities of functions and of analytically continuing outside the polycylinder of convergence; their effectiveness in numerical approximation depends on the function studied. Three types of approximant are, in general, of roughly equal efficiency:

(a) Diagonal, S.O.D. and G.O.D. approximants (equal weights);

(b) Diagonal, S.O.D. and G.O.D. approximants (PETCH weights);

(c) Chisholm-Roberts approximants.

Not quite so efficient generally is the fourth type

(d) Diagonal, S.O.D. and G.O.D. approximants (SCINCH weights).

It is very encouraging that all of these approximants, defined to satisfy specific mathematical properties, do in fact approximate! However, we must remember that the effectiveness of Padé approximants has depended to a considerable extent on the definition of generalizations suited to the solution of particular problems. I see the Chisholm approximants and their generalizations in the same light, as a basic structure which will have some immediate applications, but which will need adaptation and further generalization in order to be effective in treating other problems. Fisher's recent work [6] on the calculation of certain types of 2-variable functions arising in critical phenomena suggests one approach to generalizing 2-variable approximants. Since the number of 1-variable generalizations of Padé approximants is already quite large, we can anticipate a great variety of generalizations of 2-variable and N-variable approximants.

References

1 Chisholm, J.S.R., Rational approximants defined from double power series, Math. Comp. 27, (1973), 841.

2 Chisholm, J.S.R. and J. McEwan, Rational approximants defined from power series in N variables, Proc. Roy. Soc. A336, (1974), 421.

3 Chisholm, J.S.R. and R. Hughes Jones, Relative scale covariance of N-variable approximants, Proc. Roy. Soc. A344, (1975), 365.

4 Chisholm, J.S.R. and D.E. Roberts, Proc. Roy. Soc. A351, (1976), 585.

5 Common, A.K. and P.R. Graves-Morris, Some properties of Chisholm approximants, J. Inst. Math. Appl. 13, (1974), 229.

6 Fisher, M.E., Series expansion approximants for singular functions of many variables, Cornell University preprint.

7 Gammel, J.L., private communication.

8 Graves-Morris, P.R., R. Hughes Jones and G.J. Makinson, The calculation of some rational approximants in two variables, J. Inst. Math. Appl. 13, (1974), 311.

9 Graves-Morris, P.R., R. Hughes Jones and G.J. Makinson, A procedure for successive calculation of Chisholm approximants, Comp. J. 18, (1975), 81.

10 Graves-Morris, P.R. and R. Hughes Jones, An analysis of two variable rational approximants, J. Comp. App. Maths. 2, (1976), 41.

11 Graves-Morris, P.R. and D.E. Roberts, A subroutine and procedure for the rapid calculation of simple off-diagonal rational approximants, Comp. Phys. Comm. 9, (1975), 46.

12 Hughes Jones, R. and G.J. Makinson, The generation of Chisholm rational polynomial approximants to power series in two variables, J. Inst. Math. Appl. 13, (1974), 299.

13 Hughes Jones, R., General rational approximants in N variables, J. Approx. Theory 16, (1976), 3.

14 Roberts, D.E., D.W. Wood and H.P. Griffiths, The analysis of double power series using Canterbury approximants, J. Phys. A 8, (1975), 9.

15 Roberts, D.E., An analysis of double power series using rotationally covariant approximants, submitted for publication (1977).

J.S.R. Chisholm
Mathematics Institute
University of Kent
Canterbury
England

ASYMPTOTIC BEHAVIOR OF THE ZEROS OF SEQUENCES OF PADÉ POLYNOMIALS

Albert Edrei

In this note we present results on the behavior of the zeros of Padé polynomials which extend the classical results of Jentzsch, Szegö, Carlson, and Rosenbloom on the zeros of partial sums of a power series.

1 Introduction

Let

$$(1)\quad f(z) = \sum_{m=0}^{\infty} a_m z^m \qquad (a_0 \neq 0),$$

have a radius of convergence σ_0 $(0 < \sigma_0 \leq +\infty)$.

The entries of the Padé table of (1) are ratios of polynomials which may be represented explicitly in terms of the Hankei determinants introduced below.

Let (m,n) be a pair of nonnegative integers; put

$$a_{-j} = 0 \qquad (j = 1, 2, 3, \ldots)$$

and consider the determinants

$$A_m^{(n)} = \begin{vmatrix} a_m & a_{m-1} & \cdots & a_{m-n+1} \\ a_{m+1} & a_m & \cdots & a_{m-n+2} \\ \cdots & \cdots & \cdots & \cdots \\ a_{m+n-1} & a_{m+n-2} & \cdots & a_m \end{vmatrix}, \quad A_m^{(0)} = 1,$$

and the polynomials

$$D_{mn}(z) = \begin{vmatrix} 1 & z & z^2 & \cdots & z^n \\ a_{m+1} & a_m & a_{m-1} & \cdots & a_{m-n+1} \\ a_{m+2} & a_{m+1} & a_m & \cdots & a_{m-n+2} \\ \cdots & \cdots & \cdots & \cdots & \cdots \\ a_{m+n} & a_{m+n-1} & a_{m+n-2} & \cdots & a_m \end{vmatrix}, \quad D_{m0}(z) \equiv 1.$$

Whenever $A_m^{(n)} \neq 0$, we introduce the normalized Padé denominator of the entry (m,n) of the table of (1):

$$Q_{mn}(z) = \frac{D_{mn}(z)}{A_m^{(n)}} = 1 + q_1(m,n)z + \ldots + q_n(m,n)z^n.$$

The corresponding, normalized, Padé numerator is given by

$$P_{mn}(z) = \sum_{j=0}^{m} p_j(m,n)z^j,$$

with

$$p_j(m,n) = \frac{1}{A_m^{(n)}} \begin{vmatrix} a_j & a_{j-1} & a_{j-2} & \cdots & a_{j-n} \\ a_{m+1} & a_m & a_{m-1} & \cdots & a_{m-n+1} \\ a_{m+2} & a_{m+1} & a_m & \cdots & a_{m-n+2} \\ \cdots & \cdots & \cdots & \cdots & \cdots \\ a_{m+n} & a_{m+n-1} & a_{m+n-2} & \cdots & a_m \end{vmatrix}$$

The constant term and leading coefficient of $P_{mn}(z)$ play a dominant role in this investigation and their explicit values:

$$p_0(m,n) = a_0, \qquad p_m(m,n) = \frac{A_m^{(n+1)}}{A_m^{(n)}},$$

are important.

It is convenient to express the fundamental property of the approximant P_{mn}/Q_{mn} in terms of a Cauchy integral involving an arbitrary auxiliary polynomial of suitable degree: if $V_{mn}(z)$ is a given polynomial of degree $\leq n$, we have

$$(2)\quad \{f(z)Q_{mn}(z)-P_{mn}(z)\}V_{mn}(z) = \frac{z^{m+n+1}}{2\pi i}\int_C \frac{f(\xi)Q_{mn}(\xi)V_{mn}(\xi)}{\xi^{m+n+1}(\xi-z)}\,d\xi,$$

where we may take for contour C the circumference $|\xi| = r$ $(0 < r < \sigma_0)$ described in the positive sense. [A simple deduction of this elementary formula will be found in [3; p. 436].]

The representation (2) readily yields precise upper bounds for $|P_{mn}(z)|$ in disks which may contain poles of f(z). It is used for this purpose in my proofs.

The Padé polynomials P_{mn}, Q_{mn} are obvious generalizations of the notion of partial sum (or section) of the power series (1). One may expect that the distribution of the zeros of the Padé polynomials will be described by theorems analogous to the classical results of Jentzsch, Szegö, Carlson and Rosenbloom.

2 Main Results

My first result (Theorem 1 below) shows that, if $\sigma_0 < +\infty$, that is if the series (1) has a finite radius of convergence, the analogy with the theorems of Jentzsch and Szegö is complete.

THEOREM 1. Let f(z), defined by (1), be a non-rational function whose radius of meromorphy is τ $(\sigma_0 \leq \tau \leq +\infty)$. This means that f(z) is meromorphic in the disk $|z| < \tau$ and also that, if $\tau < +\infty$, there is a non-polar singularity of f(z) on the circumference $|z| = \tau$.

Let $n \geq 0$ be a given integer and let σ_n $(\leq +\infty)$ be the radius of the largest disk $|z| < \sigma_n$ which contains no more than n poles of f(z) and none of its non-polar singularities.

Then, if $\sigma_n < +\infty$, it is possible to find an unbounded sequence S(n), of strictly increasing positive integers, which behaves as follows.

I. $A_m^{(n)} A_m^{(n+1)} \neq 0 \quad (m \in S(n))$.

II. *Consider the normalized Padé numerator* $P_{mn}(z)$. *Given* ε $(0 < \varepsilon < 1)$ *and* $m \in S(n)$, *there are, in the annulus*

$$(1-\varepsilon)\sigma_n \le |z| \le (1+\varepsilon)\sigma_n,$$

$m(1-\eta_m)$ *zeros of* $P_{mn}(z)$, *where*

$$0 < \eta_m, \qquad \eta_m \to 0 \qquad (m \to \infty).$$

III. *If* $N(m;\phi_1,\phi_2)$ *denotes the number of zeros of* $P_{mn}(z)$ *whose arguments lie in the interval*

$$(3) \quad \phi_1 \le \arg z \le \phi_2 \qquad (\phi_1 < \phi_2 < \phi_1 + 2\pi),$$

we have

$$\frac{N(m;\phi_1,\phi_2)}{m} \to \frac{\phi_2 - \phi_1}{2\pi} \qquad (m \to \infty,\ m \in S(n)).$$

For $n = 0$, Theorem 1 coincides with Szegö's remarkably precise form of Jentzsch's theorem [8].

If $f(z)$ is entire, the preceding result does not apply because $\sigma_n = +\infty$ for all n. On the other hand, the results of Carlson [1] and Rosenbloom [6] suggest very specific conjectures concerning the behavior of the zeros of the Padé polynomials associated with entire functions of finite, positive order.

I have recently obtained the following

THEOREM 2. *Let* (1) *represent an entire function of order* λ $(0 < \lambda < +\infty)$ *and mean type* ω $(0 < \omega < +\infty)$. *Then, with every integer* $n \ge 0$, *it is possible to associate an unbounded sequence* $S(n)$ *of strictly increasing, positive integers which behaves as follows.*

I. *As*

$$(4) \quad m \to \infty, \qquad m \in S(n),$$

we have

$$(5) \quad m^{1/\lambda}\left|\frac{A_m^{(n+1)}}{A_m^{(n)}}\right|^{1/m} \longrightarrow (\lambda e\omega)^{1/\lambda}.$$

II. Put

$$R_m = \left(\frac{m}{\omega\lambda}\right)^{1/\lambda}.$$

Given $\varepsilon > 0$, there are, under the conditions (4), no more than $o(m)$ zeros of $P_{mn}(z)$ in the region

$$|z| \geq (1+\varepsilon)R_m.$$

III. Let ϕ_1 and ϕ_2 be given as in (3), and let

$$\Psi = \phi_2 - \phi_1.$$

It is then possible to determine two positive constants $K_0 = K_0\ (\lambda,\omega,\Psi)$ and $\mu = \mu\ (\lambda,\omega,\Psi)$ (both independent of m) such that, for $m > m_0$, $m \in S(n)$, there are at least μm zeros of $P_{mn}(z)$ in the sector

$$\Delta = \{z : |z| \geq K_0 R_m,\ \phi_1 \leq \arg z \leq \phi_2\}.$$

The condition that $f(z)$ is of mean type may be relaxed. At the cost of slight complications, it is possible to restate Theorem 2 so as to cover all entire functions of mean type of the Lindelöf order $(\lambda, \lambda_1, \lambda_2,\ldots,\lambda_k)$ $(\lambda > 0)$. This means that the maximum modulus

$$M(r) = \max_{|z| = r} |f(z)|,$$

satisfies the condition

$$\limsup_{r\to\infty} \frac{\log M(r)}{r^{\lambda}(\log r)^{\lambda_1}(\log_2 r)^{\lambda_2}\cdots(\log_k r)^{\lambda_k}} = \omega \quad (0<\omega<+\infty),$$

where $\log_j r$ indicates an iterated logarithm; the quantities

$$\lambda_1, \lambda_2, \ldots, \lambda_k$$

are real (not necessarily positive) and $\lambda > 0$. The above generalization permits the treatment of functions such as $1/\Gamma(z)$ (not covered by Theorem 2).

I hope that further study will enable me to extend Theorem 2 to:

(i) functions of infinite order;

(ii) all entire functions of positive order without any reference to type.

It is possible to assert a good deal more about the zeros of the Padé polynomials if one is prepared to assume more about $f(z)$. Results which are interesting and suggestive may be derived from the close study of some special choices of $f(z)$. Among them, the choice

$$f(z) = e^z$$

stands out as particularly important: it was the foundation of Padé's original work and, quite recently it has led Saff and Varga [7] to results which are a model of elegance and precision.

My proofs of Theorems 1 and 2 are too long to be included here and I shall content myself with some brief indications concerning the methods and auxiliary results which lead to the above statements.

As might be expected, my proof of Theorem 1 relies heavily on Hadamard's results on polar singularities [2; pp. 329-335].

The results of Hadamard do not suffice for my purpose. This may be seen in the following important, simple case.

Assume that $f(z)$ is regular in the unit disk and has a nonpolar singularity on the circumference $|z| = 1$. Under these hypotheses, Hadamard's theorem asserts nothing more than

$$\limsup_{m\to\infty} |A_m^{(n)}|^{1/m} = 1 \quad (n = 1,2,3,\ldots).$$

On the other hand, the proof of Theorem 1 requires the existence of an infinite sequence S(n) such that

$$(6) \quad \left|\frac{A_m^{(n+1)}}{A_m^{(n)}}\right|^{1/m} \to 1 \quad (m \to \infty,\ m \in S(n)).$$

The proof of (6) is not entirely trivial; it may have some independent interest because ratios of Hankel determinants, such as those in (6), appear in many questions concerning the Padé table.

In the final step of my proof of Theorem 1, I use an important result of Erdös and Turán [4; p. 106, Theorem 1]. It is also possible to use simpler and less precise results than the Erdös-Turán theorem, in particular Rosenbloom's Theorem XIII [6; p.25]. Since Rosenbloom's remarkable thesis was never published some readers may find my reference [4] more readily accessible.

The proof of Theorem 2 has the same structure as the proof of Theorem 1. It is, in general, more delicate and a new approach is needed to establish the limit relation (5) [(5) plays, in Theorem 2, the same role as (6) in Theorem 1]. To complete the proof of Theorem 2, I require the results of Carlson [1]. The simplest published result, embodying Carlson's main idea, and directly applicable in my proof, is due to Ganelius [5; Theorem, p. 39].

References

1 Carlson, F., Sur les fonctions entières, Arkiv för mat., astr., och fys., 35A (1948), 1-18.

2 Dienes, P., The Taylor Series, Oxford, 1931.

3 Edrei, A., The Padé table of functions having a finite number of essential singularities, Pacific J. Math., 56 (1975), 429 - 453.

4 Erdös, P. and P. Turán, On the distribution of roots of polynomials, Annals of Math., 51 (1950), 105-119.

5 Ganelius, T., Sequences of analytic functions and their zeros, Arkiv för Matematik, 3 (1954), 1-50.

6 Rosenbloom, P., Sequences of polynomials, especially sections of power series, Thesis, Stanford, 1943.

7 Saff, E.B., and R.S. Varga, On the zeros and poles of Padé approximants to e^z, Numer. Math. 25 (1975), 1-14.

8 Szegö, G., Über die Nullstellen von Polynomen die in einem Kreise gleichmässig konvergieren, Sitzungsber, Berlin Math. Gesellschaft, 21 (1922), 59-64.

Albert Edrei
Department of Mathematics
Syracuse University
Syracuse, New York

The research of the author was supported by a grant from the National Science Foundation MCS 72-04539 A04.

Added March 19, 1977

The hope, expressed above, that it should be possible to eliminate from Theorem 2 references to type and restrictions about the finiteness of the order, has now been realized.

With obvious modifications of the relation (5), and of the definition of R_m, Theorem 2 holds unrestrictedly for all functions of positive order, finite or $+\infty$.

CONFIRMING THE ACCURACY OF PADÉ TABLE APPROXIMANTS

Carl H. FitzGerald

The objective is to estimate values of a function f using limited information. It is supposed that bounds on $|f|$, certain analyticity properties of f, and approximate values of the first few terms of a series expansion for f are known. Estimates of the function at a point are found by averaging modified Padé approximants on circles centered at that point. The convergence and stability of these estimates are studied. The results suggest using these averages to confirm the values obtained by direct evaluation of the Padé approximants.

1 Introduction

During the past decade it has been shown that under quite general hypotheses the Padé approximants are arbitrarily accurate approximations on all except arbitrarily small sets. The first such results involved assumptions on the type of singularities of the function being approximated, on the size of the set of its singularities, and on its Padé approximants [2,4,6]. Nuttall [8] showed that it is not necessary to have a hypothesis directly on the approximants. Subsequently, Pommerenke [10] strengthened Nuttall's theorem by eliminating the restriction on the type of singularities allowed. A different presentation of all these results can be found in [1]. Pommerenke's formulation of the result and an indication of the proof are presented in the next section.

For a near-diagonal approximant of large degree, the set on which it may be inaccurate is small, but the union of these exceptional sets can cover a large area. In fact Wallin [13] has shown that there is an entire function such that its diagonal Padé approximants about the origin are unbounded everywhere except the origin. The difficulty of the exceptional sets is overcome here in an indirect way.

It will be shown how the Nuttall-Pommerenke Theorem can be used to obtain values of the function everywhere. The technique is to modify the Padé approximants. The new approximations converge to the function uniformly on compact subsets of the domain of the function. Furthermore, slight changes in the coefficients of the expansion of the function do not appreciably change the values of the new approximations. The proof of this stability requires an examination of a proof for the Nuttall-Pommerenke Theorem to determine how the exceptional sets arise.

2 Definitions and Background

The function to be approximated is assumed to be analytic in a neighborhood of infinity. Consequently it has an expansion there of the form $f(z) = a_0 + a_{-1}z^{-1} + \ldots$. The Padé approximants of f are denoted by $P_{mn}[f(z)]$ where m and n are nonnegative integers. Specifically, $P_{mn}[f(z)]$ is the unique rational function $p_{mn}(z)/q_{mn}(z)$ such that $q_{mn}(z)f(z) - p_{mn}(x)$ $O(1/z^{m+n+1})$ as z tends to infinity. The functions $p_{mn}(z)$ and $q_{mn}(z)$ are polynomials in $1/z$ with the degree of $p_{mn} \le n$ and the degree of $q_{mn} \le m$ and $q_{mn}(z)$ not identically zero. Discussions of this definition can be found in [9,12].

The various proofs that these approximants tend to f involve some lemma showing that a polynomial is small only on a small set. The version originally due to Cartan can be found in [11]. The following one is Pommerenke's [10].

LEMMA 2.1 *Let* r *and* ϵ *be positive numbers with* $\epsilon < 1/3$. *Consider a polynomial* $g(z)$ *of degree less than or equal* m *such that* $\max|g(z)| \ge 1$ *where the maximum is taken over the circle* $\{z : |z| = r\}$. *If* $S = \{z : |z| \le r$ *and* $|g(z)| \le \epsilon^m\}$, then the *capacity of* S *is less than or equal* $3r\epsilon$.

The notion of capacity will be used freely in the following. An adequate discussion of the concept can be found in [7]. In particular, if $\{E_k\}$ is a sequence of compact sets in the plane such that as k tends to infinity cap E_k tends to zero, then the Lebesgue measure of the area of E_k also tends to zero. Also, for such a sequence $\{E_k\}$, if C is a fixed circle, then the Lebesgue measure on the circle of $E_k \cap C$ tends to zero.

THEOREM 2.1 (Nuttall, Pommerenke [8,10]) Consider a (single-valued) function $f(z)$ that is analytic on a domain in the extended complex plane. Suppose that the complement E of the domain of f is a set in the finite plane and has zero capacity. Let r and λ be numbers greater than 1. If $\epsilon > 0$ and $\eta > 0$, there exists m_0 such that $|P_{mn}[f(z)] - f(z)| < \epsilon^m$ for all integers m and n for which $m > m_0$ and $1/\lambda \leq n/m \leq \lambda$ and for all z such that $|z| \leq r$ and $z \notin E_{mn}$ where E_{mn} is a set of capacity $\leq \eta$.

A sketch of a proof is presented. The objective is to indicate the origin of the estimate on the size of the exceptional set E_{mn}. Only the case when $m \leq n \leq \lambda m$ will be considered.

The number r can be assumed greater than 4 and large enough that $E \subset \{z : |z| < r - 4\}$. By picking ϵ still smaller if necessary, it can be assumed that $0 < \epsilon < 1/3$ and $3r\epsilon^{1/3} < \eta$. Since cap $E = 0$, there exists a positive integer k and a polynomial $h(z)$ of the form $h(z) = z^k + \ldots$ of degree k such that E is separated from infinity by the system of curves $\Gamma \equiv \{z : z \in C$ and $|h(z)| = (\epsilon^{\lambda+4}/r^{\lambda})^k\}$. (If Γ contains a zero of $h'(z)$, the curves may not be simple. By making ϵ still smaller, the curves of Γ can be assumed to be simple.) Orient the curves in the usual sense for the boundary of a domain including infinity.

Let $E_0 \equiv \{z : |h(z)| \leq \epsilon^k\}$. Note that $E \subset E_0 \subset \{z : |z| < r\}$. Let $P_{mn}[f(z)]$ have the representation $p_{mn}(z)/q_{mn}(z)$. As a normalization condition require that the maximum of $|z^m q_{mn}(z)|$ on $\{z : |z| = r\}$ be one. Define $g_{mn}(z) \equiv z^m q_{mn}(z)$. Let s be the greatest integer in m/k, so that $m - k < s\,k \leq m$.

By examining the analyticity of $\zeta^n h(\zeta)^s [q_{mn}(\zeta) f(\zeta) - P_{mn}(\zeta)]$ and its behavior at infinity, the Cauchy integral formula can be shown to apply. If z is in the infinite component bounded by Γ, then

$$z^n h(z)^s [q_{mn}(z) f(z) - p_{mn}(z)] = \frac{1}{2\pi i} \int_\Gamma \frac{\zeta^{n-m} h(\zeta)^s g_{mn}(\zeta) f(\zeta)}{z-\zeta} \, d\zeta .$$

Estimation of the integral gives

$$(2.1) \quad |z^n h(z)^s [q_{mn}(z) f(z) - p_{mn}(z)]| \le M\, r^{\lambda m} (\epsilon^{\lambda+4}/r^{\lambda})^{ks}$$
$$< M\, \epsilon^{(\lambda+4)(m-k)} .$$

To bound $z^n h(z)^s q_{mn}(z)$ away from zero, it is required that i) $|z| \ge \epsilon$, ii) z be away from Γ, specifically $z \notin E_0$, iii) z not be in the set where $g_{mn}(z)$ has small magnitude. By Lemma 2.1, the last requirement involves only a small set. If z satisfies i), ii) and iii), then $|z^n h(z)^s q_{mn}(z)| = |z^{n-m} h(z)^s g_{mn}(z)| > \epsilon^{(\lambda-1)m}\, \epsilon^{sk}\, \epsilon^m \ge \epsilon^{(\lambda+1)m}$. Dividing into inequality (2.1) gives

$$|f(z) - P_{mn}[f(z)]| < M\, \epsilon^{(\lambda+4)(m-k)-(\lambda+1)m} < M\, \epsilon^{3m-4} .$$

There exists some m_0 such that if $m > m_0$ then $|f(z) - P_{mn}[f(z)]| < \epsilon^m$.

(Note that if $n = m$, the requirement that z be away from the origin is not necessary.)

The principal point is that the estimate of the size of the exceptional set is determined by the tolerance ϵ, and the bound on $|f(z)|$ on Γ. A small change in the coefficients of the expansion to the coefficients of a function having the same set of singularities as f and satisfying the same bound of $|f(z)|$ for z on Γ would not change the *size* of the exceptional set although the location might be substantially altered.

3 Estimation Using Padé Approximation

Consider a function $f(z)$ as described in the Nuttall-

Pommerenke Theorem 2.1. Let the same notation be used for the set E of singularities of f and the Padé approximants.

Using Theorem 2.1 and its proof, it is possible to estimate f. Consider the following representation formula. If z_0 is a complex number and ρ is positive and $\{z : |z - z_0| < \rho\}$ is contained in the domain of analyticity of f, then

$$(3.1)\quad f(z_0) = \frac{1}{\pi\rho^2} \iint_{|z-z_0|\le\rho} f(z)\,dA_z$$

where the integral is with respect to area. Equation (3.1) follows from the Gauss mean value property for harmonic functions.

Thus to estimate $f(z_0)$, it suffices to estimate $f(z)$ on a disc centered at z_0. According to Theorem 2.1, the Padé approximants could be used on all but a small set. Since the values of the Padé approximants on the small sets could differ from those of $f(z)$ by an arbitrarily large amount and could contribute significantly to the areal integral, a modification is necessary.

In stipulating that the function f(z) is analytic in domain, it is to be understood that on each compact subset of the domain a bound B for $|f(z)|$ is determined [3]. Once this bound B is obtained, it is unreasonable to use $P_{mn}[f(z)]$ as an estimate of $f(z)$ where the Padé approximant has magnitude greater than 2B. It would be more accurate to use zero as the estimate at such points. This discussion motivates defining the truncated Padé approximants as follows:

$$\hat{P}_{mn}[f(z)] \equiv \begin{cases} 0 & \text{if } |P_{mn}[f(z)]| \ge 2B \\ P_{mn}[f(z)] & \text{otherwise.} \end{cases}$$

Then it is the truncated Padé approximants which should be used in the integral of Equation (3.1).

The correct order of picking the sets and numbers is now indicated. Specify the compact set K on which f is to be estimated. Of course K must be contained in the domain of analyticity of f. Let ρ be a positive number such that all points within ρ of K are in the domain of analyticity of f. Define $K_\rho \equiv \{z$: the distance from z to K does not exceed $\rho\}$. Then K_ρ is a compact subset of the domain of f, and there is a bound $B > 1$ of $|f(z)|$ on K_ρ. Fix $\lambda > 1$ and consider pairs m,n such that $m \leq n < \lambda m$.

Let $\frac{1}{3} > \epsilon > 0$ be the desired tolerance. By Theorem 2.1 for sufficiently large m, the area of the set for which $|P_{mn}[f(z)] - f(z)| > \epsilon/2$ is less than $\epsilon\pi\rho^2/(6B)$. Hence

$$\left|\frac{1}{\pi\rho^2}\iint_{|z-z_0|\leq\rho} f(z)dA_z - \frac{1}{\pi\rho^2}\iint_{|z-z_0|\leq\rho} \hat{P}_{mn}[f(z)]dA_z\right|$$

$$\leq \frac{1}{\pi\rho^2}(3B)\left(\frac{\epsilon}{2}\frac{1}{3B}\pi\rho^2\right) + \frac{1}{\pi\rho^2}\left(\frac{\epsilon}{2}\right)(\pi\rho^2) = \epsilon.$$

The first term after the inequality sign estimates the contribution on the small set where $P_{mn}[f(z)]$ differs from $f(z)$ by more than $\epsilon/2$. Even on that set $\hat{P}_{mn}[f(z)]$ cannot differ from $f(z)$ by more than $3B$.

Recalling equality (3.1),

$$\left|f(z_0) - \frac{1}{\pi\rho^2}\iint_{|z-z_0|\leq\rho} \hat{P}_{mn}[f(z)]dA_z\right| < \epsilon$$

uniformly for z_0 on K.

An examination of the proof shows a certain stability of the estimate. The location of the exceptional sets is now of no importance, it is necessary to know only that they are of small size. Only general information about f is required in the proof of the Nuttall-Pommerenke Theorem and the preceding discussion: specifically, that the set of singularities has capacity zero and is away from the set on which convergence is being proved, and a bound for $|f(z)|$ on the system of curves separating the singularities of f

from infinity. If the bound B on $|f|$ is chosen larger than necessary, then there are other functions satisfying the conditions for the estimates. Any coefficients arising from one of them would also have Padé approximants which when truncated and averaged would give an approximation to f. Considering such functions. it is clear that all of the coefficients in the series for f could be changed slightly without significantly altering the estimate obtained for $f(z)$. (Related observations concerning noise in the coefficients for a special case were made by Gammel [5].) The following statement has been proved.

THEOREM 3.1 Let f be a function as described in Theorem 2.1, and let $\lambda > 1$. The areal means of trancated Padé approximants having $m \leq n \leq \lambda m$ converge normally to $f(z)$. The approximations to f are stable under small changes in the coefficient expansion.

In place of the areal average of equality (3.1), the average over a circle can be taken:

$$(3.2)\quad f(z_0) = \frac{1}{2\pi\rho} \int_{|z-z_0|=\rho} f(z)\,|dz|$$

Again the truncated Padé approximants can be used. Theorem 2.1 shows that for sufficiently large m and n the Padé approximants are accurate estimates on all but a small portion of the circle. Also the same remarks concerning stability hold.

THEOREM 3.2 The previous theorem holds if areal means are replaced by circular means.

Other weightings on the circle would give values of f elsewhere within the circle or coefficients of the power series expansion of f about z_0. The same remarks about convergence and stability hold for any of these functionals.

4 Illustrative Example

Suppose $f(z)$ is known to be meromorphic in the extended plane with up to five poles in $\{z : 3 < |z| < 4\}$ with no two poles being closer than one unit. Suppose the sum of the magnitudes of the residues is less than three. And suppose the coefficients of the expansion about infinity are known accurately but not precisely. For z near infinity

$$f(z) \doteq 1 + 1/z^2 - 0.15/z^3 + 10/z^4 - 0.15/z^5 + 100.015/z^6 - 1.5/z^7 + 1{,}000.015/z^8 - 15.015/z^9 + 10{,}000.15/z^{10} + \ldots .$$

The problem is to estimate $f(0)$. Clearly $|f(z)| < 5$ on $\{z : |z| \leq 2\}$, but it will not be necessary to use this bound. Using the coefficients as given,

$$P_{3,4}[f(z)] = \frac{1 - 80.713/z + 1.8968/z^2 + 726.56/z^3 + 23.0038/z^4}{1 - 80.713/z + .8968/z^2 + 807.423/z^3}$$

$$P_{5,5}[f(z)] = \frac{1 - 9/z^2 - 1/(20z^3) + 9/(20z^5)}{1 - 10/z^2 + 1/10z^3 - 1/z^5} .$$

z:	1	i	-1	-i	0(averaging)
Re $P_{3,4}$:	.92196	.90896	.8988	.92196	.8988
Re $P_{5,5}$:	.76768	.89558	1.037	.89558	.8990
z:	2	2i	-2	-2i	0(averaging)
Re $P_{3,4}$:	.86374	.92850	.80124	.92850	.8805
Re $P_{5,5}$:	.8179	.92853	.8492	.92835	.88095

Averaging $P_{5,5}$ over eight points equally spaced about $\{z : |z| = \sqrt{2}\}$ gives $f(0) \doteq .90197$. Direct evaluation of $P_{5,5}$ at $z = 0$ gives $-.45$. This must be a poor estimate of f since at least some of the values of $P_{3,4}$ on $\{z : |z| = 1\}$

must be fairly accurate. The good agreement among the averaged values confirms that $f(0) \triangleq .89$.

The structure of the example becomes clear by factoring $P_{5,5}$

$$P_{5,5}[f(z)] = \frac{1 - 9/z^2}{1 - 10/z^2} \cdot \frac{1 - 1/20z^3}{1 + 1/10z^3} .$$

Thus the actual function might be the first factor, and the second factor the source of the noise. In that case .90 is the exact value of $f(0)$ and the averaging of $P_{3,4}$ and $P_{5,5}$ gave good estimates.

5 Conclusions

For functions satisfying the hypotheses of the Nuttall-Pommerenke Theorem, if one considers averages of the truncated Padé approximants, then (i) there is no exceptional set arising from the poles of the Padé approximants, (ii) the convergence is normal, (iii) the estimates are stable under small changes in the coefficients consistent with the bounds for $|f|$ and the geometry of the domain of analyticity of f. These observations suggest testing estimates of f at a point by averaging the truncated approximants on circles about the point of interest.

Acknowledgements: Helpful conversations with Professors Gragg and Henrici are gratefully acknowledged.

References

1. Baker, G.A., Jr., Essentials of Padé Approximants, Academic Press, New York, (1975).

2. Beardon, A.F., On the convergence of Padé approximants, J. Math. Anal. Appl. 21 (1968), 344-346.

3. Bishop, E., Foundations of Constructive Analysis, McGraw Hill, New York, (1967).

4. Chisholm, J.S.R., Approximation by sequences of Padé approximants in regions of meromorphy, J. Math. Phys 7 (1966), 39-44.

5. Gammel, J.L., Effects of random errors (noise) in the terms of a power series on the convergence of the Padé approximants, Padé Approximants edited by P.R. Graves-Morris, The Institute of Physics, London, (1973).

6. Gammel, J.L. and J. Nuttall, Convergence of Padé approximants to quasianalytic functions beyond natural boundaries, J. Math. Anal. Appl., 43 (1973), 694-696.

7. Goluzin, G.M. Geometric Theory of Functions of a Complex Variable. Trans. of Math. Mono. Vol. 26, Am. Math. Soc., (1969).

8. Nuttall, J. The convergence of Padé approximants of meromorphic functions, J. Math. Anal. Appl. 31 (1970), 147-153.

9. Perron, O. Die Lehre von den Kettenbrüchen, Chelsea Publ., New York, (1950).

10. Pommerenke, Ch., Padé approximants and convergence in capacity, J. Math. Anal. Appl. 41 (1973), 775-780.

11. Veech, W.A., A Second Course in Complex Analysis, W.A. Benjamin, New York, (1967).

12. Wall, H.S., Analytic Theory of Continued Fractions, Van Nostrand-Reinhold, Princeton, (1948).

13. Wallin, H., On the convergence theory of Padé approximants. Linear Operators and Approximation (Proc. Conf., Oberwolfach, 1971), pp. 461-469. Internat. Ser. Numer. Math., Vol. 20, Birkhauser, Basel, 1972. Math. Rev. 51, #10645.

Carl H. FitzGerald*
Department of Mathematics
University of California, San Diego
San Diego, California 92093

*Research supported in part by NSF grant MCS 76-07277.

LAURENT, FOURIER, AND CHEBYSHEV-PADÉ TABLES

William B. Gragg

In [6, 9] the Padé table of a formal power series was extended to (doubly infinite) formal Laurent series with a view toward algorithmic "near best" uniform rational approximation of functions on intervals. In this paper we summarize and extend the results of [9], and comment briefly on further extensions.

1 Laurent-Padé fractions

Let

$$\phi(\zeta) = \sum_{-\infty}^{\infty} c_k \zeta^k$$

be a complex formal Laurent series, and let m and n be non-negative integers. Put $\ell \equiv \max\{m, n\}$, and consider the two linear systems

$$L^{\pm}_{m,n} : \begin{cases} \alpha^{\pm}_i = 2 \sum_{j=0}^{i}{}' c_{\pm(i-j)} \beta^{\pm}_j, & 0 \le i \le \ell, \\ \sum_{j=0}^{n} c_{\pm(i-j)} \beta^{\pm}_j = 0, & m < i \le m+n, \end{cases}$$

for the determination of polynomials

$$p^{\pm}_{m,n}(\zeta) = \sum_0^{\ell} \alpha^{\pm}_i \zeta^i, \quad q^{\pm}_{m,n}(\zeta) = \sum_0^{n} \beta^{\pm}_j \zeta^j \ne 0.$$

Here "$'$" means that c_0 is to be replaced by $c_0/2$, and we put $\beta^{\pm}_j \equiv 0$ for $j > n$. In [9] it is shown that the rational functions, $r^{\pm}_{m,n} \equiv p^{\pm}_{m,n}/q^{\pm}_{m,n}$, determined by $L^{\pm}_{m,n}$ are unique, and also that

$$r_{m,n}(\zeta) \equiv \frac{1}{2}\left[r^{+}_{m,n}(\zeta) + r^{-}_{m,n}(\zeta^{-1})\right] \equiv \frac{p_{m,n}(\zeta)}{q_{m,n}(\zeta)}$$

with Laurent polynomials $p_{m,n}$, $q_{m,n}$ of degrees at most m and n, respectively. The rational function $r_{m,n}$ is the Laurent-Padé fraction of type (m, n) for ϕ; it depends on only the coefficients c_k with $|k| \le m + n$.

The construction is based on the <u>additive splitting</u>

$$\phi(\zeta) = \frac{1}{2}\left[\phi^{+}(\zeta) + \phi^{-}(\zeta^{-1})\right]$$

with formal <u>power</u> series

$$\phi^{\pm}(\zeta) \equiv c_0 + 2\sum_1^{\infty} c_{\pm k}\zeta^k.$$

If $m \geq n$ then $r^{\pm}_{m,n}$ is the classical Padé fraction for $\phi^{\pm}$. If $m < n$ the osculatory interpolation property of the Padé fraction is preserved by $r^{\pm}_{m,n}$; the additional coefficients in $p^{\pm}_{m,n}$ are used to insure that $\deg p_{m,n} \leq m$. When ϕ itself is a formal power series, $r_{m,n}$ <u>is</u> the classical Padé fraction $r^{P}_{m,n}$.

The c-<u>table</u> of ϕ consists of the <u>Laurent determinants</u>

$$c_{m,n} \equiv \det\left(c_{m+i-j}\right)^{n}_{i,j=1}, \quad |m| < \infty, \quad n \geq 0,$$

with $c_{m,0} \equiv 1$. The c-tables of $\phi(\zeta)$ and $\phi(\zeta^{-1})$ are related by reflection in the line $m = 0$. ϕ is <u>normal</u> if all $c_{m,n} \neq 0$. An interesting class of normal Laurent series is known [4]. The prototype,

$$\phi(\zeta) = \sum_{-\infty}^{\infty} I_k(a)\zeta^k = e^{a(\zeta+\zeta^{-1})/2},$$

is normal for real $a \neq 0$. I_k is the modified Bessel function of the first kind. See also [17]. The c-table of a normal ϕ can be constructed recursively, from the coefficients $c_{m,1} = c_m$, using the classical identity

$$c_{m-1,n}c_{m+1,n} + c_{m,n-1}c_{m,n+1} = c^2_{m,n}.$$

This identity also forms the basis for the qd- and $\pi\zeta$-algorithms [19, 13, 7], which now may be extended to negative indices m. The two extended tables are related by reflection in the line $m = 0$. Day [3] has given an explicit formula for $c_{0,n}$ when ϕ represents a rational function with <u>simple</u> zeros and poles. The corresponding result for $c_{m,n}$, with <u>multiplicities</u> permitted, as in [20] and the related work [18], will provide extensions of the Hadamard theory [11, 8].

As in the classical case, in general, sets of vanishing determinants $c_{m,n}$ occur in square blocks in the c-table. The block structure theorem in [7] also appears to extend, with appropriate modifications. When $c_{m,n} \neq 0$ the linear system $L^{+}_{m,n}$ provides determinant representations for $q^{+}_{m,n}$ and $p^{+}_{m,n}$. Likewise for $q^{-}_{m,n}$ and $p^{-}_{m,n}$ if $c_{-m,n} \neq 0$. If both these conditions are satisfied it follows that $r_{m,n}(\zeta)$, when measured in the Riemann metric, is a continuous function of the coefficients $\{c_k\}$.

The identities of Frobenius type, given in [7] for the Padé table, extend to the general case, provided the indices m involved are all nonnegative. In particular when ϕ is normal and we normalize all $q^{+}_{m,n}(0) \equiv 1$, then

$$p^{+}_{m+1,n}q^{+}_{m,n} - p^{+}_{m,n}q^{+}_{m+1,n} = (-1)^{n}2\,\rho_{m,n}\zeta^{m+n+1},$$
$$p^{+}_{m,n+1}q^{+}_{m,n} - p^{+}_{m,n}q^{+}_{m,n+1} = (-1)^{n}2\,\rho_{m,n}\zeta^{m+n+1},$$
$$p^{+}_{m+1,n+1}q^{+}_{m,n} - p^{+}_{m,n}q^{+}_{m+1,n+1} = (-1)^{n}2\,\rho_{m,n}\zeta^{m+n+1},$$

with $\rho_{m,n} \equiv c_{m+1,n+1}/c_{m,n}$. In special cases these can be used to prove convergence of the columns, rows, or diagonals of the Laurent-Padé table. Finally, Wynn's identity [21],

$$\frac{1}{r^{\pm}_{m+1,n} - r^{\pm}_{m,n}} + \frac{1}{r^{\pm}_{m-1,n} - r^{\pm}_{m,n}} = \frac{1}{r^{\pm}_{m,n+1} - r^{\pm}_{m,n}} + \frac{1}{r^{\pm}_{m,n-1} - r^{\pm}_{m,n}}$$

is valid for normal ϕ, with the external boundary conditions

$$r^{\pm}_{m,-1}(\zeta) \equiv \infty, \qquad r^{\pm}_{-1,n}(\zeta) + r^{\mp}_{0,n}(\zeta^{-1}) \equiv 0.$$

This permits the (pointwise) construction of the Laurent-Padé table, starting with the partial sums

$$r^{\pm}_{m,0}(\zeta) = c_0 + 2\sum_{1}^{m} c_{\pm k}\zeta^{k}.$$

2 Fourier and Chebyshev-Padé fractions

ϕ is Hermitian if $c_{-1} \equiv \overline{c}_k \equiv a_k + ib_k$. The partial sums of the real Fourier series

$$\tau(\theta) \equiv a_0 + 2\sum_1^\infty (a_k \cos k\theta + b_k \sin k\theta)$$

are then the Cauchy partial sums of $\phi(e^{i\theta})$, $p_{m,n}(e^{i\theta})$ and $q_{m,n}(e^{i\theta})$ are real trigonometric polynomials of degree at most m and n, respectively, and $q_{m,n}(e^{i\theta}) \geq 0$ for real θ. If $q_{m,n}(e^{i\theta})$ has no real zeros then, by construction, the Fourier expansion of the Fourier-Padé fraction of type (m, n) for τ, $r^F_{m,n}(\tau;\theta) \equiv r_{m,n}(\phi; e^{i\theta})$, agrees with τ through as many initial terms as possible.

For real θ we have $r^F_{m,n}(\theta) = \mathrm{Re}\left[r^+_{m,n}(e^{i\theta})\right]$. With the introduction of horizontal and vertical differences in the $r^\pm$-table there results a compact "code" for the pointwise evaluation of Fourier-Padé fractions for real θ:

```
for n = 0, 1, 2, ...
| h_n ← ∞,
| if n = 0 then
| | ζ = e^{iθ}, z ← 1; r_0 ← v_0 ← c_0;
| otherwise
| | z ← ζz; v_n ← 2c_n z;
| | r_n ← r_{n-1} + Re v_n;
| | for k ← n - 1, n - 2, ... , 1
| | | h_k ← 1/(1/h_k + 1/v_{k+1} - 1/v_k);
| | | v_k ← v_k + h_k - h_{k-1};
| | | r_k ← r_{k-1} + Re v_k;
| | h_0 ← 1/(1/h_0 + 1/v_1 - 1/v_0/2);
| | r_0 ← v_0 ← v_0 + Re h_0.
```

At the completion of the nth stage we have $r_k = r^F_{k,n-k}(\theta)$, $0 \leq k \leq n$. A related idea is in [22].

ϕ is symmetric if $c_{-1} \equiv c_k$. The partial sums of the Cheybshev series of the first kind,

$$f(z) = c_0 + 2\sum_1^\infty c_k T_k(z),$$

are then the Cauchy sums of $\phi(\zeta)$, where $z \equiv (\zeta + \zeta^{-1})/2$,

$\zeta \equiv z + \sqrt{z^2 - 1}$, and we take $|\zeta| > 1$ for $z \notin [-1, 1]$. The symmetry implies that the T-Padé fraction for f, $r^T_{m,n}(f; z) \equiv r_{m,n}(\phi; \zeta)$, is a rational function in z of type (m, n). If $r^T_{m,n}$ has no poles in $[-1, 1]$ then, by construction, its T-expansion agrees with f through as many initial terms as possible. The T-Padé fractions are distinct from Maehly fractions $r^M_{m,n}$ [14], which depend on c_k, $|k| \le m + 2n$, and do not possess the generalized interpolation property. They were mentioned in [6, 5] for $m \ge n - 1$. The extension to $m \le n - 2$ is due essentially to Clenshaw and Lord [2, 1], where interesting numerical results are also given. Our treatment [9] is somewhat more algebraic.

ϕ is skew-symmetric if $c_{-k} + c_k \equiv 0$. The Chebyshev series of the second kind,

$$g(z) = \sum_0^\infty c_{k+1} U_k(z),$$

is then related, in the Cauchy manner, to $\phi(\zeta)/(\zeta - \zeta^{-1})$. The skew-symmetry implies that the U-Padé fraction for g, $r^U_{m,n}(g; z) \equiv r_{m+1,n}(\phi; \zeta)/(\zeta - \zeta^{-1})$, is a rational function in z of type (m, n). If $r^U_{m,n}$ has no poles in $[-1, 1]$ then its U-expansion agrees with g through as many initial terms as possible. Minor notational changes occur if, as in the next section, we denote the U-coefficients by $\{c_k\}_0^\infty$ instead of $\{c_{k+1}\}_0^\infty$.

We can also work with Chebyshev series adjusted to any bounded interval. For instance, let

$$f(z) = \sum_0^\infty c_k(a) t_k(z; a)$$

be holomorphic on $[-a, a]$, with monic T-polynomials $t_k(z; a)$ adjusted to $[-a, a]$. When $a = 0$ this is the Maclaurin expansion of f. For $a > 0$ the function $f_a(z) \equiv f(az)$ has an ordinary T-expansion, with coeffieients $2(a/2)^k c_k(a)$, and we define $r^{T,a}_{m,n}(f; z) \equiv r^T_{m,n}(f_a; z/a)$. Let $c_{m,n}(a)$ be the Laurent determinant formed from the symmetric sequence $\{c_k(a)\}_{-\infty}^\infty$. Then

$a \to +0$ and $c_{m,n}(0) \neq 0$ imply that $r^{T,a}_{m,n}(z) \to r^{P}_{m,n}(z)$ in the Riemann metric. This uses very elementary determinant manipulations and the continuity of Laurent-Padé fractions when $c_{m,n}c_{-m,n} \neq 0$. Likewise for U-Padé fractions $r^{U,a}_{m,n}$ adjusted to $[-a, a]$. A study of the Chebyshev-Padé fractions for e^{-z} on $[0, a)$, as $a \to +\infty$, would seem to be of interest.

3 Convergence results

In special cases there are connections between certain sequences of polynomials $q^{+}_{m,n}$ and orthogonal polynomials. See [10] for the latter. Polynomials orthogonal on the circle and line now appear in the same unified context. In what follows ν and μ are bounded nondecreasing functions, with infinitely many points of increase in the respective intervals $[-\pi, \pi]$, $[-1, 1]$, and $\tau = \nu'$, $\omega = \mu'$, are their derivatives, which exist almost everywhere.

The $\{c_n\}_{-\infty}^{\infty}$ are trigonometric moments,

$$c_n = \frac{1}{2\pi}\int_{-\pi}^{\pi} e^{-in\theta} d\nu(\theta),$$

if and only if all $c_{0,n} > 0$. Then, taking $q^{+}_{0,n}(0) \equiv 1$, we have

$$q^{+}_{0,n}(\zeta) = \left(\frac{c_{0,n+1}}{c_{0,n}}\right)^{\frac{1}{2}} \psi^{*}_{n}(\zeta)$$

where $\{\psi_n\}_0^{\infty}$ are the polynomials orthonormal on $|\zeta| = 1$ with respect to $\tau/2\pi$ and, for any polynomial ψ of degree n, $\psi^{*}(\zeta) \equiv \zeta^{n}\overline{\psi}(\zeta^{-1})$. The zeros of ψ_n are in $|\zeta| < 1$, so $q_{0,n}(e^{i\theta}) > 0$ for real θ. Moreover, if $\log \tau$ is integrable, then $1/\psi^{*}_{n}(\zeta) \to \sqrt{\gamma_{\tau}(\zeta)}$ for $|\zeta| < 1$, where

$$\gamma_{\tau}(\zeta) \equiv \exp\left\{\frac{1}{2\pi}\int_{-\pi}^{\pi} \log \tau(\theta)\, \frac{e^{i\theta} + \zeta}{e^{i\theta} - \zeta}\, d\theta\right\}$$

is an analytic extension of the geometric mean $\gamma_{\tau}(0) > 0$ of τ. Refined results of this type, as in [10, 12], will be useful for establishing the convergence and asymptotic behavior of sequences $\{r_{m,n}(\zeta)\}_{n=0}^{\infty}$, both on $|\zeta| = 1$ and in annuli $\rho^{-1} < |\zeta| < \rho$.

The $\{c_n\}_0^{\infty}$ are Hausdorf moments,

$$c_n = \int_{-1}^{1} t^n d\mu(t),$$

if and only if all Hankel determinants $c_n^{(0)} \equiv (-1)^{n(n-1)/2} c_{n-1,n}$ are positive. Taking $q_{n-1,n}^{+}(0) \equiv 1$ we then have

$$q_{n-1,n}^{+}(\zeta) = \left(\frac{c_{n+1}^{(0)}}{c_n^{(0)}}\right)^{\frac{1}{2}} q_n^*(\zeta),$$

where $\{q_n\}_0^{\infty}$ are the polynomials orthonormal with respect to ω. The classical identity

$$\sqrt{2\pi}\left(1 + \frac{\psi_{2n}(0)}{\psi_{2n}^*(0)}\right)^{\frac{1}{2}} q_n(z) = \zeta^{-n}\psi_{2n}(\zeta) + \zeta^{n}\psi_{2n}(\zeta^{-1}),$$

with $\zeta = z + \sqrt{z^2 - 1}$, then relates the $\{q_n\}$ to polynomials $\{\psi_n\}$, where $\tau(\theta) \equiv \omega(\cos\theta)|\sin\theta|$. Let D be the domain of the z-plane obtained by deleting the real points $z = x$ with $|x| \geq 1$, and put

$$\rho(z) \equiv z/\left(1 + \sqrt{1 - z^2}\right), \quad G_\omega(z) \equiv \gamma_\tau(\rho(z)).$$

Then G_ω is real since τ is even, and $|\rho(z)| < 1$ for $z \in D$. The condition $G_\omega(0) = \gamma_\tau(0) > 0$ is equivalent with the integrability of $\log \omega(t)/\sqrt{1 - t^2}$, which will henceforth be assumed. We thus have

$$\frac{\left(1 + \sqrt{1 - z^2}\right)^n}{q_n^*(z)} \to \sqrt{2\pi G_\omega(z)}, \qquad z \in D,$$

together with the Frobenius identity

$$\frac{1}{2}\left[r_{n,n+1}^{+}(z) - r_{n-1,n}^{+}(z)\right] = \frac{q_{n+1}^*(0)}{q_n^*(0)} \; \frac{z^{2n}}{q_n^*(z) q_{n+1}^*(z)}.$$

The Hausdorf moments $\{c_n\}_0^\infty$ are the Maclaurin coefficients of

$$h(z) \equiv \int_{-1}^{1} \frac{d\mu(t)}{1 - zt} .$$

Since $r^-_{n,n+1} \equiv c_0$ we conclude, by majorization and the Padé property, that

$$h(z) - r^P_{n-1,n}(h;\ z) \sim \frac{2\pi}{\sqrt{1 - z^2}}\, G_\omega(z)[\rho(z)]^{2n}, \qquad z \in D.$$

The $\{c_n\}_0^\infty$ are also the T-coefficients for

$$f(z) \equiv \int_{-1}^{1} \frac{1 - t^2}{1 - 2zt + t^2}\, d\mu(t).$$

Note the partial fraction decomposition of the kernel, as a function of t, when $z = (\zeta + \zeta^{-1})/2$. Now $\zeta = z + \sqrt{z^2 - 1}$ and $|\zeta| > 1$ for $z \in D$, $z \neq \overline{z}$; in this case it follows that $|\rho(\zeta^{-1})| < |\rho(\zeta)|$. We may apply the above results to get, for $z \in D$,

$$\begin{aligned} f(z) - r^T_{n-1,n}(f;\ z) &\sim \frac{2\pi}{\sqrt{1 - \zeta^2}}\, G_\omega(\zeta)[\rho(\zeta)]^{2n}, & z \neq \overline{z}, \\ &\sim \mathrm{Re}\left\{\frac{4\pi}{\sqrt{1 - \zeta^2}}\, G_\omega(\zeta)[\rho(\zeta)]^{2n}\right\}, & z = \overline{z}. \end{aligned}$$

Finally, $\{c_n\}_0^\infty$ are the U-coefficients for

$$g(z) \equiv \int_{-1}^{1} \frac{d\mu(t)}{1 - 2zt + t^2} .$$

The replacement $c_{n+1} \leftarrow c_n$ means that we must replace the Frobenius identity by

$$\frac{1}{2}\left[r^+_{n+1,n+1}(\zeta) - r^+_{n,n}(\zeta)\right] = \frac{q^*_{n+1}(0)}{q^*_n(0)}\ \frac{\zeta^{2n+1}}{q^*_n(\zeta)\, q^*_{n+1}(\zeta)} .$$

There results, for $z \in D$,

$$g(z) - r^U_{n-1,n}(g;\ z) \sim \frac{2\pi\zeta}{(\zeta - \zeta^{-1})\sqrt{1 - \zeta^2}} G_\omega(\zeta)[\rho(\zeta)]^{2n}, \quad z \neq \bar{z}$$
$$\sim \frac{4\pi i}{\zeta - \bar{\zeta}} \operatorname{Im}\left\{\frac{\zeta}{\sqrt{1 - \zeta^2}} G_\omega(\zeta)[\rho(\zeta)]^{2n}\right\}, \quad z = \bar{z}.$$

Consequently, for $z \in D$,

$$|h(z) - r^P_{n-1,n}(h;\ z)|^{1/2n} \to |\rho(z)|,$$

$$|f(z) - r^T_{n-1,n}(f;\ z)|^{1/2n} \to |\rho(\zeta)|, \qquad z \neq \bar{z},$$

and likewise for g. For $z = x \in (-1, 1)$ a comparison of the convergence factors $|\rho(x)|$ and $|\rho(\zeta)|$ shows that the former is smaller for $x^2 < 2/(1 + \sqrt{5})$ and larger outside this subinterval.

If we adjust the T-series for f to $[-a, a]$, $0 < a < 1$, then we must consider $f(az)$, which can be written

$$f(az) = \int_{-\alpha}^{\alpha} \frac{1 - t^2}{1 - 2zt + t^2}\, d\mu_a(t),$$

with $\alpha = \rho(a)$ and a related measure μ_a. We have $\alpha < 1$ so the asymptotic theory cannot be applied directly. However, we may set $s = t/\alpha$, use the partial fraction decomposition in s, and then replace z by z/a, to get the corresponding results. In particular,

$$\overline{\lim}\ |f(z) - r^{T,a}_{n-1,n}(f;\ z)|^{1/2n} \leq \left|\rho\left(\frac{z + \sqrt{z^2 - a^2}}{1 + \sqrt{1 - a^2}}\right)\right|, \qquad z \in D.$$

The right side tends to $|\rho(z)|$ as $a \to 0$. Likewise for the adjusted U-Padé fractions for g.

Meinardus [15] has conjectured that, for best uniform approximations $r^*_{m,n}$ on $[-1, 1]$,

$$\|e^x - r^*_{m,n}\|_\infty \sim \varkappa_{m,n} \equiv \frac{m!n!}{2^{m+n}(m+n)!(m+n+1)!}, \qquad m + n \to \infty.$$

The T-coefficients for e^z are trigonometric moments,

$$c_n = \frac{1}{2\pi}\int_{-\pi}^{\pi} e^{-in\theta}e^{\cos\theta}d\theta = I_n(1) \sim \frac{1}{2^n n!}, \qquad n \to \infty.$$

The T-Padé fractions may be used, as Meinardus used the Maehly fractions, to prove the conjecture for $n = 1$, $m \to \infty$. The asymptotic theory can presumably be applied to study $\|e^x - r^T_{0,n}\|_\infty$ similarly. Numerical results reported in [9] indicate that

$$\|e^x - r^T_{n,n}\|_\infty / \varkappa_{n,n} \to \sqrt{e}.$$

In this connection we conjecture that

$$c_{n,n}/e_{n,n}(1/2) \to e^{1/8},$$

where $e_{m,n}(\gamma) = \gamma^{mn} e_{m,n}(1)$ are the well known Laurent determinants for $e^{\gamma z}$; see [7]. Finally we note, with Newman [16], that e^z has a multiplicative splitting: $e^z = e^{\zeta/2} e^{\zeta^{-1}/2}$. Hence we may put $r^N_{m,n}(z) \equiv r^P_{m,n}(\zeta/2) r^P_{m,n}(\zeta^{-1}/2)$, with $r^P_{m,n}$ the Padé fraction for e^z. The known asymptotic behavior of $r^P_{n,n}$ can then be used to show that

$$\|e^x - r^N_{n,n}\|_\infty / \varkappa_{n,n} \to e.$$

The multiplicative splitting idea has obvious extensions to Laurent series ϕ. The practical difficulty is that of factoring ϕ into a product of ascending and descending power series.

References

1 Clenshaw, C.W. and K. Lord, Rational approximations from Chebyshev series, Studies in Numerical Analysis (B.K.P. Scaife, editor), Academic Press, London, 1974, pp. 95-113.

2 Clenshaw, C.W., Rational approximations for special functions, Software for Numerical Mathematics (D.J. Evans, editor), Academic Press, London, 1974, pp. 275-284.

3 Day, K.M., Toeplitz matrices generated by an arbitrary rational function, Trans. Amer. Math. Soc., 206 (1975), 224-245.

4 Edrei, A., On the generating function of a doubly infinite, totally positive sequence, Trans. Amer. Math. Soc., 74 (1953), 367-383.

5 Fleischer, J., Generalizations of Padé approximants, Padé Approximants (P.R. Graves-Morris, editor), The Institute of Physics, London, 1973, pp. 126-131.

6 Frankel, A.P., and W.B. Gragg, Algorithmic almost best uniform rational approximation with error bounds (abstract). SIAM Rev., 15 (1973), 418-419.

7 Gragg, W.B., The Padé table and its relation to certain algorithms of numerical analysis, SIAM Rev., 14 (1972), 1-62.

8 Gragg, W.B., On Hadamard's theory of polar singularities, Padé Approximants and Their Applications, (P.R. Graves-Morris, editor), Academic Press, London, 1973, pp. 117-123.

9 Gragg, W.B., and G.D. Johnson, The Laurent-Padé table, Information Processing 74, Proc. IFIP Congress 74, North-Holland, Amsterdam, 1974, pp. 632-637.

10 Grenander, U. and G. Szegö, Toeplitz Forms and Their Applications, University of California, Press, Berkeley, 1958.

11 Hadamard, J., Essai sur l'étude des fonctions données par leur développement de Taylor, J. Math. Pures Appl., 8 (1892), 101-186.

12 Hartwig, R.E., and M.E. Fisher, Asymptotic behavior of Toeplitz matrices and determinants, Arch. Rational Mech. Anal., 32 (1969), 190-225.

13 Henrici, P., Some applications of the quotient-difference algorithm, Proc. Symposium Appl. Math., vol. 15, Amer. Math. Soc., Providence, 1963, pp. 159-183.

14 Kogbetliantz, E.G., Generation of elementary functions, Mathematical Methods for Digital Computers, vol. 1 (A. Ralston and H.S. Wilf, editors), Wiley, New York, 1967, pp. 5-35.

15 Meinardus, G., Approximation of Functions: Theory and Numerical Methods, Springer-Verlag, New York, 1967.

16 Newman, D.J., Super good rational approximation to e^x, J. Approximation Theory, to appear.

17 Norman, E., A discrete analogue of the Weierstrass transform, Proc. Amer. Math. Soc., 11 (1960), 596-604.

18 Parlett, B., Global convergence of the basic QR algorithm on Hessenberg matrices, Math. Comp., 22 (1968), 803-817.

19 Rutishauser, H., Der Quotienten-Differenzen Algorithmus, Mitt. Inst. Angew. Math. Zürich, 7 (1957), 74pp.

20 Wilson, R., Determinantal criteria for meromorphic functions, Proc. London Math. Soc., 4 (1954), 357-374.

21 Wynn, P., Upon systems of recursions which obtain among the quotients of the Padé table, Numer. Math., 8 (1966), 246-269.

22 Wynn, P., Transformations to accelerate the convergence of Fourier series, Blanch Anniversary Volume, Aerospace Research Lab., U.S. Air Force, Washington, D.C., 1967, pp. 339-379.

William B. Gragg*
Department of Mathematics
University of California, San Diego
La Jolla, California 92093

*Research supported in part by the Air Force Office of Scientific Research under Grant AFOSR-76-2910.

GENERALISATIONS OF THE THEOREM OF DE MONTESSUS USING CANTERBURY APPROXIMANTS

P. R. Graves-Morris

The theorem of Chisholm and Graves-Morris which generalises de Montessus' theorem to two variables is reviewed and revised in the light of recent progress. Weighting schemes for the approximants are reviewed in §2 and an outline proof is given in §3.

1 Introduction

After the advent of Chisholm approximants [1], a quantity of collaborative work in Canterbury followed which showed the viability of the original scheme and suggested some generalisations. Off-diagonal N-variable approximants were defined, culminating in the Hughes Jones (general off-diagonal) approximants [4]. Numerical work in model applications was encouraging, and so a convergence theorem seemed to Chisholm and me to be a natural target. De Montessus' theorem is a natural theorem for rows of the Padé table, and we hoped for a straight-forward generalisation to two variables. After the theorem [2] was proved, weighting schemes for the symmetrising equations evolved, leading to important invariance and stability properties. In this article, the weighting schemes are briefly reviewed, and our theorem is stated in this up-to-date context. An error in our original paper is also corrected here.

Padé approximants are defined by

$$(1)\quad [m/n] = \sum_{i=0}^{m} a_i z^i \Big/ \sum_{i=0}^{n} b_i z^i$$

with $b_0=1$ and the formal requirement that

$$(2)\quad [m/n] - f(z) = O(z^{m+n+1}) \quad .$$

This is the Baker definition, based on the accuracy-through-order principle; existence of the approximants is not guaranteed a priori. From (2),

$$(3)\quad \sum_{i=0}^{m} a_i z^i - \left(\sum_{i=0}^{\infty} c_i z^i\right)\left(\sum_{i=0}^{n} b_i z^i\right) = O(z^{m+n+1}) .$$

The coefficients $\{b_i , i=1,2,\ldots,n\}$ are located in fig.1.

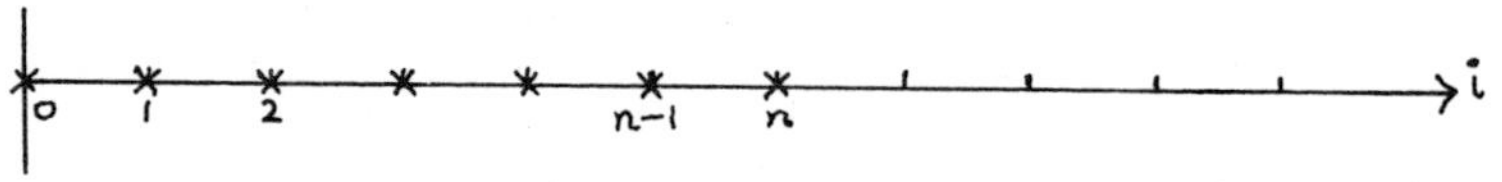

Fig.1. The denominator lattice

The values of b_i are obtained from $b_0=1$ and (3) using the equalities of fig.2.

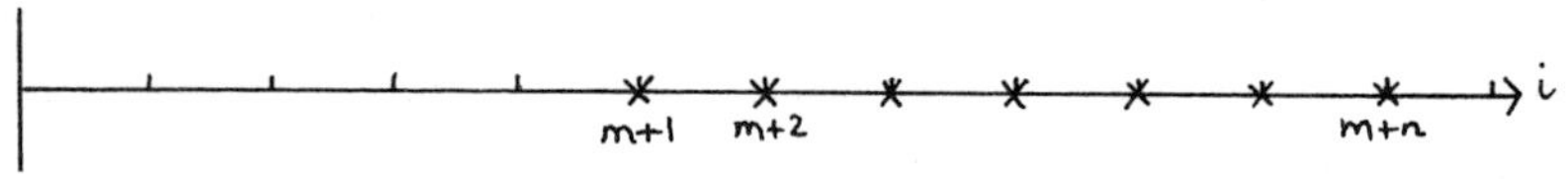

Fig.2. The equality lattice

For present purposes, the generalisations to two dimensions are indicated by figs 3 and 4.

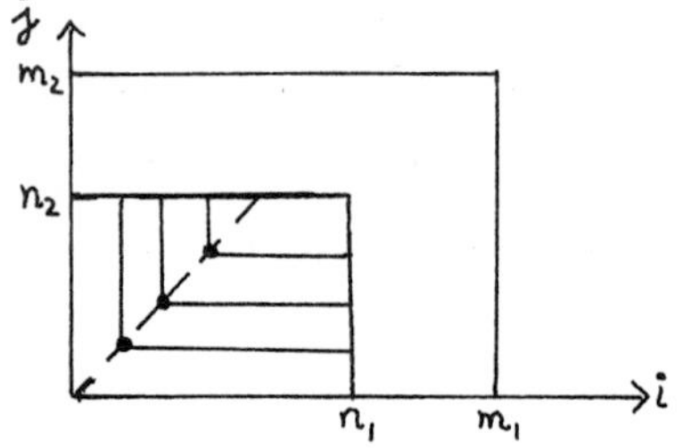

Fig.3

The numerator is an "$m_1 \times m_2$" lattice
The denominator is an "$n_1 \times n_2$" lattice

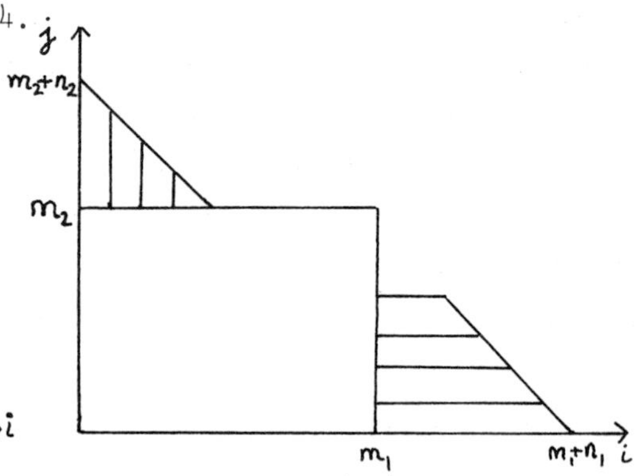

Fig.4

The equality lattice

The two variable approximants originating in Canterbury all take the general form

$$(4)\quad [\underline{m}/\underline{n}] \;=\; \sum_{i=0}^{m_1}\sum_{j=0}^{m_2} a_{ij} z_1^{\,i} z_2^{\,j} \Bigg/ \sum_{i=0}^{n_1}\sum_{j=0}^{n_2} b_{ij} z_1^{\,i} z_2^{\,j} \,.$$

They are constructed with $b_{00}=1$ and by defining

$$(5)\quad \left(\sum_{i=0}^{n_1}\sum_{j=0}^{n_2} b_{ij} z_1^{\,i} z_2^{\,j}\right)\left(\sum_{i=0}^{\infty}\sum_{j=0}^{\infty} c_{ij} z_1^{\,i} z_2^{\,j}\right) =$$

$$\sum_{i=0}^{m_1}\sum_{j=0}^{m_2} a_{ij} z_1^{\,i} z_2^{\,j} + \text{as high order terms as possible.}$$

The prong method of solution is an organisation of the denominator coefficients and their defining equations as indicated in fig.5 and 6.

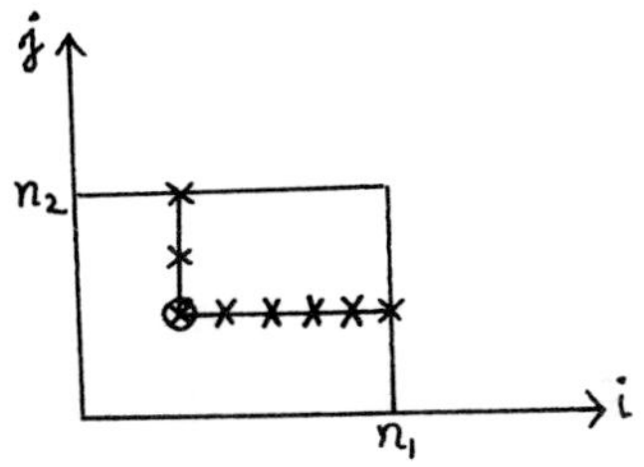

Fig.5 The denominator lattice

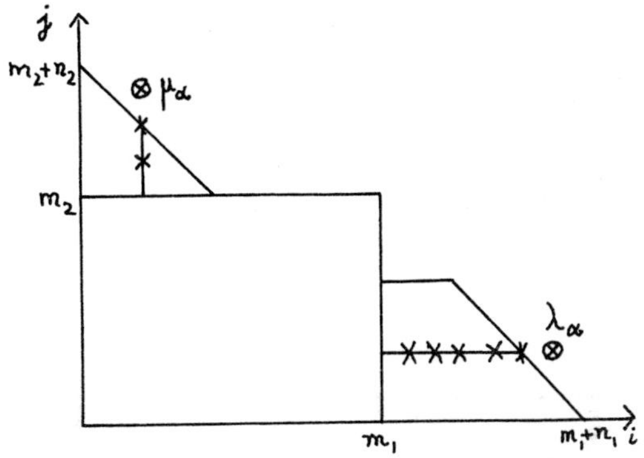

Fig.6 The equality lattice

The equation derived from (5) at the end of the horizontal branch is multiplied by λ_α and added to μ_α times the equation at the end of the vertical branch of the αth prong, to form a single weighted symmetrised equation. The denominator coefficients of the αth prong $(1 \leq \alpha \leq \min(n_1,n_2))$ are

$$\underline{b}_\alpha = (b_{n_1,\alpha}, b_{n_1-1,\alpha}, \ldots, b_{\alpha+1,\alpha} \;;\; b_{\alpha,n_2}, b_{\alpha,n_2-1}, \ldots, b_{\alpha,\alpha+1} \;;\; b_\alpha)^T.$$

The structure of the equations for the αth prong is

$$(6) \quad D_{\alpha\alpha}\, \underline{b}_\alpha + \sum_{\beta=0}^{\alpha-1} G_{\alpha\beta}\, \underline{b}_\beta = 0 ,$$

where

$$(7) \quad D_{\alpha\alpha} = \begin{pmatrix} C_\alpha^{(1)} & 0 & \underline{X}_\alpha^{(1)} \\ 0 & C_\alpha^{(2)} & \underline{X}_\alpha^{(2)} \\ \lambda_\alpha X^{(1)T} & \mu_\alpha X^{(2)T} & Y_\alpha \end{pmatrix}$$

and $(C_\alpha^{(1)})_{ij} = c_{m_1-n_1+i+j-1,0}$; $(\underline{X}^{(1)})_j^T = c_{m_1-\alpha+j-1,0}$

and $Y_\alpha = \lambda_\alpha c_{m_1+n_1-2\alpha+1,0} + \mu_\alpha c_{0,m_2+n_2-2\alpha+1}$ etc.. [3] gives further details in complete generality.

2 Weighting schemes

The block lower triangular structure of (1.6) allows the denominator coefficients to be determined sequentially in prongs. Furthermore, we must analyse $D_{\alpha\alpha}$ completely. Notice the block factorisation

$$(1) \quad D_{\alpha\alpha} = \begin{pmatrix} S^{11} & 0 \\ 0 & S^{22} \\ 0,0,\ldots 0,\lambda_\alpha & 0,0,\ldots 0,\mu_\alpha \end{pmatrix} \begin{pmatrix} E^{11} & 0 & E^{13} \\ 0 & E^{22} & E^{23} \end{pmatrix} ,$$

where

$$S^{11} = \begin{pmatrix} 1 & 0 & \ldots & 0 & 0 \\ 0 & 1 & \cdots & \cdot & \cdot \\ \cdot & \cdot & & \cdot & \cdot \\ \cdot & \cdot & & \cdot & \cdot \\ 0 & 0 & \ldots & 1 & 0 \end{pmatrix}, E^{11} = \begin{pmatrix} c_{m_1-n_1+1,0} & c_{m_1-n_1+2,0} & \cdots & c_{m_1-\alpha,0} \\ c_{m_1-n_1+2,0} & & & \vdots \\ \vdots & & & \vdots \\ c_{m_1-\alpha+1,0} & \cdots & \cdots & c_{m_1+n_1-2\alpha,0} \end{pmatrix}$$

and $E = (c_{m_1-\alpha+1,0}\, , c_{m_1-\alpha+2,0}, \ldots, c_{m_1+n_1-2\alpha+1,0})^T$ etc..

Notice that the notation of (1.7) and (1),

$$C^{(1)}_{\alpha-1} = \begin{pmatrix} C^{(1)}_{\alpha} & X^{(1)}_{\alpha} \\ X^{(1)}_{\alpha} & Y^{(1)}_{\alpha} \end{pmatrix} = (E^{11}\ E^{13}) \quad ,$$

is entirely self-consistent. Application of the Cauchy-Binet theorem to (1) shows that

$$(2)\ \det D_{\alpha\alpha} = \lambda_{\alpha} \det C^{(1)}_{\alpha-1} \det C^{(2)}_{\alpha} + \mu_{\alpha} \det C^{(1)}_{\alpha} \det C^{(2)}_{\alpha-1} \quad .$$

We must choose $\lambda_{\alpha}:\mu_{\alpha}$ to best advantage. Chisholm's original choice of $\lambda_{\alpha}=\mu_{\alpha}$ is appropriate for symmetric functions for which $f(z_1,z_2) = f(z_2,z_1)$. It is clear that there is a choice of $\lambda_{\alpha}:\mu_{\alpha}$ rendering $\det D_{\alpha\alpha}=0$; avoidance of such degeneracies and maximum stability of the defining equations motivates the weighting scheme known as PETCH, the pessimistic choice. We maximise $|\det D_{\alpha\alpha}|$ subject to $|\lambda_{\alpha}|^2 + |\mu_{\alpha}|^2 = 1$ to find

$$\lambda_{\alpha} : \mu_{\alpha} = [\det C^{(1)}_{\alpha-1} \det C^{(2)}_{\alpha}]^* : [\det C^{(1)}_{\alpha} \det C^{(2)}_{\alpha-1}]^* \ .$$

The alternative approach is to require as much homographic covariance of the approximants as possible in the hope that the diagonal sequence converges in as large a domain of $\mathbb{C}^2$ as possible. If

$$(2)\quad z_1 = Aw_1/(1+Bw_1) \quad \text{and} \quad z_2 = Cw_2/(1+Dw_2)$$

and $g(w_1,w_2) \equiv f(z_1,z_2)$, then $[n,n/n,n]_g(w_1,w_2) = [n,n/n,n]_f(z_1,z_2)$ with the scale invariant choice SCINCH defined by

$$\lambda_{\alpha} : \mu_{\alpha} = \det C^{(1)}_{\alpha} \det C^{(2)}_{\alpha-1} : \det C^{(1)}_{\alpha-1} \det C^{(2)}_{\alpha} \quad .$$

It turns out that PETCH has only covariances under the transformations (2) when $|A| = |C|$.

Several interesting problems are associated with the weighting problem, which might shed light on an aesthetic choice for $\lambda_{\alpha} : \mu_{\alpha}$. Which degeneracies are avoidable and unavoidable? Are there recurrence relations for the approximants, their numerators and denominators? Does the explicit determinantal formula yield a lower order Canterbury approximant in the presence of degeneracy? Some answers are given in [3] .

3 A convergence theorem

THEOREM <u>Let</u> $f(z_1,z_2) \equiv f(\underline{z})$ <u>be analytic in some polydisc</u> $|z_1| \leq \rho_1$, $|z_2| \leq \rho_2$. <u>Let there exist a polynomial</u>

$$(1) \quad B(\underline{z}) = \sum_{i=0}^{n_1} \sum_{j=0}^{n_2} B_{ij} z_1^{\,i} z_2^{\,j} \quad \text{such that}$$

(2) $A(\underline{z}) \equiv B(\underline{z}) f(\underline{z})$ <u>is analytic in the larger polydisc</u> $|z_1| < R_1$, $|z_2| < R_2$, <u>and further let</u> $B(z_1,0)$ <u>have</u> n_1 <u>zeros at</u> $z_1 = p_\nu^{(1)}$, $\nu = 1,2,\ldots,n_1$ <u>and</u> $B(0,z_2)$ <u>have</u> n_2 <u>zeros at</u> $z_2 = p_\nu^{(2)}$, $\nu = 1,2,\ldots,n_2$. <u>These zeros have their multiplicity counted and are ordered so that</u>

$$(3) \quad \begin{array}{l} 0 < |p_1^{(1)}| \leq |p_2^{(1)}| \leq \ \ldots \ \leq |p_{n_1}^{(1)}| < R_1 \\ 0 < |p_1^{(2)}| \leq |p_2^{(2)}| \leq \ \ldots \ \leq |p_{n_2}^{(2)}| < R_2 \ . \end{array}$$

<u>Also, let no distinct zeros of</u> $B(z_1,0)$ (or $B(0,z_2)$) <u>be equidistant from the origin: to wit, if</u> $|p_{\nu-1}^{(i)}| = |p_\nu^{(i)}|$, <u>then</u> $p_{\nu-1}^{(i)} = p_\nu^{(i)}$ <u>for</u> $\nu = 1,2,\ldots,n_i$ <u>and</u> $i = 1,2$. <u>If</u> $m_1,m_2 \to \infty$ <u>along a ray such that</u> $|m_2 - m_1\theta|$ <u>is bounded, define</u>

$$f_{\theta,\alpha} = \max\{|p_{n_1-\alpha+1}^{(1)} / p_1^{(1)}| \ , |p_{n_2-\alpha+1}^{(2)} / p_1^{(2)}|^{\theta}\} \ .$$

$$(4) \quad \begin{array}{l} \text{Provided } R_1 > |p_{n_1}| \, (r_1/\rho_1) \, (r_2/\rho_2)^{\theta} \prod_{\alpha=1}^{n} f_{\theta,\alpha} \\ \text{and} \quad R_2 > |p_{n_2}| \, (r_1/\rho_1)^{1/\theta} \, (r_2/\rho_2) \prod_{\alpha=1}^{n} f_{\theta,\alpha} \end{array}$$

where $n = \min(n_1, n_2)$, then the $[m_1, m_2/n_1, n_2]$ Canterbury approximants with SCINCH or PETCH weights converge to $f(\underline{z})$ on any compact subset of

$$\mathcal{D} = \{\underline{z} : |z_1| < r_1 \ , \ |z_2| < r_2 \ , \ B(\underline{z}) \neq 0\} \ .$$

Outline Proof. Just as (2.6) is a representation of (1.5) using prongs, so (2) is represented in prongs by

$$(5) \quad D_{\alpha\alpha} \underline{B}_\alpha + \sum_{\beta=0}^{\alpha-1} G_{\alpha\beta} \underline{B}_\beta = \underline{A}_\alpha \ .$$

Let $\underline{\Delta}_\alpha = \underline{B}_\alpha - \underline{b}_\alpha$, $\Delta_{ij} = B_{ij} - b_{ij}$, so that $\Delta_{00} = 0$, and for $1 \leqslant \alpha \leqslant n$, subtraction of (2.6) from (5) gives

$$(6) \quad D_{\alpha\alpha} \underline{\Delta}_\alpha + \sum_{\beta=0}^{\alpha-1} G_{\alpha\beta} \underline{\Delta}_\beta = \underline{A}_\alpha \ .$$

We must prove that each $\underline{\Delta}_\alpha \to 0$ sufficiently fast as $m_1, m_2 \to \infty$. Hence we need bounds on $(D_{\alpha\alpha})^{-1}$, and we use formulae of type (2.2), which entail det $C_\alpha^{(i)}$. Omitting the case where $f(z_1, 0)$ has multipoles, the conditions of the theorem ensure that the coefficients $c_{i,0}$ have a dominant part

$$c_{i,0} \simeq \sum_{\nu=1}^{n} r_\nu^{(1)} \, p_\nu^{(1)-i} \ ,$$

where $r_\nu^{(1)}$, $\nu = 1, 2, \ldots, N$ are non-zero residues. Then, omitting the superscript (1) for brevity, we find the dominant part of $C_\alpha^{(1)}$ to be given by

$$\begin{pmatrix} 1 & 1 & \cdots & 1 \\ p_1^{-1} & p_2^{-1} & \cdots & p_N^{-1} \\ \vdots & \vdots & & \vdots \\ p_1^{1-N} & p_2^{1-N} & \cdots & p_N^{1-N} \end{pmatrix} \begin{pmatrix} r_1 p_1^{-m_1+\alpha} & & & \\ & r_2 p_2^{-m_1+\alpha} & & \\ & & \ddots & \\ & & & r_N p_N^{-m_1+\alpha} \end{pmatrix} \begin{pmatrix} p_1^{N-1} & \cdots & p_1 & 1 \\ p_2^{N-1} & \cdots & p_2 & 1 \\ \vdots & & \vdots & \vdots \\ p_N^{N-1} & \cdots & p_N & 1 \end{pmatrix}$$

where $N = n_1 - \alpha$. Hence we deduce that

$$\det C_\alpha^{(1)} \simeq (\text{non zero constant}) \,.\, \prod_{i=1}^{n_1-\alpha} r_i^{(1)} \, [p_i^{(1)}]^{-m_1} \,.$$

Using (6), this and similar formulae lead to a bound for each Δ_α successively, and condition (4) of the theorem is designed to ensure the bound

$$(7) \quad |\Delta_\alpha| \simeq O(\rho_1/r_1)^{m_1} \, O(\rho_2/r_2)^{m_2}$$

Having established that

$$B^{[m_1 m_2/n_1 n_2]}(z_1,z_2) \to B(z_1,z_2) \text{ as } m_1,m_2 \to \infty \,,$$

it remains to establish that the approximants converge in the designated region $\mathfrak{D}$. This region, as Wallin observed, [5], is not the polydisc of analyticity of $A(z_1,z_2)$, as was originally stated. Define the notation $[.]_0^{\underline{m}}$ for the Maclaurin polynomial by

$$[A(\underline{z})]_0^{\underline{m}} = \sum_{i=0}^{m_1} \sum_{j=0}^{m_2} a_{ij} \, z_1^{\,i} \, z_2^{\,j} \,.$$

Then $$f(\underline{z}) - \frac{A^{[\underline{m}/\underline{n}]}(\underline{z})}{B^{[\underline{m}/\underline{n}]}(\underline{z})} = f(\underline{z}) - \frac{[A(\underline{z})]_0^{\underline{m}}}{B^{[\underline{m}/\underline{n}]}(\underline{z})} - P(\underline{z}) \,,$$

where $P(\underline{z}) = [(B^{[\underline{m}/\underline{n}]}(\underline{z}) - B(\underline{z}))\, f(\underline{z})]_0^{\underline{m}}$

is a polynomial. Eq. (7) shows that each coefficient of $P(\underline{z})$ tends to zero; knowing that $c_{ij} = O(\rho_1^{-i}\,\rho_2^{-j})$, each of the $(m_1+1)\,(m_2+1)$ terms of $P(\underline{z})$ may be bounded in $|z_1| < r_1$, $|z_2| < r_2$ and convergence of the row sequence, according to the theorem, now follows.

COROLLARY 1 Let $f(z_1,z_2) \equiv f(\underline{z})$ be symmetric, so that $f(z_1,z_2) = f(z_2,z_1)$, and let $f(\underline{z})$ be analytic in the polydisc $|z_1| < \rho$, $|z_2| < \rho$. Let there exist a polynomial

$$B(\underline{z}) = \sum_{i=0}^{n} \sum_{j=0}^{n} B_{ij} z_1^{\,i} z_2^{\,j}$$

such that $A(\underline{z}) = B(\underline{z}) f(\underline{z})$ is analytic in the larger polydisc $|z_1| < R$, $|z_2| < R$, and let $B(w,0) = B(0,w)$ have its n zeros at $w = p_\nu$, $\nu=1,2,\ldots,n$. As before, these zeros have their multiplicity counted and may not be equidistant from the origin unless they are multipoles. Provided

$$R > |p_n| \left(\frac{r}{\rho}\right)^2 \prod_{\alpha=1}^{n-1} |p_{n-\alpha+1} / p_1| ,$$

then the $[m,m/n,n]$ Canterbury approximants converge to $f(\underline{z})$ on any compact subset of

$$\mathcal{D} = \{\underline{z} : |z_1| < r , |z_2| < r , B(\underline{z}) \neq 0 \} .$$

COROLLARY 2 Let $f(\underline{z})$ be analytic at $\underline{z} = \underline{0}$, and let there exist a polynomial

$$B(\underline{z}) = \sum_{i=0}^{n_1} \sum_{j=0}^{n_2} B_{ij} z_1^{\,i} z_2^{\,j}$$

such that $B(\underline{z}) f(\underline{z})$ is analytic. Provided $B(w,0)$ has degree n_1 and $B(0,w)$ has degree n_2 and neither has zeros equidistant from the origin, except for multipoles, then the $[m_1,m_2 / n_1,n_2]$ Canterbury approximants with SCINCH or PETCH weighting converge to $f(\underline{z})$ on any compact subset of

$$\mathcal{D} = \{\underline{z} : B(\underline{z}) \neq 0\} .$$

Acknowledgement

I am grateful to Professor J. S. R. Chisholm and Dr. R. Hughes Jones for engaging in the collaborative work underlying this review. I am also grateful to numerous colleagues in Canterbury for helpful discussions and to Professor H. Wallin for [5] .

References

1 Chisholm, J. S. R., Rational approximants defined from double power series, Math. Comp. 27 (1973) 841-848.

2 Chisholm, J. S. R. and Graves-Morris, P. R., Generalizations of the theorem of de Montessus to two variable approximants, Proc. Roy. Soc. Lond. A 342 (1975) 341-372.

3 Graves-Morris, P. R. and Hughes Jones, R., An analysis of two variable rational approximants, J. Comp. Appld. Math. 2 (1976) 41-48.

4 Hughes Jones, R., General rational approximants N-variables J. Approx. 16 (1976) 201-233.

5 Wallin, H., private communication; Karlsson J. and Wallin, H., University of Umeå mathematical preprint (1976) No.8.

P. R. Graves-Morris,
Mathematical Institute,
University of Kent,
Canterbury,
Kent,
England.

RATIONAL APPROXIMATION BY AN INTERPOLATION PROCEDURE IN SEVERAL VARIABLES

J. Karlsson and H. Wallin

We study the convergence properties of certain rational approximation schemes in several variables. These schemes are defined by interpolation conditions generalizing those that define Padé approximants in one variable.

0 Introduction

If f is a (formal) power series in one complex variable we define the $[n,\nu]$ Padé approximant to f as the (unique) rational function P/Q of type $[n,\nu]$, i.e. with denominator of degree $\leq \nu$, and numerator of degree $\leq n$, such that the Taylor expansion of $Qf-P$ starts with terms of degree $\geq n+\nu+1$. Padé approximants have been widely studied in recent years due to their usefulness in various applications (cf [1]). In many applications the need for approximants to functions of several variables arises naturally. This paper is concerned with generalizing the definition of Padé approximants to higher dimensions and studying properties of the approximants obtained. Seemingly, the first investigations in this line were carried out by Chisholm´s group in Canterbury (cf. Chisholm [3]).

1 Definitions and some properties

Throughout this paper, we will restrict attention to two complex variables z_1 and z_2 for simplicity. We write $(z_1,z_2) = z \in C^2$. Most results carry over directly to any number of variables and sometimes the appropriate changes will be indicated.

A rational approximant to a power series f

$$f(z_1,z_2) = \sum_{j,k} c_{jk} z_1^j z_2^k$$

will be of the form P/Q where P and Q are polynomials from given classes such that $Qf-P$ satisfies given conditions on its Taylor coefficients with $Q \not\equiv 0$. Schemes of this type have previously been described by Chisholm et. al. [2], [5], [10] and Lutterodt [14]. These authors consider

$$P = \sum_{j=0}^{n} \sum_{k=0}^{n} \alpha_{jk} z_1^j z_2^k \quad \text{and}$$

$$Q = \sum_{j=0}^{\nu} \sum_{k=0}^{\nu} \beta_{jk} z_1^j z_2^k .$$

They also look at the more general setting when the sums range from 0 to n_1, from 0 to n_2, from 0 to ν_1, and from 0 to ν_2. We refer to [10] and [14] for details.

In Lutterodt the conditions on the coefficients of

$$Qf-P = \sum d_{jk} z_1^j z_2^k$$

are

$$d_{jk} = 0 \quad (j,k) \in S$$

where S, <u>the interpolation set</u>, is a subset of $N \times N$ with $(n+1)^2 + (\nu+1)^2 - 1$ elements, containing the set $\{(j,k) | (j,k) \leq (n,n)\}$, and satisfying the <u>rectangle rule</u>: $(j,k) \in S \Rightarrow (\ell,m) \in S \ \forall (\ell,m) < (j,k)$ under the usual partial order.

In Chisholm's case the interpolation set is smaller

$$S = \{(j,k) | (j,k) \leq (n,n) \quad \text{or} \quad j+k \leq \nu+n\}$$

and in addition he requires

$$d_{\nu+n+1-p,p} + d_{p,\nu+n+1-p} = 0 \quad \text{for} \quad p=1, \ldots, n+\nu.$$

The latter conditions may be changed to arbitrary linear conditions [5].

We propose rational approximants where

$$P = \sum_{j+k \leq n} \alpha_{jk} z_1^j z_2^k$$

$$Q = \sum_{j+k\leq\nu} \beta_{jk} z_1^j z_2^k .$$

Such approximants are said to be of <u>type</u> $[n,\nu]$ or $[n,\nu]$ approximants. We require that the interpolation set S contains $\{(j,k)|j+k\leq n\}$, obeys the rectangle rule and has $(n+1)(n+2)/2 + (\nu+1)(\nu+2)/2 - 1$ elements.

The above three definitions are all chosen so that in analogy with Padé approximants P is the appropriate partial sum of the Taylor series to Qf. Furthermore the remaining conditions give rise to linear equations in the coefficients of Q, obtained by equating coefficients of Qf or linear combinations of them to zero. The number of equations is one less than the number of unknowns. By elementary linear algebra a nontrivial solution then always exists. It is clear that existence is independent of the actual choice of the interpolation set, only depending on the number of elements in it.

Uniqueness of the approximants is a trickier matter, however. If we do not require that P be the partial sum of Qf we are in trouble because then a valid approximant to $f \equiv 0$ would be e.g. $z_1/1$ provided we do not ask for interpolation in the point $(1,0)$, i.e. require that $(1,0) \in S$.

THEOREM 1.1. <u>If</u> P/Q <u>is a rational approximant of type</u> $[n,\nu]$ <u>with an interpolation set</u> S <u>containing, in addition to</u> $\{(j,k)|j+k\leq n\}$ <u>as many points as possible from some given enumeration of the points in</u> $N\times N$, <u>then</u> P/Q <u>is unique</u>.

<u>Remark</u>. With the rectangle rule this gives in one variable an equivalent definition of Padé approximant (see [18], Prop. 1).

<u>Proof</u>. Suppose Q_1 and Q_2 are two linearly independent polynomials giving the same (maximal) interpolation set. Then $aQ_1 + bQ_2$ gives at least the same interpolation set and we can in fact choose a and b to get an extra interpolation point contradicting maximality.

Remark. If no finite maximal S exists, it can be shown by the methods in section 3 that f is rational provided f is analytic at the origin.

THEOREM 1.2. Whenever an interpolation set satisfies the rectangle rule and P/Q is an approximant to f, $f(0,0) \neq 0$, then Q/P is an approximant with the same interpolation set to $1/f$. (Lutterodt's, Chisholm's, or our definition)(c.f. [2], [14])

Proof. By definition $Qf-P = \sum d_{jk} z_1^j z_2^k$ with $d_{jk} = 0$ for $(j,k) \in S$. Multiplying by $1/f$: $Q-P/f = \sum e_{jk} z_1^j z_2^k$ with $e_{jk} = 0$ for $(j,k) \in S$ by the rectangle rule. For the proof in Chisholm's case we refer to [2]. Note that this works also with the approximants of Th. 1.1 because if we do not get maximal interpolation to $1/f$ we would get more interpolation to f by taking the reciprocal of the approximant to $1/f$.

In the same way, under the same conditions, for all three definitions, but provided $n = \nu$ we get: If P/Q is an approximant to f and $ad - bc \neq 0$ then $(a\,P/Q + b)/(c\,P/Q + d)$ is an approximant with the same interpolation set to $(af + b)/(cf + d)$ (c.f. [2], [14]).

Furthermore, provided $n = \nu$, approximants under all three definitions are invariant under certain changes of variables, i.e. if w_k are functions of z_k, $k = 1, 2$, then the approximant to $f(w_1,w_2)$ as a function of (z_1,z_2) is $P(w_1,w_2)/Q(w_1,w_2)$ where P/Q is the approximant to f.

The relevant variable changes are: In Lutterodt's case: [14]

$$w_k = \frac{A_k z_k}{1 + B_k z_k}, \quad k = 1, 2.$$

In Chisholm's case: [2]

$$w_k = \frac{A z_k}{1 + B_k z_k}, \quad k = 1, 2.$$

(By changing the second set of defining conditions different constants in the numerator can be allowed [5].) For our case the relevant variable changes are

$$w_k = \frac{A_k z_k}{1 + Bz_1 + Cz_2}, \quad k = 1, 2.$$

This invariance is in all cases established by first noting that $P(w_1,w_2)/Q(w_1,w_2)$ will have correct kinds of numerator and denumerator and then checking the interpolation.

Chisholm's and Lutterodt's schemes (with suitable interpolation sets) give the approximants the property that on setting z_1 or $z_2 = 0$ we get Padé-approximants in the remaining variable (projection property [2]). Furthermore a product of two functions $g_1(z_1)$ and $g_2(z_2)$ will have as a Chisholm (or Lutterodt) type approximant the product of the Padé approximants to g_1 and g_2 [6]. Neither of these properties is in general satisfied by our definition. The first will be satisfied if the interpolation set contains $\{(0,k)|0 \leq k \leq n+\nu\} \cup \{(j,0)|0 \leq j \leq n+\nu\}$.

In comparison: for all types of approximants one can establish invariance under certain transformations both of functions and of arguments. The approximants of Chisholm or Lutterodt type have nice relationships to functions on the axes ($z_1 = 0$, $z_2 = 0$). Our approximants have the nice property that, projected onto any complex line $c_1z_1 + c_2z_2 = 0$ we get rational approximants of type $[n,\nu]$, while the Chisholm-Lutterodt approach gives type $[n,\nu]$ on the axes and type $[2n,2\nu]$ on all other lines.

2 Convergence of [n,n] approximants

The fact that our approximants have the same type on all complex lines makes it natural to prove results by projecting onto complex lines and using standard techniques from one-variable Padé theory on every line. Our approximants will not be Padé approximants when projected but throughout this section our approximants are chosen so that S contains $\{(j,k)|j+k\leq[n\sqrt{2}]\}$, the square brackets denoting the integer part. This will give sufficient interpolation to use the same techniques. This was first exploited by Gončar [8] who proved the following analog of a theorem of Nuttall's [15] for functions analytic at $z = 0$.

THEOREM 2.1. If $f(z_1,z_2)$ is meromorphic in C^2 then the rational approximants P_n/Q_n of type $[n,n]$ (for a class of definitions including ours) converge in measure to f on compact subsets of C^2.

By the same method we get the following theorem, for the notation of which we refer to [12] and [8].

THEOREM 2.2. Suppose $f \in R_0(\lambda(n))$. Then on any complex line

$$\text{cap}\{|f - P_n/Q_n| > \varepsilon^n\} \le \text{const.}\ \lambda(n)^{\sqrt{2}-1}/\varepsilon.$$

Note that the degree of convergence in capacity obtained here is not the same as in the one variable case (Karlsson [8]). In general the exponent $\sqrt{2}-1$ is $2^{1/N}-1$ where N is the complex dimension of the space studied.

The following generalizes a theorem of Wallin's [19].

THEOREM 2.3. Suppose that the power series $f = \sum c_{jk} z_1^j z_2^k$ has coefficients that satisfy

$$\sum_\nu \left(\max_{[n_\nu\sqrt{2}]-n_\nu<j+k\le[n_\nu\sqrt{2}]} |c_{jk}| \right)^{\alpha/n_\nu} < \infty$$

for some $\alpha > 0$ and some sequence $n_\nu \to \infty$. Then if f_k denotes the Taylor polynomial of degree k to f, we have on every complex line

$$f_{[n_\nu\sqrt{2}]} - P_{n_\nu}/Q_{n_\nu} \to 0 \quad \text{as} \quad \nu \to \infty$$

except on a set of α-dimensional Hausdorff measure zero.

Remark on proof. Projecting on a complex line we get a function, the nth Taylor coefficient of which we can majorize by $\max_{|j+k|=n} |c_{jk}| \cdot \binom{2+n-1}{n}$. (In N variabels the 2 is changed to N.) Our approximants interpolate to this function of order $[n_\nu\sqrt{2}]$. If we carry out the same calculations as in Wallin [19] we get the desired result.

The reason for the surprisingly neat formulation of theorem 2.1 is that once we have proved convergence in (2-dimensional Lebesgue) measure on every line it follows that we have convergence in (2N-dimensional Lebesgue) measure. In theorems 2.2 and 2.3 it is not easy to summarize the results in this way.

We also want to note the following result (cf. Baker [1, p. 170], Jones, Thron [11] for the one-variable case; in several variables cf. Lutterodt [13]).

THEOREM 2.4. Let P_ν/Q_ν be rational approximants to f such that the sum of the degrees of P_ν and Q_ν tends to infinity with ν and such that P_ν/Q_ν is uniformly (in ν) bounded in an open set, containing the origin, where f is analytic. Then P_ν/Q_ν converges uniformly to f on compact subsets. This theorem is true for any of the definitions of rational approximants.

Proof. Take any subsequence of P_ν/Q_ν. By the Montel theorem in several variables there exists a subsequence uniformly convergent on compact subsets. By the interpolation properties (use e.g. Cauchy's estimates) we get convergence to f in a polydisc about the origin, and by analytic continuation the convergence is to f everywhere. Thus every subsequence has a subsequence converging to f. Hence the whole sequence converges to f.

3 Convergence of $[n,\nu]$ approximants, ν fixed

The purpose of this section is to analyze the possibility to generalize the Montessus de Ballore theorem (Theorem 3.2) from one to several variables. However, we also give a simple unified treatment in the one-variable case of some more or less well-known results. The calculations are based on Cauchy's estimates.

3.1 The one-variable case, $z \in \mathbb{C}$.

THEOREM 3.1. *Let f be a meromorphic function of one complex variable in $|z| < R$ with poles $\xi_1, \ldots, \xi_\mu$ - counted with their multiplicities - where $0 < |\xi_j| < R$. Let P_n/Q_n, $Q_n \not\equiv 0$, be a rational approximant of type $[n,\nu]$, $0 \leq \mu \leq \nu$, ν fixed, such that*

$$(3.1) \quad (Q_n f - P_n)(z) = \sum_{j>n+\mu} d_j^{(n)} z^j.$$

Then there exist $\nu-\mu$ points $\xi_{\mu+1}, \ldots, \xi_\nu$ and a subsequence of the sequence $\{P_n/Q_n\}$ converging uniformly to f, and even with geometric degree of convergence (see (3.5)), on compact subsets of $\{|z| < R\} \setminus \{\xi_j | 1 \leq j \leq \nu\}$.

Proof. We put $Q(z) = (z-\xi_1) \ldots (z-\xi_\mu)$ and get by (3.1)

$$(3.2) \quad (Q\, Q_n f - Q\, P_n)(z) = \sum_{j>n+\mu} e_j^{(n)} z^j.$$

Normalize P_n and Q_n so that $\max\{|Q_n(z)| : |z| = R\} = 1$. We choose $R_1 < R$. Since degree $Q\, P_n \leq n+\mu$, Cauchy's estimates and the normalization of Q_n give

$$|e_j^{(n)}| \leq \max_{|z|=R_1} |(Qf)(z)| R_1^{-j}, \quad \text{for } j > n+\mu.$$

By summing a geometric series this and (3.2) give, with some constant $M(R_1)$

$$(3.3) \quad |(Q\, Q_n f - Q\, P_n)(z)| \leq M(R_1)(|z|/R_1)^{n+\mu} \quad \text{for } |z| < R_1.$$

Since $\{Q_n\}$ is a sequence of polynomials of degree $\leq \nu$ which are uniformly bounded in $|z| \leq R$ we can choose a subsequence converging uniformly on compact sets, $Q_{n_j} \to q \not\equiv 0$, q a polynomial of degree $\leq \nu$. By (3.3) the analogous subsequence of $\{P_n\}$ converges uniformly to a function p on compact subsets of $|z| < R_1$ and hence on $|z| < R$. Also

$$(3.4) \quad Qqf - Qp = 0 \quad \text{in } |z| < R.$$

If z is a pole of f, then $Q(z) = 0$, $(Qf)(z) \neq 0$, and hence $q(z) = 0$ by (3.4). Consequently, the zeros of q are completely determined by the zeros of Q if $\mu = \nu$. If $\mu < \nu$, let $\xi_{\mu+1}, \ldots, \xi_\nu$ be the rest of the zeros of q.

Let E be a compact subset of $\{|z| \leq R_2\} \setminus \{\xi_j | 1 \leq j \leq \nu\}$ where $R_2 < R_1$. From (3.3) we get

$$\limsup_{n\to\infty} \max_{z\in E} |(Q\,Q_n f - Q\,P_n)(z)|^{1/n} \leq R_2/R_1.$$

If we combine this with the fact that

$$Q\,Q_{n_j} \to Qq \quad \text{uniformly on } E, \quad Qq \neq 0 \quad \text{on } E,$$

we finally get what we want to prove:

$$\limsup_{j\to\infty} \max_{z\in E} |(f - P_{n_j}/Q_{n_j})(z)|^{1/n_j} \leq R_2/R_1 < 1. \tag{3.5}$$

Remark. With the same assumptions and proof we get that every subsequence of $\{P_n/Q_n\}$ contains a convergent subsequence of the kind mentioned in the theorem. As a corollary we also get:

THEOREM 3.2. (Montessus de Ballore) Suppose that f is analytic at the origin and meromorphic with ν poles in $|z| < R$. Then the $[n,\nu]$ Padé approximant to f, P_n/Q_n, is, when n is sufficiently large, the unique rational function of type $[n,\nu]$ which interpolates to f at the origin of order at least $n+\nu+1$. The poles of P_n/Q_n converge to the poles of f and P_n/Q_n converges uniformly to f, with geometric degree of convergence, in those compact subsets of $|z| < R$ which do not contain any poles of f.

Proof. With the same notation as in the proof of Theorem 3.1 q and Q have the same zeros. By multiplying Q_n with a suitable constant we can make $q \equiv Q$. According to the above remark this is true even if we start from a subsequence of $\{P_n/Q_n\}$. Consequently $Q_n \to Q$ uniformly on compact sets which gives the required convergence. As $Q_n(0) \to Q(0) \neq 0$, we do not lose, for large n, any interpolation when dividing by Q_n in the definition of the Padé approximant. This gives the desired interpolation and the uniqueness follows from the uniqueness of the Padé approximant.

Remarks. a) The method with the normalization of Q_n to give a simple proof of Montessus de Ballore's theorem is due to Harold Shapiro (personal communication). In fact, our proof is only a technical modification of Shapiro's proof. Other treatments are in [17] and [8].

b) Theorem 3.1 has a connection to a conjecture by Baker and Graves-Morris (see [9] pp. 61-62) saying that if f is analytic at the origin and meromorphic with μ poles, $\mu \le \nu$, in $|z| < R$, then there exists a subsequence of $[n,\nu]$ Padé approximants (ν fixed) converging to f in $|z| < R$ (except at the poles of f). Theorem 3.1 shows that the conjecture is true in the weaker form allowing $\nu-\mu$ exceptional points where convergence is not required.

c) From (3.4) follows that if z_0, $|z_0| < R$, is not a pole of f, then $p(z_0) = 0$ iff $q(z_0) = 0$ or $f(z_0) = 0$, i.e. z_0 is, according to a standard theorem in complex analysis, an accumulation point of zeros of Padé approximants iff $f(z_0) = 0$ or z_0 is accumulation point of poles of Padé approximants.

3.2 The many-variable case, $z = (z_1,z_2)$. In this case the transition from (3.1) to (3.3) does not work (see the discussion after Theorem 3.4) and the situation is more complicated. Among other things we shall give counterexamples to the natural generalization of the Montessus de Ballore theorem. However, if we in this theorem change the assertion that for all large n there exist unique rational functions of type $[n,\nu]$ interpolating to f of order $n+\nu+1$ to the assertion that there exist rational functions of type $[n,\nu]$ interpolating to f of order at least $n+1$ at the origin, then we get a convergence theorem also in several variables (for the one-variable case see [20]):

THEOREM 3.3. Let $f = F/Q$ where F is analytic in the polydisc $\{z = (z_1,z_2): |z_i| < R_i,\ i = 1,2\}$ and Q is a polynomial of degree ν, $Q(0) \neq 0$. Then there exist rational functions $R_{n\nu}$

of type $[n,\nu]$ interpolating to f at the origin of order at least $n+1$ (i.e. having the same Taylor polynomial of degree n as f) and converging uniformly, as $n \to \infty$, to f, with geometric degree of convergence, in compact subsets of $\{z \mid |z_i| < R_i\} \setminus \{z \mid Q(z) = 0\}$.

Proof. Let π_n be the Taylor polynomial of degree n of fQ. Choose $\varepsilon_i > 0$ such that $(R_1-\varepsilon_1)/R_1 = (R_2-\varepsilon_2)/R_2$. By replacing, if necessary, R_i by smaller numbers, we may assume that F is analytic in $|z_i| \le R_i$. Consequently, Cauchy's estimates give, for some constant $M(R_1,R_2)$

$$\max_{|z_i| \le R_i - \varepsilon_i} |(fQ-\pi_n)(z)| \le \max_{|z_i| \le R_i - \varepsilon_i} \sum_{j+k>n} \frac{M(R_1,R_2)}{R_1^j \, R_2^k} |z_1|^j |z_2|^k \le$$

$$\le M(R_1,R_2) \sum_{s=n}^{\infty} (s+1)\left(\frac{R_1-\varepsilon_1}{R_1}\right)^s = O(n)\left(\frac{R_1-\varepsilon_1}{R_1}\right)^n .$$

If E is a compact subset of $\{|z_i| \le R_i - \varepsilon_i\} \setminus \{Q(z) = 0\}$ we conclude

$$\limsup_{n\to\infty} \max_{z \in E} |(f - \pi_n/Q)(z)|^{1/n} \le \frac{R_1-\varepsilon_1}{R_1} < 1.$$

This gives the desired convergence property for $R_{n\nu} = \pi_n/Q$. The interpolation follows from the choice of π_n since $Q(0) \ne 0$.

The estimates in this proof work if we replace Q by a polynomial Q_n of degree $\le \nu$ and π_n by the Taylor polynomial of degree n of $Q_n f$ if we normalize Q_n to have maximum 1 in a suitable polydisc. If we combine this with the method from section 3.1 of choosing a subsequence of $\{Q_n\}$ converging to q, we obtain, as in the proof of Theorem 3.1, the following generalization to several variables of Theorem 3.1 in the case $\mu = 0$.

THEOREM 3.4. Let f be analytic in $\{|z_i| < r_i, \ i = 1,2\}$. Let $\{P_n/Q_n\}$, $Q_n \not\equiv 0$, be a sequence of rational functions of type $[n,\nu]$, $\nu \ge 0$, ν fixed, such that P_n is the Taylor polynomial of degree n of $Q_n f$. Then there exists a polynomial

q, $q \neq 0$, <u>of degree</u> $\leq \nu$ <u>and a subsequence of</u> $\{P_n/Q_n\}$ <u>converging uniformly to</u> f, <u>with geometric degree of convergence, on compact subsets of</u> $\{|z_i| < r_i\} \setminus \{z|q(z) = 0\}$.

The last Theorem gives a convergence result only in the polydisc where f is analytic. Now we assume that $f = F/Q$ where Q is a polynomial of degree $\leq \nu$, $Q(0) \neq 0$, f is analytic in the polydisc $|z_i| \leq r_i$, and F in the polydisc $|z_i| \leq R_i$. Let P_n/Q_n be a rational function of type $[n,\nu]$ such that P_n is the Taylor polynomial of degree n of $Q_n f$. Then we have an expansion

$$(3.6)\quad (Q_n f - P_n)(z) = \sum_{j+k>n} d_{jk}^{(n)} z_1^j z_2^k .$$

In order to be able to use the fact that $F = Qf$ is analytic in $|z_i| \leq R_i$ we multiply by Q and get

$$(3.7)\quad (Q\, Q_n f - Q\, P_n)(z) = \sum_{j+k>n} e_{jk}^{(n)} z_1^j z_2^k .$$

Suppose that $\max\{|Q(z)| \,\big|\, |z_i| \leq R_i\} = 1$. Since $Q\, P_n$ has degree $\leq n+\nu$ we get by Cauchy´s estimates

$$(3.8)\quad |e_{jk}^{(n)}| \leq \frac{M(R_1,R_2)}{R_1^j R_2^k} \quad \text{for } j+k \geq n+\nu+1.$$

The coefficients $e_{jk}^{(n)}$ with $n+1 \leq j+k \leq n+\nu$ depend also on $Q\, P_n$ and can not be estimated in this way. There are

$$\sum_{s=2}^{\nu+1} (n+s) = \nu n + \nu(\nu+3)/2$$

such coefficients. We can impose $\nu(\nu+3)/2$ conditions on them by choosing the coefficients of Q_n so that, for instance, $\nu(\nu+3)/2$ of the $d_{jk}^{(n)}$-coefficients, $n+1 \leq j+k \leq n+\nu$ become zero. However, we cannot - as in the one-dimensional case - choose interpolation conditions which guarantee that all the coefficients $e_{jk}^{(n)}$, $n+1 \leq j+k \leq n+\nu$ become zero. This leads to counterexamples to a general analogue of the Montessus de Ballore theorem (see section 4). An estimate of $e_{jk}^{(n)}$, $n+1 \leq j+k \leq n+\nu$, or of the corresponding coefficients $d_{jk}^{(n)}$ is given by the following lemma where we use the same notation and assumptions as above.

LEMMA 3.5. For some constant M and $n+1 \leq j+k \leq n+\nu$,

$$|d_{jk}^{(n)}| \leq M \max_{|z_i| \leq r_i} |Q_n(z) - Q(z)| r_1^{-j} r_2^{-k} + M R_1^{-j} R_2^{-k}.$$

The same inequality holds for $e_{jk}^{(n)}$.

Proof. Cauchy's integral formula gives

$$(2\pi i)^2 d_{jk}^{(n)} = \int_{|z_i|=r_i} \frac{(fQ_m)(z)}{z_1^{j+1} z_2^{k+1}} dz =$$

$$= \int_{|z_i|=r_i} \frac{f(Q_m-Q)(z)}{z_1^{j+1} z_2^{k+1}} dz + \int_{|z_i|=R_i} \frac{(Qf)(z)}{z_1^{j+1} z_2^{k+1}} dz$$

and this gives the desired estimate.

Remark. Chisholm and Graves-Morris [4] have studied the problem of generalizing the Montessus de Ballore theorem to functions of several variables for the Chisholm type approximants. In their estimates they do not seem to have taken into consideration their problem corresponding to those coefficients in the right member of (3.7) which are influenced by $Q P_n$ (see the fourth line after (4.57) in [4]). It is not clear how this gap in their proof can be filled.

4 Convergence of [n,1] approximants

When $\nu = 1$ explicit calculations are easy. We give a convergence result and two counterexamples. The results, which should be compared to the case $\nu = 1$ in Montessus de Ballore's theorem, show that the several variable case is more complicated than the one variable case.

THEOREM 4.1. Let $f = F/Q$ where $Q(z) = 1-b_1z_1-b_2z_2$, $b_1 \neq 0$, f is analytic in $|z_i| < r_i$, $i = 1,2$, with Taylor coefficients c_{jk}, F is analytic in $|z_i| < R_i$ with Taylor coefficients a_{jk}, $F(b_1^{-1},0) \neq 0$, and $0 < r_1 \leq 1/|b_1| < R_1$, $0 < r_2 \leq R_2$. Let

P_n/Q_n be a rational approximant of type $[n,1]$ satisfying at least the interpolation conditions determined by

$$(4.1)\quad (f\,Q_n - P_n)(z) = \sum_{j+k>n} d_{jk}^{(n)} z_1^j z_2^k, \quad d_{n+1,0}^{(n)} = d_{n,1}^{(n)} = 0.$$

Then P_n/Q_n converges uniformly to f, with geometric degree of convergence, in compact subsets of $\{|z_i| \leq s_i,\ i=1,2\} \setminus \{z \mid Q(z)=0\}$ if

$$s_1 < |b_1| R_1 r_1, \quad s_2 < |b_1| R_1 r_2, \quad s_2 < R_2.$$

Example 1. $R_1 = R_2 = \infty$ gives convergence everywhere except where $Q = 0$.

Example 2. $r_1 = r_2 = r$, $R_1 = R_2 = R$ gives convergence in the polydisc $|z_i| < |b_1| Rr$ which is larger than the polydisc $|z_i| < r$ where f is analytic - we have got an expansion factor $|b_1|R$. By choosing the interpolation conditions differently it is possible to get other expansion factors, for instance $|b_2|R$.

Proof of Theorem 4.1. We first observe that $P_n(z_1,0)/Q_n(z_1,0)$ is the one-dimensional Padé approximant to $f(z_1,0)$. Because of that $Q_n(0) \to Q(0) \neq 0$, by Montessus de Ballore's theorem. For large n we may therefore normalize Q_n so that $Q_n(0) = 1$ and put $Q_n(z) = 1 - \beta_1^{(n)} z_1 - \beta_2^{(n)} z_2$.

The condition $d_{n+1,0}^{(n)} = 0$ gives, since

$$(fQ_n)(z) = (F \cdot 1/Q \cdot Q_n)(z) = \sum a_{jk} z_1^j z_2^k \sum (b_1 z_1 + b_2 z_2)^j Q_n(z),$$

that

$$\sum_{j=0}^{n+1} a_{j0} b_1^{n+1-j} - \beta_1^{(n)} \sum_{j=0}^{n} a_{j0} b_1^{n-j} = 0, \quad \text{i.e.}$$

$$(4.2)\quad \beta_1^{(n)} - b_1 = \frac{a_{n+1,0}}{b_1^n A_n}, \quad \text{where } A_n = \sum_0^n a_{j0} b_1^{-j}.$$

Note that $A_n \to F(b_1^{-1},0) \neq 0$ by the assumptions. Analogously, the condition $d_{n,1}^{(n)} = 0$ gives after simplification with the same A_n as in (4.2)

$$(4.3)\quad \beta_2^{(n)} - b_2 = \frac{a_{n1}}{b_1^n A_n} + (b_1 - \beta_1^{(n)}) \frac{\sum_0^{n-1} a_{j1} b_1^{n-1-j} + b_2 \sum_0^{n-1} a_{j0}(n-j) b_1^{n-1-j}}{b_1^n A_n} .$$

Replacing, if necessary, r_i and R_i by smaller numbers, we may assume that f is analytic in $|z_i| \leq r_i$ and F in $|z_i| \leq R_i$. Hence $a_{jk} = O(R_1^{-j} R_2^{-k})$ and (4.2) gives

$$(4.4)\quad \beta_1^{(n)} - b_1 = O(1)(|b_1|R_1)^{-n}.$$

Analogously, a simple calculation and (4.3) give

$$(4.5)\quad \beta_2^{(n)} - b_2 = O(n)(|b_1|R_1)^{-n}.$$

The last two formulas give an estimate of $Q_n - Q$. We can consequently use Lemma 3.5. Combined with (3.6) - (3.8) this gives in the usual way after some calculation, if we put

$$\delta_1 = \max\left(\frac{s_1}{r_1}, \frac{s_2}{r_2}\right) \quad \text{and} \quad \delta_2 = \max\left(\frac{s_1}{R_1}, \frac{s_2}{R_2}\right),$$

$$\max_{|z_i| \leq s_i} |(Q\, Q_n f - Q\, P_n)(z)| = O(n)(|b_1|R_1)^{-n} (n+2)\delta_1^{n+1} + O(1) \sum_{n+1}^{\infty} (j+1)\delta_2^j .$$

Since $Q_n \to Q$ this gives in the usual way the desired convergence if $\delta_1 < R_1|b_1|$ and $\delta_2 < 1$ and the theorem follows.

Counterexample 1. We construct an example where

$$f(z) = F(z)/(1 - b_1 z_1), \quad b_1 \neq 0, \quad F(z) = \sum a_{jk} z_1^j z_2^k ,$$

and F is analytic in C^2. We choose a_{0k} so that $F(0,z_2)$ is an entire function in z_2 having $[n,1]$ Padé approximants with poles clustering everywhere in the complex plane, which is possible according to an example by Perron [16, p. 270]. We then choose a_{jk}, $j > 0$, so that F becomes analytic in C^2. Let

P_n/Q_n be a rational approximant of type [n,1] determined by

$$(f\,Q_n - P_n)(z) = \sum_{j+k>n} d_{jk}^{(n)} z_1^j z_2^k\,, \quad d_{n+1,0}^{(n)} = d_{0,n+1}^{(n)} = 0.$$

By these interpolation conditions $P_n(0,z_2)/Q_n(0,z_2)$ becomes the [n,1] Padé approximant of $f(0,z_2) = F(0,z_2)$ and its poles cluster everywhere in C by the construction.

Hence, P_n/Q_n cannot converge to f in any polydisc around $z = 0$. A comparison between this counterexample and Theorem 4.1 (Example 1) shows the importance of the choice of the interpolation conditions. Furthermore, it is easy to see that $Q_n(z)$ does not converge to $Q(z) = 1 - b_1 z_1$, i.e. the singularities of the approximants do not even converge to the singularities of f. We now give a more complicated counterexample in a case where $Q_n \to Q$.

Counterexample 2. We use the same notation and assumptions as in Theorem 4.1 and determine $f = F/Q$ by putting

$$F(z) = \sum_{k=0}^{\infty} z_2^k + \sum_{j=0}^{\infty} (\rho b_1)^j z_1^j z_2 \text{ where } 0<\rho<1,\ \rho|b_2|>1.$$

(4.2) and (4.3) give with $Q_n(z) = 1 - \beta_1^{(n)} z_1 - \beta_2^{(n)} z_2$

$$(4.6)\quad \beta_1^{(n)} = b_1 \quad\text{and}\quad \beta_2^{(n)} - b_2 = a_{n1}/b_1^n = \rho^n \quad\text{for } n \geq 1.$$

The coefficients $d_{jk}^{(n)}$ are given by (4.1) and we see that

$$d_{0,n+1}^{(n)} = \sum_0^{n+1} a_{0j}\, b_2^{n+1-j} - \beta_2^{(n)} \sum_0^n a_{0j}\, b_2^{n-j} =$$

$$= (b_2 - \beta_2^{(n)}) \sum_0^n b_2^{n-j} + 1.$$

Since $\rho|b_2| > 1$, this gives by means of (4.6), with some constant $M > 0$,

$$|d_{0,n+1}^{(n)} z_2^{n+1}| \geq M(\rho^n|b_2|^n - 1)|z_2|^{n+1} \to \infty$$

if $|z_2| > \rho^{-1}|b_2|^{-1}$. If we use (3.6) - (3.8) and note that $d_{0,n+1}^{(n)} = e_{0,n+1}^{(n)}$ this gives that

$$(Q\,f\,Q_n - Q\,P_n)(0,z_2) = d_{0,n+1}^{(n)}\,z_2^{n+1} + \sum_{n+2}^{\infty} e_{0k}^{(n)}\,z_2^k$$

diverges for $\rho^{-1}|b_2|^{-1} < |z_2| < 1 = R_2$, since the last sum converges for these values of z_2. But $Q_n \to Q$ and we get

$$(f - P_n/Q_n)(0,z_2) \to \infty \quad \text{for} \quad \rho^{-1}|b_2|^{-1} < |z_2| < 1 = R_2.$$

This example shows that we can not get convergence in the z_2-plane in the whole disc $|z_2| < R_2$ where $Q(0,z_2) \neq 0$. It is also easy to see that this counterexample shows that Theorem 4.1 is sharp.

References

1 Baker, G.A. Jr., Essentials of Padé Approximants, Academic Press, New York, 1975.

2 Chisholm, J.S.R., Rational approximants defined from double power series, Math. Comp., 27 (1973), 841-848.

3 Chisholm, J.S.R., Rational polynomial approximants in N variables, Lecture Notes in Physics 47 (1976), 33-54.

4 Chisholm, J.S.R. and P.R. Graves-Morris, Generalization of the theorem of de Montessus to two-variable approximants, Proc. Royal Society Ser. A, 342 (1975), 341-372.

5 Chisholm, J.S.R. and R. Hughes Jones, Relative scale covariance of N-variable approximants, U. of Kent preprint, Canterbury (1974).

6 Common, A.K. and P.R. Graves-Morris, Some properties of Chisholm approximants, J. Inst. Maths. Applics, 13 (1974), 229-232.

7 Gončar, A.A., A local condition for the single-valuedness of analytic functions of several variables, Math. USSR Sbornik, 22 (1974), 305-322, (Russian original Mat. Sb., 93 (1974).)

8 Gončar, A.A., On the convergence of generalized Padé approximants to meromorphic functions, Mat. Sb., 98 (140) (1975), 564-577, (Russian).

9 Graves-Morris, P.R., Convergence of rows of the Padé table, Lecture Notes in Physics 47 (1976), 55-68.

10 Hughes Jones, R., General rational approximants in N variables, J. Approximation Theory, 16 (1976), 201-233.

11 Jones, W.B. and W.J. Thron, On convergence of Padé approximants, SIAM J. Math. Anal., 6 (1975), 9-16.

12 Karlsson, J., Rational interpolation and best rational approximation, J. Math. Anal. Appl., 53 (1976), 38-52.

13 Lutterodt, C.H., A two-dimensional analogue of Padé Approximant theory, U. of Birmingham preprint (1973).

14 Lutterodt, C.H., Rational approximants to holomorphic functions in n dimensions, J. Math. Anal. Appl., 53 (1976), 89-98.

15 Nuttall, J., The convergence of Padé approximants of meromorphic functions, J. Math. Anal. Appl., 31 (1970), 147-153.

16 Perron, O., Die Lehre von den Kettenbrüchen, Band II, Teubner, Stuttgart, 1957.

17 Saff, E.B., An extension of Montessus de Ballore's theorem on the convergence of interpolating rational functions, J. Approximation Theory, 6 (1972), 63-68.

18 Wallin, H., On the convergence theory of Padé approximants, Linear Operators and Approximation, Proceedings of a conference in Oberwolfach 1971, Birkhäuser, Stuttgart, 1972, pp. 461-469.

19 Wallin, H., The convergence of Padé approximants and the size of the power series coefficients, Applicable Analysis, 4 (1974), 235-252.

20 Walsh, J.L., The convergence of sequences of rational functions of best approximation, Math. Annalen, 155 (1964), 252-264.

J. Karlsson
Department of Mathematics
University of Umeå
S-901 87 Umeå, Sweden

H. Wallin
Department of Mathematics
University of Umeå
S-901 87 Umeå, Sweden

THE CONVERGENCE OF PADÉ APPROXIMANTS TO FUNCTIONS WITH BRANCH POINTS

J. Nuttall

This paper describes the result of work on the convergence of diagonal Padé approximants to a class of functions with an even number of branch points with principal singularities of square root type. Convergence in capacity is shown away from a set of arcs whose location is completely determined by the location of the branch points. A conjecture about the possible form of extensions of this work is presented.

1 Introduction

Suppose, for example, that we have a function of the form

$$(1.1) \quad \psi(t) = \int_L dt' \; (t'^2-1)^{-\frac{1}{2}} \mu(t')(t'-t)^{-1}$$

where $\mu(t)$ is a polynomial and L a finite arc joining the points -1, 1. Thus $\psi(t)$ is a function having a square root singularity at $t = \pm 1$. In this representation, $\psi(t)$ is analytic in the complex plane cut along L. The diagonal Frobenius Padé approximants (P.A.'s) derived from the expansion of $\psi(t)$ about $t = \infty$ are determined from the moments

$$\int_L dt' \; (t'^2-1)^{-\frac{1}{2}} \mu(t')t'^k \quad , \; k = 0,1,\dots \; .$$

These moments are unchanged if L is distorted in the finite plane, provided its ends remain fixed. Thus the P.A.'s are independent of the location of L.

In this example, the convergence of the sequence of diagonal P.A.'s can be discussed in an elementary manner from the properties of orthogonal polynomials $p_n(t)$ of degree n satisfying

$$(1.2)\quad \int_S dt(t^2-1)^{-\frac{1}{2}}\,\mu(t)\,p_n(t)t^k \quad = 0,\ k = 0,1,\ldots,n-1.$$

Since their definition is also independent of the choice of L, we have chosen it to be S, the line segment joining -1,1. The behavior of $p_n(t)$ for large n can be worked out with the help of Christoffel's formula [11] and the properties of Jacobi polynomials, and the convergence of the P.A.'s obtained from

$$(1.3)\quad [n/n] = p_n^{-1}(t)\int_S dt'(t'^2-1)^{-\frac{1}{2}}\,\mu(t')[p_n(t)-p_n(t')](t'-t)^{-1}.$$

It follows, even for $\mu(t)$ which are complex on S, provided that $\mu(t) \neq 0$ on S, that the diagonal P.A.'s converge uniformly to the $\psi(t)$ of (1.1) with L = S, in any closed bounded domain in the complex plane not intersecting S.

This is the most favorable situation that a Padé enthusiast could have imagined. The P.A.'s fail to converge on an arc joining the two branch points, and not in some larger region. The P.A.'s have chosen this arc S, which is the same for all weight functions $\mu(t)$ of the type considered.

It has been shown [7], generalizing the proof of Szegö [12], that the same convergence behavior holds, with L = S, for an extended class of weight functions, namely those satisfying a certain Lipschitz condition. The earlier results of Baxter [2] imply the same result for an overlapping class of weight functions.

The choice made by the P.A. for the cut joining two branch points is hardly surprising, but it is by no means so obvious what the corresponding result will be for more than two branch points. The aim of our work has been to begin to answer this question.

Historically, the first clue is probably that provided by the work of Dumas [3]. Starting from the analysis of Jacobi [5], he constructed explicitly the diagonal P.A.'s for a function which is the square root of a fourth degree polynomial plus a second degree polynomial chosen to make the whole $O(t^{-1})$ at ∞. Dumas showed that, as $n \to \infty$, all, except for at most one, of the poles of $[n/n]$ approach a certain locus S containing the four branch points. Away from S and the remaining pole, which moves about the complex plane, $[n/n]$ approaches the function being approximated. It may be shown that there is convergence in capacity outside S. Achyeser's [1] later work is related to a special case of Dumas' results.

The result described below was first suggested by an empirical analysis of the formula giving $p_n(t)$ as an n-dimensional integral. In Szegö's book [10] this formula is referred to as not being suitable for the derivation of the properties of orthogonal polynomials, but perhaps this is not the case.

2 An Even Number of Branch Points

Now suppose that we have an even number of distinct branch points d_i, $i = 1,\ldots, 2\ell$. For a certain class of functions with these branch points, we have shown that the diagonal P.A.'s converge in capacity away from a set S, which is uniquely determined by the branch points. The set S may be described in terms of the function $\phi(t)$, where

$$(2.1)\quad \phi(t) = \int_{d_1}^{t} dt'\, X^{-\frac{1}{2}}(t')\, Z(t') .$$

Here $X(t) = \prod_{i=1}^{2\ell} (t-d_i)$ and $Z(t)$ is a monic polynomial of degree $\ell-1$ which is determined uniquely [9] by the requirement that the periods of the hyperelliptic integral of the third kind $\phi(t)$ are all pure imaginary. The set S is the locus $\mathrm{Re}\,\phi(t) = 0$, which consists of ℓ analytic Jordan arcs joining pairs of the branch points. In general, no zero of $Z(t)$ will lie on S, and in this case, the ℓ arcs are non-intersecting.

Consider a function f(t) given by

$$(2.2)\quad f(t) = \int_S dt'\; X_+^{-\frac{1}{2}}(t')\;\sigma(t')(t'-t)^{-1}$$

where $X_+^{-\frac{1}{2}}$ means the limit from a particular side of S. Suppose that, for t, t' ε S,

$$(2.3)\quad |\sigma(t') - \sigma(t)| < C\;(\ell n|t'-t|)^{-1-\lambda}\quad ,\qquad C\;,\;\lambda > 0\;,$$

and

$$(2.4)\quad A > |\sigma(t)| > B > 0\;.$$

(The smoothness condition may be relaxed somewhat near the ends of S). We have proved the following

THEOREM [8]. <u>If</u> S <u>consists</u> <u>of</u> ℓ <u>non-intersecting</u> <u>arcs</u>, <u>the</u> <u>sequence</u> <u>of</u> [n/n] <u>P.A.'s</u> <u>to</u> f(t) <u>converges</u> <u>in</u> <u>capacity</u> <u>to</u> f(t) <u>as</u> $n \to \infty$, <u>in</u> <u>any</u> <u>closed</u>, <u>bounded</u> <u>domain</u> <u>not</u> <u>intersecting</u> S.

It is not expected that the restriction on the form of S is necessary, provided σ obeys suitable conditions near the points where arcs of S intersect.

It may be verified that, for four points, the set S is the same as that found by Dumas [3].

3 Method of Proof

The method of proof of the theorem is a generalization of that used by Szegö to obtain the asymptotic behavior of polynomials orthogonal on the interval (-1,1) with real positive weight function satisfying (2.3) and (2.4). We find an approximation to $\sigma(t)$ of the form $\rho_m^{-1}(t)$, where ρ_m is a degree $(m-\ell+1)$ polynomial. We construct explicitly the orthogonal polynomials $q_{n,m}(t)$ of degree $n \geq m$ for the approximate weight.

Suppose that R is the Riemann surface $y^2 = X(t)$, which consists of two sheets joined together at S. Let the zeroes of ρ_m be r_i, $i = 1,\ldots, s = m-\ell+1$, which we shall take to be on the second sheet. The Jacobi inversion theorem [5] tells us that there is at least one set of points α_i, $i = 1,\ldots, \ell-1$, $\alpha_i \varepsilon R$, such that the following equivalence relation between divisors holds:

$$(3.1) \quad \alpha_1\cdots\alpha_{\ell-1}r_1\cdots r_s(\infty_2)^{n-m} \sim (\infty_1)^n .$$

It follows that there is a function F(t), meromorphic on R, that has zeroes at the points of the divisor on the left and a pole of order n at ∞ on the first sheet. It may be shown that, on S,

$$(3.2) \quad q_{n,m}(t) = F_+(t) + F_-(t) ,$$

where $F_\pm(t)$ means the limit of F(t) as t approaches one or other side of S from the first sheet of R. It may also be shown that, on the first sheet, F(t) is an analytic function in the plane cut along S, with zeroes only at those points α_i lying on the first sheet, which is $O(t^n)$ at ∞, and which satisfies on S

$$(3.3) \quad \rho_m^{-1}(t)F_+(t)F_-(t) = \text{const.} \prod_{i=1}^{\ell-1} (t-\alpha_i)$$

Equation (3.3) may be solved, using the methods of Muskhelishvili [6], and it is found, on the first sheet away from S, that a factor $\phi^n(t)$ dominates the behavior of F(t).

We then show that $p_n(t)$ satisfies the integral equation

$$(3.4) \quad \mu_n p_n(t) = q_{n,n}(t) + \int_S dt' \, X_+^{-\frac{1}{2}}(t')[\sigma(t')-\rho_n^{-1}(t')]K(t,t')(t'-t)^{-1}p_n(t') ,$$

where

$$(3.5) \quad K(t,t') = q_{n,n}(t)q_{n+1,n}(t') - q_{n+1,n}(t)q_{n,n}(t') ,$$

and μ_n is a constant depending on the points α_i. For an infinite sequence of values of n, with successive members differing by no more than ℓ, we show that (3.4) may be solved by iteration and in this way the asymptotic behavior of $p_n(t)$ is determined. The gaps in the sequence are filled in with the help of a generalization of the usual three-term recurrence relation, which may no longer hold when there are blocks in the Padé table. An equation corresponding to (1.3) is used to study the behavior of the P.A.'s.

A possible alternative method of proof would be to replace ρ_m^{-1} in (3.3) by σ, giving $F'(t)$ say, and use Faber polynomials [14] $Q_n(t)$, with

$$(3.6) \quad Q_n(t) = \int_\Gamma dt' F'(t')(t'-t)^{-1} ,$$

where Γ is a large circle including t and S. An integral equation similar to (3.4) may be obtained.

4 Possible Further Developments

The functions that we have studied so far have had branch points where the strongest singularity was of the form $(t-d)^{-\frac{1}{2}}$. We expect that it should be possible to remove this limitation to some extent so that singularities such as $(t-d)^{\nu}$, $\nu > -1$, can be dealt with. As evidence for this, we point to the work of Szegö [13] and Geronimus [4], who study real weight functions with singularities of this nature. The leading term in the asymptotic form of $p_n(t)$ away from the segment (-1,1) is given by a formula identical to that found for the previous class of weight functions, although on S there is a difference. The same thing happens in the case of Jacobi polynomials, where, with $X = (t^2-1)$, we have $\sigma(t) = (1-t)^{\alpha} (1+t)^{\beta}$, even for α, β complex. The behavior of $p_n(t)$ away from S is all that is required to prove the convergence of [n/n]. Unfortunately, our method of proof involves a knowledge of $p_n(t)$ on S, so that, to study other types of singularity, we expect that the approximate polynomials of the previous section will have to be modified.

Let us suppose that the extension of the theorem to more singular weights $\sigma(t)$ has been accomplished. What could we then learn about the convergence of P.A.'s to functions with singularities of type other than square root? We shall illustrate our ideas by considering the function $T(t) = \tau^{-1/3}(t)$, where τ is a degree 3 polynomial with zeroes at d_1, d_2, d_3. If we could find a set S of the type described in Sect. 2, containing d_1, d_2, d_3, such that the discontinuity of T(t) across any arc of S is not identically zero, we could write T in the form (2.2), where $\sigma(t)$ had at most isolated zeroes, and apply the extended version of our theorem.

We therefore look at sets S that are obtained from four points, three of which we wish to be d_1, d_2, d_3. For the fourth point d_4 in general position, S will contain an arc joining d_4 to say d_1, and another disconnected arc joining d_2, d_3. Such an S will not satisfy the requirement above, since T(t) has no singularity at d_4 and its discontinuity across the arc d_4 d_1 will be zero. The only satisfactory S is one for which the zero of Z(t) (a polynomial of degree 1) coincides with d_4. In this case, S will consist of three arcs radiating from d_4 to d_1, d_2, and d_3. Using the Riemann period relations [9], we have shown that, given d_1, d_2, and d_3, there exists a unique point d_4 for which S has the above character. Our conjecture is that the diagonal P.A.'s to T(t) will converge in capacity away from this locus. It seems likely that similar results will hold for functions with more branch points.

An interesting property of the set S corresponding to an even number of points d_i is the following. It can be shown that S is the unique set of minimum capacity among all sets whose connected components contain an even number of the points d_i. Also,the set described above in connection with T(t) is the unique set of minimum capacity of all sets such that the three points d_i lie in the same connected component. We expect similar results to be true for more points. No application of this property to the theory of orthogonal polynomials or P.A.'s is known.

The extension of these results to functions of several variables might eventually prove interesting. This would presumably be facilitated by defining a generalized P.A. that was related to a polynomial with certain orthogonality properties, which appears to be possible. A function of two complex variables, such as the square root of a polynomial, will be singular on a manifold of real dimension 2, and it is tempting to speculate that a sequence of generalized P.A.'s might converge away from a manifold of real dimension 3 bounded by the manifold of singularity. Perhaps this manifold is the solution of a minimum problem analogous to the one above.

Acknowledgements

The author thanks Dr. S.R. Singh for many helpful discussions. He is also grateful to Prof. A. Edrei for his comments and for bringing attention to the work of Dumas.

References

1 Achyeser, N., Über eine Eigenschaft der elliptischen Polynome, Comm. de la Soc. Math. de Kharkof (Zapiski Inst. Mat. Mech.) (4) 9 (1934), 3-8.

2 Baxter, G., A convergence equivalence related to polynomials on the unit circle, Trans. Amer. Math. Soc. 99 (1961), 471-487; A norm inequality for a finite-section Wiener-Hopf equation, Illinois Jour. Math. 7 (1963), 97-103.

3 Dumas, S., Sur le développement des fonctions elliptiques en fractions continues, Thesis, Zurich (1908).

4 Geronimus, Y.L., Polynomials Orthogonal on a Circle and Interval, Pergamon, New York, 1960.

5 Jacobi, C.G.J., Note sur une nouvelle application de l'analyse des fonctions elliptiques à l'algèbre, Crelle Journal für die reine und angewandte Mathematik, 7 (1830), 41-43.

6 Muskhelishvili, N.I., Singular Integral Equations, Noordhoff, Groningen, 1953.

7 Nuttall, J., Orthogonal polynomials for complex weight functions and the convergence of related Padé approximants, unpublished (1972).

8 Nuttall, J. and S.R. Singh, Orthogonal polynomials and Padé approximants associated with a system of arcs, submitted to J. Approximation Theory (1976).

9 Siegel, C.L., Topics in Complex Function Theory, Vol. 2, Chap. 4, Wiley-Interscience, New York, 1971.

10 Szegö, G., Orthogonal Polynomials, American Mathematical Society, New York, 1959, p27.

11 Ref. 10, p29.

12 Ref. 10, Chaps. 10, 11, 12.

13 Ref. 10, p296.

14 Widom, H., Extremal polynomials associated with a system of curves in the complex plane, Advances in Math. 3 (1969), 127-232.

J. Nuttall,*
Department of Physics,
University of Western Ontario,
London, Ontario, Canada, N6A 3K7.

* Research supported in part by the National Research Council Canada.

PADÉ APPROXIMANTS AND INDEFINITE INNER PRODUCT SPACES

H. van Rossum

Convergence theory in the Padé table for a Stieltjes series can be based on the convergence of a sequence of self adjoint operators to the multiplication operator in Hilbert space. In the more general case of a Padé table for a normal power series, Hilbert space has to be replaced by an indefinite inner product space where the multiplication operator satisfies a condition QN (for a definition see section 1).

1 Introduction

Let V denote an infinite dimensional non-degenerate indefinite inner product space with inner product $\langle .,. \rangle$. Let A denote a symmetric operator $V \to V$. Let $u_o \in V$ be non-neutral such that all powers of A on u_o exist. We generate the sequence $u_o, u_1 = Au_o, \ldots, u_n = A^n u_o, \ldots$. From the symmetry of A it follows:

$$\langle u_i, u_j \rangle = \langle u_k, u_l \rangle \quad \text{if } i+j = k+l,\ i,j,k,l \in \mathbb{N}_o .$$

Moreover $\langle u_i, u_j \rangle \in \mathbb{R}$ $(i,j \in \mathbb{N}_o)$. Thus we obtain a sequence of real numbers $(c_n)_{n=o}^{\infty}$, with $c_n = \langle A^n u_o, u_o \rangle$. From now on we impose on A the extra <u>condition QN</u>

$$D_n = |c_{i+j}|_{i,j=o}^{n} \neq 0 \;, \quad (n \in \mathbb{N}_o).$$

Remark. The condition QN implies the non-neutrality of u_o. The sequence $(c_n)_{n=o}^{\infty}$ satisfying QN is called <u>quasi normal</u>.

THEOREM 1.1. <u>The set</u> $\{u_o, u_1, \ldots, u_n\}$ <u>is linearly independent</u>.

<u>Proof</u>. Follows immediately from condition QN satisfied by the operator generating the sequence.

THEOREM 1.2. <u>For every</u> $n \in \mathbb{N}$, <u>the space</u> $V_n = L_{\mathbb{R}}(u_o, u_1, \ldots u_{n-1})$ <u>is non-degenerate</u>.

Proof. (We show that u_n has a unique projection on V_n).
Put $u'_n = -\alpha_0 u_0 - \alpha_1 u_1 - \ldots - \alpha_{n-1} u_{n-1}$; $\alpha_0, \alpha_1, \ldots, \alpha_{n-1}$ can be uniquely determined such that $u_n - u'_n$ is orthogonal to $u_0, u_1, \ldots, u_{n-1}$. Indeed, forming the inner product of $u_n - u'_n$ with $u_0, u_1, \ldots, u_{n-1}$ respectively gives

$$
\begin{array}{lllll}
\langle u_o,u_o\rangle\alpha_o & +\langle u_o,u_1\rangle\alpha_1 & +\ldots+\langle u_o,u_{n-1}\rangle\alpha_{n-1} & +\langle u_o,u_n\rangle & =0 \\
\langle u_1,u_o\rangle\alpha_o & +\langle u_1,u_1\rangle\alpha_1 & +\ldots+\langle u_1,u_{n-1}\rangle\alpha_{n-1} & +\langle u_1,u_n\rangle & =0 \\
\cdot & \cdot & \cdots \quad \cdot & \cdot & \cdot \\
\langle u_{n-1},u_o\rangle\alpha_o & +\langle u_{n-1},u_1\rangle\alpha_1 & +\ldots+\langle u_{n-1},u_{n-1}\rangle\alpha_{n-1} & +\langle u_{n-1},u_n\rangle & =0
\end{array}
$$

Condition QN ensures the determinant of this system to be non-vanishing. Hence u_n has u'_n as its unique projection on V_n □.

We now consider the space $V_u = L_{\mathbb{R}}(u_0,u_1,\ldots,u_n,\ldots)$ and apply the Gram-Schmidt orthogonalization process to the sequence $(u_n)_{n=0}^{\infty}$. Put $u_0 = p_0$, next $p_1 = u_1 - \alpha_1 u_0$ and choose α_1, such that $\langle p_1,u_o\rangle = 0$. In general $p_k = u_k - \alpha_1 u_{k-1} - \ldots - \alpha_k u_0$, where we require $\langle p_k,u_i\rangle = 0$ $(i = 0,1,\ldots,k-1)$. We have the matrix equation

$$
(1.1)\qquad \begin{pmatrix}
\langle u_{k-1},u_o\rangle & \langle u_{k-2},u_o\rangle & \cdots & \langle u_o,u_o\rangle \\
\langle u_{k-1},u_1\rangle & \langle u_{k-2},u_1\rangle & \cdots & \langle u_o,u_1\rangle \\
\vdots & \vdots & & \vdots \\
\langle u_{k-1},u_{k-1}\rangle & \langle u_{k-2},u_{k-1}\rangle & \cdots & \langle u_o,u_{k-1}\rangle
\end{pmatrix}
\begin{pmatrix} \alpha_1 \\ \alpha_2 \\ \vdots \\ \alpha_k \end{pmatrix}
=
\begin{pmatrix} \langle u_k,u_o\rangle \\ \langle u_k,u_1\rangle \\ \vdots \\ \langle u_k,u_{k-1}\rangle \end{pmatrix}.
$$

There is a unique solution since $D_{k-1} \neq 0$. Hence $L_{\mathbb{R}}(p_0,p_1\ldots,p_k) = L_{\mathbb{R}}(u_0,u_1,\ldots,u_k)$, $(k \in \mathbb{N}_0)$. Adjoining to (1.1) the equation

$$\langle p_k,p_k\rangle = \langle u_k,u_k\rangle - \Sigma_{j=1}^{k}\alpha_j\langle u_{k-j},u_k\rangle ,$$

we find, solving by Cramer's rule $\langle p_k,p_k\rangle = D_k/D_{k-1}$, $(k=1,2,\ldots)$. Recurrence relations for three consecutive elements of the orthogonal sequence $(p_n)_{n=0}^{\infty}$ are formed in the usual way to obtain:

$$p_0=u_0,\ p_1=(A-\frac{c_1}{c_0}I)p_0,\ p_{n+1}=(A-a_nI)p_n - b_{n-1}p_{n-1}\ (n=1,2,\ldots),$$

with $a_n = \frac{\langle Ap_n,p_n\rangle}{\langle p_n,p_n\rangle}$, $b_{n-1} = \frac{\langle p_n,p_n\rangle}{\langle p_{n-1},p_{n-1}\rangle}$, $(n = 1,2,\ldots)$.

Because of condition QN we have $b_n \neq 0$, $(n \in \mathbb{N}_o)$.

2 The multiplicationoperator and its matrix

Consider the space $P = L_{\mathbb{R}}(1,x,\ldots,x^n,\ldots)$, vector space of polynomials in an indeterminate x with real coefficients. Let Ω be a linear functional on P defined by $\Omega(x^n) = c_n$ $(n \in \mathbb{N}_o)$ where $(c_n)_{n=0}^{\infty}$ is the quasi normal sequence introduced in section 1. We introduce an inner product $(.,.)$ in P as follows: For any two elements Q_n,Q_m of P,

$$(Q_n,Q_m) := \Omega(Q_n(x)Q_m(x)). \quad \text{Let } \phi(x^n) = u_n, \ (n \in \mathbb{N}_o).$$

The mapping $\phi : P \to V_u$ is easily seen to be an isometric isomorphism. Let $\hat{\Phi} : \mathrm{Hom}(P,P) \to \mathrm{Hom}(V_u,V_u)$ be the mapping assigning to $\mathfrak{A} \in \mathrm{Hom}(P,P)$ the mapping $\hat{\mathfrak{A}} \in \mathrm{Hom}\,(V_u,V_u)$ such that the following diagram is commutative

$$\begin{array}{ccc} P & \overset{\phi}{\to} & V_u \\ \mathfrak{A}\downarrow & & \downarrow\hat{\mathfrak{A}} \\ P & \overset{\phi}{\to} & V_u \end{array} \qquad \phi\,\mathfrak{A} = \hat{\mathfrak{A}}\,\phi .$$

Now let $\mathfrak{A}$ be the multiplication operator in P, i.e., $\mathfrak{A}Q(x) = xQ(x)$ for every element $Q \in P$. Then we have

$$\hat{\mathfrak{A}}(u_n) = \hat{\mathfrak{A}}(\phi(x^n)) = \phi(\mathfrak{A}(x^n)) = \phi(x^{n+1}) = u_{n+1} \quad (n \in \mathbb{N}_o).$$

Hence $\hat{\mathfrak{A}} = A$, where A is the operator introduced in section 1.

Put $\phi^{-1}(p_n) = B_n(x)$. Then $(B_n(x))_{n=0}^{\infty}$ is a sequence of polynomials orthogonal in $P(.,.)$.

We derive the matrix representation of $\mathfrak{A}$ with respect to the orthonormal sequence $(C_n(x))_{n=0}^{\infty}$ where

$$C_n(x) = \left|\frac{D_{n-1}}{D_n}\right|^{\frac{1}{2}} B_n(x) \ , \ (n = 1,2,\ldots).$$

The recurrence relations are

$$(2.1) \quad xC_n(x) = a_{n,n+1}C_{n+1}(x) + a_{n,n}C_n(x) + a_{n,n-1}C_{n-1}(x) .$$

From (2.1):

(2.2) $a_{n,n-1} = \Omega(xC_n(x)C_{n-1}(x))\varepsilon_{n-1}$, where $\varepsilon_{n-1} = \pm 1$.

Substituting in (2.2), $xC_{n-1}(x) = a_{n-1,n}C_n(x) + R_{n-1}(x)$ where $R_{n-1}(x)$ is a polynomial in x of degree n-1 at most,

$$|a_{n,n-1}| = |\Omega(C_n(x)\{a_{n-1,n}C_n(x) + R_{n-1}(x)\})| = |\Omega(a_{n-1,n}C_n(x)C_n(x))| = $$
$$= |a_{n-1,n}|.$$

Moreover, comparing leading coefficients in (2.1) yields:

$$a_{n,n+1} = \frac{|D_{n-1}D_{n+1}|^{\frac{1}{2}}}{D_n} \neq 0$$

The matrix $(a_{i,j})$ for Ω satisfies $|a_{n-1,n}| = |a_{n,n-1}|$, n=1,2,... . We call it quasi definite. Conversely, given a quasi definite tri-diagonal matrix $(a_{i,j})$ of the type considered above there corresponds to it a unique sequence of moments $(c_n)_{n=0}^{\infty}$ that is quasi normal. For a proof compare [6].

Remark. Let $B_n(x) = \beta_0 + \beta_1 x + \ldots + \beta_n x^n$ be an arbitrary element of P, then applying the operator ϕ defined above:

$$p_n = \phi(B_n(x)) = \sum_{k=0}^{n}\beta_k\phi(x^k) = \sum_{k=0}^{n}\beta_k u_k = \sum_{k=0}^{n}\beta^k A^k u_o = B_n(A)u_o.$$

3 Orthogonality properties of Padé approximants

Let $f(x) = \sum_{n=0}^{\infty} c_n x^n$ be a formal power series for which all determinants $\left|c_{i+j+k}\right|_{i,j=0}^{n}$ $(k,n \in \mathbb{N}_0)$ differ from zero. The series as well as the sequence $(c_n)_{n=0}^{\infty}$ are called <u>semi normal</u>. If also $(c'_n)_{n=0}^{\infty}$ is semi normal where $\sum_{n=0}^{\infty} c'_n x^n$ is reciprocal to $\sum_{n=0}^{\infty} c_n x^n$, then $(c_n)_{n=0}^{\infty}$ is called <u>normal</u>; $\sum_{n=0}^{\infty} c_n x^n$ is called normal power series. For the present, let $\sum_{n=0}^{\infty} c_n x^n$ be semi normal. For any fixed integer $k \in \mathbb{N}_0$ we have

(3.1) $Q_{n,k+n}(x)f(x) - P_{n,k+n}(x) = d_{n,k+n}x^{k+2n+1} + Q(x^{k+2n+2})$

where $P_{n,k+n}(x)/Q_{n,k+n}(x)$ is the (n,n+k) Padé approximant to f(x). Normality of the sequence $(c_n)_{n=0}^{\infty}$ is equivalent to:

degree $Q_{n,k+n}(x) = n$, degree $P_{n,k+n}(x) = k+n$ and $d_{n,k+n} \neq 0$.
From (3.1) it can easily be deduced that the sequence

$$(x^n Q_{n,k+n}(x^{-1}))_{n=o}^{\infty}$$

is orthogonal with respect to the inner product $(.,.)_k$ defined on P as follows: For any two elements $Q_n, Q_m \in P$

$$(Q_n, Q_m)_k := \Omega_k(Q_n(x)Q_m(x))$$

where the linear functional Ω_k is defined by $\Omega_k(x^n) = c_{k+n+1}$ $(n \in \mathbb{N}_o)$.
The sequence $(x^n Q_{n,n-1}(x^{-1}))_{n=1}^{\infty}$ is orthogonal with respect to the inner product introduced in section 2, if $(c_n)_{n=o}^{\infty}$ is normal.
In the space P(.,.) considered in section 2, we introduce a new inner product, the $\mathfrak{A}$-inner product $(.,.)_{\mathfrak{A}}$ as follows: For any two elements $Q_n, Q_m \in P$

$$(Q_n, Q_m)_{\mathfrak{A}} = (\mathfrak{A} Q_n, Q_m)$$

It is clear what is meant by $\mathfrak{A}$-orthogonal, $\mathfrak{A}$-neutral, etc.

$$(Q_n, Q_m)_{\mathfrak{A}} = (\mathfrak{A} Q_n, Q_m) = \Omega_o(Q_n(x)Q_m(x)), \text{ where } \Omega_o(x^n) = c_{n+1} .$$

Returning for a moment to the matrix $(a_{i,j})$ with respect to the sequence $(C_n(x))_{n=0}^{\infty}$ we have

$$(C_n, C_n)_{\mathfrak{A}} = (xC_n, C_n) = a_{n,n}(C_n, C_n).$$

So if a diagonal element of the matrix is zero, the corresponding basis element is $\mathfrak{A}$-neutral. In the case of the Padé table for exp(x), the "inverse" Padé denominators $x^n Q_{nn}(x^{-1}) = x^n {}_1F_1(-n;-2n;-x^{-1})$ are $\mathfrak{A}$-neutral for all $n > 1$ as follows directly from the recurrence relations.
If $(c_n)_{n=o}^{\infty}$ is semi normal, the inverse Padé denominators $x^n Q_{n,n+k}(x^{-1})$ $(n \in \mathbb{N}_o)$ belonging to $f(x) = \Sigma_{n=o}^{\infty} c_n x^n$ are $\mathfrak{A}^{k+1}$-orthogonal. This holds for all $k \in \mathbb{N}_o$.
A characterization of semi normality in terms of the functionals $\Omega_o, \Omega_1, \Omega_2, \ldots$ is contained in the next theorem.

THEOREM 3.1. The sequence $(c_n)_{n=0}^{\infty}$ is semi normal iff the restrictions of $\Omega_0, \Omega_1, \ldots, \Omega_{n-1}$ to $L_{\mathbb{C}}(1,x,\ldots,x^{n-1}) = U_n$ are linearly independent for $n = 1,2,\ldots$.

Proof. The mapping $\alpha : U_n \to \mathbb{C}^n$ defined by $\alpha(a_0+a_1x+\ldots+a_{n-1}x^{n-1}) = (a_0,a_1,\ldots,a_{n-1})$ is an isomorphism and $\beta: \mathbb{C}^n \to (\mathbb{C}^n)^*$ defined by:

$\beta(\eta_0,\ldots,\eta_{n-1}) = f \Leftrightarrow f(\xi_0,\ldots,\xi_{n-1}) = \xi_0\eta_0+\ldots+\xi_{n-1}\eta_{n-1}$ for all $(\xi_0,\ldots,\xi_{n-1}) \in \mathbb{C}^n$, is also an isomorphism. We have: $U_n^* \overset{\alpha}{\to} \mathbb{C}^n$ and $U_n^* \overset{\alpha^*}{\leftarrow} (\mathbb{C}^n)^* \overset{\beta}{\leftarrow} \mathbb{C}^n$. Let $f_0,\ldots,f_{n-1}$ be the restrictions of $\Omega_0,\ldots,\Omega_{n-1}$ to U_n. Then $\alpha^*(\beta(c_k,c_{k+1},\ldots,c_{k+n-1})) = f_k$ $(k = 0,1,\ldots,n-1)$. If $(c_n)_{n=0}^{\infty}$ is semi normal we have $D_{n-1} \neq 0$ and hence the columns of this determinant are linearly independent. Because $\alpha^*\beta$ is an isomorphism, this implies that $f_0,\ldots,f_{n-1}$ are also linearly independent. If on the other hand the restrictions of $f_0,\ldots,f_{k+n-1}$ are linearly independent on U_{k+2n-2}, so are $f_k,f_{k+1},\ldots,f_{k+n-1}$. Because this implies that the columns of $\left|c_{i+j+k}\right|_{i,j=0}^{n-1}$ are linearly independent, the just mentioned determinant differs from zero and we are led to the semi normality of $(c_n)_{n=0}^{\infty}$.

REMARK Prof N.H. Kuiper (Bûres sur Yvette) drew my attention to this fact; the proof given above is due to E. Hendriksen (Amsterdam).

4 Convergence

The space $V_u\langle .,. \rangle$ introduced in section 2 is the span of an infinite orthogonal system $(p_n)_{n=0}^{\infty}$. Hence V_u is an infinite dimensional, non-degenerate, decomposable space. We write $V_u = V_u^+ \oplus V_u^-$, where $\oplus$ stands for orthogonal direct sum; the positive p_n belong to V_u^+ the negative p_n belong to V_u^-. The space V_u^+ can be embedded uniquely as a dense subspace in a Hilbert space H_1. Similarly V_u^- can be completed with respect to the norm $\|x^-\|=(-\langle x,x\rangle)^{\frac{1}{2}}$. This yields the space H_2, the antispace of a

Hilbert space. The space $K = H_1 \oplus H_2$ is a Krein space (see [1] for an axiomatic definition). Let P^+ and P^- be the orthogonal projectors mapping K onto H_1 and H_2 respectively. Put $J = P^+ - P^-$ then one has in K the positive definite inner product $(.,.)_J$ defined as follows

$$(x,y)_J = \langle Jx,y \rangle.$$

The J-norm $\|.\|$ is defined by $\|x\| = (x,x)_J^{\frac{1}{2}}$ $(x \in K)$. Notions like dense, closed, bounded, etc always refer to the strong topology induced by the J-norm.

If A is the symmetric operator on V_u discussed before, then because A is bounded and densely defined on K, it has a unique closure. Denote this closure by B. Since the spectrum of B is contained in the disc $|z| \leq \|B\|$ in the complex plane, we have the convergent expansion for the resolvent $R_z(B)$ for all $z \in \mathbb{C}$ satisfying $|z| > \|B\|$,

$$R_z(B) = (zI - B)^{-1} = Iz^{-1} + Bz^{-2} + \ldots + B^{n-1}z^{-n} + \ldots$$

For the leading element of the matrix $R_z(B)$ with respect to the basis $(p_n)_{n=o}^{\infty}$ we have (assuming p_o to be positive),

$$((zI-B)^{-1}p_o,p_o) = ((\lim_{m\to\infty} \Sigma_{n=o}^{m} z^{-n-1} B^n)p_o,p_o) =$$

$$= \lim_{m\to\infty} ((\Sigma_{n=o}^{m} z^{-n-1} B^n)p_o,p_o) = \lim_{m\to\infty} \Sigma_{n=o}^{m} z^{-n-1}(B^n p_o,p_o) =$$

$$= \lim_{m\to\infty} \Sigma_{n=o}^{m} c_n z^{-n-1} = \Sigma_{n=o}^{\infty} c_n z^{-n-1}.$$

A result in [1] implies that our spaces $V_n = L_{\mathbb{R}}(u_o,u_1,\ldots,u_{n-1})$ being finite dimensional as well as non-degenerate are <u>ortho-complemented</u>. Hence for every $n \in \mathbb{N}_o$ an orthogonal projector E_n exists with range V_n mapping every element of K into its projection on V_n. For each $n \in \mathbb{N}_o$ we obtain an operator $B_n = E_n B E_n$ which is symmetric, and $\|B_n\| \leq \|B\|$. The acting of B_n on V_n can be described as follows (compare [5]):

$$u_k = B_n^k u_o, \ (k=0,1,\ldots,n-2); \quad B_n u_{n-1} = E_n u_n$$

The resolvent $R_z(B_n)$ is bounded for $|z| > \|B\|$; $R_z(B_n)$ transforms V_n into itself. So

(4.1) $(zI-B_n)^{-1}p_o = \xi_o p_o + \xi_1 p_1 + \dots + \xi_{n-1}p_{n-1}$,

whence $((zI-B_n)^{-1}p_o,p_o) = \xi_o$. Also from (4.1) we obtain

(4.2) $p_o = \Sigma_{i=o}^{n-1} \xi_i(zI-B_n)p_i$.

Let (a_{ij}) be the matrix of B (of A) with respect to the orthonormal basis $(p_n)_{n=o}^{\infty}$. Solving the system of linear equations from (4.2) by Cramer's rule for ξ_o we get

$$\xi_o = \frac{\begin{vmatrix} z-a_{11} & -a_{12} & -a_{23} & \dots & 0 \\ -a_{21} & z-a_{22} & -a_{23} & \dots & 0 \\ \vdots & \vdots & \vdots & \dots & \vdots \\ 0 & 0 & 0 & -a_{n-1,n-2} & z-a_{n-1,n-1} \end{vmatrix}}{\begin{vmatrix} z-a_{00} & -a_{01} & 0 & \dots & 0 \\ -a_{10} & z-a_{11} & -a_{12} & \dots & 0 \\ \vdots & \vdots & \vdots & \dots & \vdots \\ 0 & 0 & 0 & -a_{n-1,n-2} & z-a_{n-1,n-1} \end{vmatrix}}.$$

From this we see, that the leading element $((zI-B_n)^{-1}p_o,p_o)$ of the matrix of the resolvent $R_z(B_n)$ equals the (n,n-1)-Padé approximant for $\Sigma_{n=o}^{\infty} c_n z^{-n}$, where $c_n = \langle B^n p_o,p_o\rangle = (B^n p_o,p_o)$.

The sequence $(B_n)_{n=o}^{\infty}$ converges strongly to B (see [5]).
As a consequence $(R_z(B_n))_{n=o}^{\infty}$ converges strongly to $R_z(B)$. Since strong convergence implies weak convergence we have

$$\lim_{n\to\infty} (R_z(B_n)p_o,p_o) = (R_z(B)p_o,p_o) \;, \quad |z| > \|B\|.$$

This means: The (n,n-1)-Padé approximant to $\Sigma_{n=o}^{\infty} c_n z^n$ converges to $\Sigma_{n=o}^{\infty} c_n z^n$ for all z, with $|z| < \frac{1}{\|B\|}$.

EXAMPLE The Padé table for ${}_1F_1(1;c;z)$ $(c > 0)$. The (n,n-1)-inverse Padé denominator is

$$B_n(z) = z^n {}_1F_1(-n;-c-2n+2;-z^{-1}).$$

The sequence $(B_n)_{n=1}^{\infty}$ $(B_o \equiv 1)$ is orthogonal with respect to the inner product $(.,.)_{-1}$ based on the sequence

$$1, \frac{1}{c}, \frac{1}{c(c+1)}, \ldots$$

The inner square of B_n is

$$(B_n,B_n)_{-1} = (-1)^n \frac{(c+2n-1)\Gamma(c)\Gamma(c+n-1)}{\{\Gamma(c+2n)\}^2} \quad (n \quad {}_o),$$

hence in the basis $(B_n)_{n=o}^{\infty}$ there are infinitely many positive - as well as negative elements. So the space $P(.,.)_{-1}$ becomes a Krein spaces after completion.

Apart from normalization the polynomials B_n are the generalized Bessel polynomials.

In this example, from [4], Ch VI it follows that the underlying operator is bounded so the convergence result derived above applies.

References

1 Bognár, J., Indefinite Inner Product Spaces, Springer, Berlin, 1974.

2 Bruin, M.G. de and H. van Rossum, Formal Padé approximation, Nieuw Archief voor Wiskunde (3), 23 (1975), 115-130.

3 Masson, D., Hilbert Space and the Padé Approximant in: The Padé Approximant in Theoretical Physics, ed. G.A. Baker and J.L. Gammel, Academic Press, New York, 1971.

4 Rossum, H. van, A theory of orthogonal polynomials based on the Padé table, van Gorcum, Assen, 1953.

5 Vorobyev, Yu.V., Methods of Moments in Applied Mathematics, Gordon and Breach, New York, 1965.

6 Wall, H.S., Analytic Theory of Continued Fractions, D. van Nostrand, Toronto-New York-London, 1948.

H. van Rossum
Instituut voor Propedeutische Wiskunde
Universiteit van Amsterdam
Roetersstraat 15, Amsterdam, The Netherlands

THE TRANSFORMATION OF SERIES BY THE USE OF PADÉ QUOTIENTS AND MORE GENERAL APPROXIMANTS

P. Wynn

Methods for extracting a generating function from an asymptotic series are reviewed, and classical results concerning the equivalence of such methods are outlined. A general approximant, of which the Padé quotient is a special form, is introduced. Transformations of the Euler Maclaurin series by the use of both Padé quotients and the more general approximants are considered. Methods for evaluating an integral over a semi-infinite interval in terms of the derivatives of the integrand at the finite end point are referred to. A method for approximating the sum of a series in terms of the derivatives of its first term is described.

1 Introduction and summary

Much of what has recently been written on the theory of continued fractions was known and better understood fifty years ago. A useful purpose is therefore served by providing a synopsis of the development of a selected component of the theory as originally presented, together with original references. This is done with regard to certain generalisations of continued fractions and Padé quotients arising from the transformation of asymptotic series and series of functions; new developments in this subject are mentioned. To illustrate the theory, transformations of the Euler Maclaurin series by both classical continued fractions and their generalised variants are considered. Reference is made to certain recently introduced methods for approximating an integral over a semi-infinite interval in terms of the derivatives of the integrand at the finite end point, and one method is briefly described. As a bonne bouche, it is shown that the selected method may be combined with use of the Euler-Maclaurin formula to approximate the sum of a series in terms of the derivatives of its first term.

2 Notation and conventions

The index of single summation is always ν; if the upper limit is infinity it is omitted from the summation sign, and if the lower limit is also zero, both limits are omitted: $\Sigma_1^n f_\nu, \Sigma_1 f_\nu, \Sigma f_\nu$ are $\sum_{\nu=1}^{n} f_\nu, \sum_{\nu=1}^{\infty} f_\nu, \sum_{\nu=0}^{\infty} f_\nu$ respectively. $H[t_{\tau+m}]_r$ $(0\leq m, r<\infty)$ is the value of the Hankel determinant of order $r+1$ whose $(\nu+1)^{th}$ row $(\nu=0,1,\ldots,r)$ contains the elements $t_{\tau+m+\nu}$ $(\tau=0,1,\ldots,r)$. With $-\infty\leq\alpha<\beta\leq\infty$, $\sigma\in\mathbb{B}(\alpha,\beta)$ means that the real valued function σ is bounded and nondecreasing over $[\alpha,\beta]$ and such that all moments $\int_\alpha^\beta t^\nu d\sigma(t)$ $(\nu=0,1,\ldots)$ exist. $\mathcal{D}$ is the differential operator with respect to μ: $\mathcal{D}^2\psi(\mu)$ is $d^2\psi(\mu)/d\mu^2$.

3 Asymptotic series

It was shown by Borel [6] and Carleman [7] that if a sequence f_ν $(\nu=0,1,\ldots)$ of finite complex numbers is given, and a sector $\Delta(a,b)$ containing the points λ for which $a\pi\leq\arg(\lambda)\leq b\pi$ $(-1<a\leq 1;$ $a<b<a+2)$ is prescribed, then a function $F(\lambda)$ exists such that

$$(1)\qquad \lim \lambda^n\{F(\lambda)-\Sigma_0^{n-1} f_\nu \lambda^{-\nu-1}\}=0 \qquad (n=0,1,\ldots)$$

as λ tends to infinity in $\Delta(a,b)$. If $|A|<\infty$, $0<B<\infty$ and $C<1/2(b-a)$, then

$$(2)\qquad \lim \lambda^n A\exp\{-B(e^{-\frac{1}{2}(a+b)i\pi}\lambda)^C\}=0 \qquad (n=0,1,\ldots)$$

as λ tends to infinity in $\Delta(a,b)$: the function $F(\lambda)$ in (1) may be replaced by a suitable sum of the form

$$F(\lambda)+\Sigma_1^r A_\nu \exp\{-B_\nu(e^{-\frac{1}{2}(a+b)i\pi}\lambda)^{C_\nu}\}.$$

Other functions obey limiting relationships of the form (2). In short, asymptotic series are associated not with unique functions but with classes of functions.

Nevertheless, auxiliary conditions sufficient to determine a unique member of the asymptotic class may be imposed. We consider two such sets of conditions. The first set results from theory due to Watson [33] and F. Nevanlinna [18] (for an authoritative discussion of this theory, see [20]). It is supposed that the function $F(\lambda)$ is regular in $\mathbb{D}(a,b;c)$, the union of the concave sector $(a-\frac{1}{2})\pi\leq\arg(\lambda)\leq(b+\frac{1}{2})\pi$ $(a\in(-\frac{1}{2},0), b\in(0,\frac{1}{2}))$, $|\lambda|<\infty$ and the disc $|\lambda|\leq c\in(0,\infty)$, and that, setting

$$F(\lambda)=\Sigma_0^{n-1} f_\nu \lambda^{-\nu-1}+R_n(\lambda) \qquad (n=0,1,\ldots) \tag{3}$$

we have, for some fixed $\xi\in(0,\infty)$, $f_\nu=O(\nu!\xi^{-\nu})$ for large ν and, uniformly for λ in $\mathbb{D}(a,b;c)$, $R_n(\lambda)=O(n!\xi^{-n}|\lambda|^{-n-1})$ for large n. Under these conditions $F(\lambda)$ both satisfies relationships (1) in $\Delta(a,b)$ and is uniquely determined in this convex sector: when $a\pi<\arg(\lambda)<b\pi$, the function $\tilde{F}(\lambda,u)$ defined for small u by the series

$$\Sigma f_\nu u^\nu/(\nu!\lambda^{\nu+1}) \tag{4}$$

is regular for $u\in(0,\infty)$ and $F(\lambda)=\int_0^\infty e^{-u}\tilde{F}(\lambda,u)du$.

The second set of auxiliary conditions derives from the theories of the Hamburger [11] and Stieltjes [29] moment problems. It is supposed that $F(\lambda)$ is regular for all finite nonreal λ with $\mathrm{Im}\{F(\lambda)\}\leq 0$ for $\mathrm{Im}(\lambda)>0$. Relationships (1) holding over $\Delta(\delta,1-\delta)$ for some $\delta\in(0,\frac{1}{2})$ imply that F belongs to a subclass of functions representable in the form

$$F(\lambda)=\int_{-\infty}^{\infty}(\lambda-t)^{-1}d\sigma(t) \qquad (\sigma\in\mathbb{B}(-\infty,\infty)). \tag{5}$$

To ensure that only one normalised function σ is permissible in the representation (5), i.e. that F is uniquely determined, a further condition upon the $\{f_\nu\}$ must be imposed; for example (Carleman [7]) that the series $\Sigma_1 f_{2\nu}^{-1/2\nu}$ should diverge (e.g. $f_\nu=O\{(\nu+K)!\xi^{-\nu}\}$, $0\leq K<\infty$, $0<\xi<\infty$). If $F(\lambda)$ is regular for all finite $\lambda\notin[0,\infty]$ and $\mathrm{Im}\{\lambda F(\lambda^2)\}\leq 0$ for $\mathrm{Im}(\lambda)>0$, and relationships (1) hold over $\Delta(\delta,2-\delta)$ for some $\delta\in(0,1)$, then F has a representation of the form

$$F(\lambda)=\int_0^\infty(\lambda-t)^{-1}d\sigma(t) \qquad (\sigma\in\mathbb{B}(0,\infty)) \tag{6}$$

and is uniquely determined if the series $\Sigma_1 f_\nu^{-1/2\nu}$ diverges (e.g. $f_\nu = O\{(2\nu+K)!\xi^{-\nu}\}$, $0 \le K < \infty$, $0 < \xi < \infty$). Furthermore, when $\lambda \in (-\infty, 0)$, the series $\Sigma f_\nu \lambda^{-\nu-1}$ is semi-convergent; in the notation of formula (3), $|R_n(\lambda)| < f_n |\lambda|^{-n-1}$ $(n=0,1,\ldots)$.

By way of relating the above theory directly to that of the moment problem, we remark that a representation of the form (5) for the function $F(\lambda)$ concerned exists if and only if $\sigma \in \mathbb{B}(-\infty,\infty)$ can be found such that

$$(7) \qquad \int_{-\infty}^{\infty} t^\nu d\sigma(t) = f_\nu; \qquad (\nu=0,1,\ldots)$$

the solubility of the Hamburger moment problem (7) with σ not reducing to a simple step function with a finite number of salti implies and is implied by the condition that

$$(8) \qquad H[t_\tau]_r > 0; \qquad (r=0,1,\ldots)$$

if the series $\Sigma_1 f_{2\nu}^{-1/2\nu}$ diverges, the moment problem is determinate: only one normalised function σ can be taken in formula (7). A representation of the form (6) for the function $F(\lambda)$ concerned exists if and only if the Stieltjes moment problem

$$(9) \qquad \int_0^{\infty} t^\nu d\sigma(t) = f_\nu \qquad (\nu=0,1,\ldots)$$

is soluble which, with σ not reducing to a simple step function with a finite number of salti, implies and is implied by the condition that

$$(10) \quad H[t_\tau]_r, H[t_{\tau+1}]_r > 0; \qquad (r=0,1,\ldots)$$

if the series $\Sigma_1 f_\nu^{-1/2\nu}$ diverges, the moment problem (9) is determinate.

4 <u>Methods for extracting the generating function</u>

Borel's method [6] applied to a power series $\Sigma f_\nu \lambda^{-\nu-1}$ proceeds as follows: it is assumed that for all λ belonging to a prescribed point set $\mathbb{M}$, a) the series (4) converges for small values of u, b) the function $\tilde{F}(\lambda,u)$ so defined is regular for $u\in(0,\infty)$ and c) the integral $S(\lambda)=\int_0^\infty e^{-u}\tilde{F}(\lambda,u)du$ exists; following Hardy ([12] §8.11), $\Sigma f_\nu\lambda^{-\nu-1}$ is then said to be summable (B*) to S over $\mathbb{M}$. Precisely this method of summation features in the Watson-Nevanlinna theory.

Le Roy's method [17] may be regarded as an extension of that of Borel: applied to the power series $\Sigma f_\nu\lambda^{-\nu-1}$, it is assumed that for a fixed $\chi\in(0,\infty)$ and all λ belonging to a prescribed point set $\mathbb{M}$, a) the series $\Sigma f_\nu u^\nu/((\chi\nu)!\lambda^{\nu+1})$ converges for small values of u, b) the function $F_\chi(\lambda,u)$ so defined is regular for $u\in(0,\infty)$ and c) the integral

$$S_\chi(\lambda)=\chi^{-1}\int_0^\infty \exp(-u^{1/\chi})u^{(1/\chi)-1}F_\chi(\lambda,u)du$$

exists; $\Sigma f_\nu\lambda^{-\nu-1}$ is then said to be summable (B,χ) to S_χ over $\mathbb{M}$. (B*) is, of course, (B,1).

An extraction method of quite different type, whose use in the transformation of asymptotic series of a special but large class was pioneered by Stieltjes [29], is based upon the use of continued fractions. If conditions (8) hold, the series $\Sigma f_\nu\lambda^{-\nu-1}$ generates an associated continued fraction ([23] Ch. 3) whose successive convergents have the form

$$(11) \quad C_i(\lambda) = \frac{v_1}{\lambda-w_1-}\,\frac{v_2}{\lambda-w_2-}\cdots\frac{v_i}{\lambda-w_i}. \qquad (i=1,2,\ldots)$$

If the series $\Sigma_1 f_{2\nu}^{-1/2\nu}$ diverges, $\{C_i(\lambda)\}$ converges uniformly to the function $F(\lambda)$ of formulae (5,7) in any bounded domain in the λ-plane not containing any point of the real axis. If conditions (10) hold, the series $\Sigma f_\nu\lambda^{-\nu-1}$ generates a corresponding continued fraction whose even order convergents have the form (11), and now, in particular, if the series $\Sigma_1 f_\nu^{-1/2\nu}$ diverges, uniform convergence of the sequence $\{C_i(\lambda)\}$ to the function $F(\lambda)$ of formulae (6,8) holds over any bounded domain not containing any point of the nonnegative real axis.

The above continued fraction methods are constructive in the sense that for a fixed value of λ lying in the appropriate convergence domain the $\{C_i(\lambda)\}$ offer approximations to the value of the limit function $F(\lambda)$. The methods of Borel and Le Roy merely provide analytic expressions for $F(\lambda)$ of possible use in subsequent work.

5 Equivalence of extraction methods

It was shown by Hamburger [10] and F. Bernstein [3] that, subject to certain conditions, the extraction methods of the two distinct types described above produce the same result.

Hamburger's argument was addressed to the case in which the coefficients of the series $\Sigma f_\nu \lambda^{-\nu-1}$ satisfy conditions (8) and

$$(12)\quad |f_\nu| \le K\xi^{-\nu}\nu! \qquad (\nu=0,1,\ldots)$$

for fixed $K,\xi\in(0,\infty)$. The convergents $\{C_i(\lambda)\}$ of formula (11) of the continued fraction associated with the above series may be expressed in the form

$$(13)\quad C_i(\lambda)=\Sigma_1^i M_\nu^{(i)}(\lambda-t_\nu^{(i)})^{-1} \qquad (0<M_\nu^{(i)}<\infty, -\infty<t_\nu^{(i)}<\infty)$$

the $\{t_\nu^{(i)}\}$ being distinct. Since, when $t_\nu^{(i)}$ is real

$$(\lambda-t_\nu^{(i)})^{-1}=\int_0^{-i\infty}\exp\{-u(\lambda-t_\nu^{(i)})\}du$$

for finite λ with $\mathrm{Im}(\lambda)\ge\delta>0$,

$$(14)\quad C_i(\lambda)=\int_0^{-i\infty}F_i(u)e^{-u\lambda}du$$

where

$$(15)\quad F_i(u)=\Sigma_1^i M_\nu^{(i)}\exp(t_\nu^{(i)}u).$$

Hamburger showed firstly that $\{F_i(u)\}$ converges for $|u|\le\xi-\delta$, where $\delta\in(0,\infty)$ is arbitrarily small, to $\hat{F}(u)$, the function defined for small u by the series

(16) $\Sigma f_\nu u^\nu/\nu!$,

that the $\{F_i(u)\}$ remain bounded for all $Re(u)\in[-\xi+\delta,\xi-\delta]$, and that in consequence $\{F_i(u)\}$ converges to $\hat{F}(u)$, and $\hat{F}(u)$ is bounded, for all such u. The convergence of the $\{C_i(\lambda)\}$ to the function

(17) $F(\lambda)=\int_0^{-i\infty}\hat{F}(u)e^{-u\lambda}du$

for all λ as restricted above now follows directly, and it is easily snown that the functions $F(\lambda)$ of formulae (5) and (17) are identical. In short, Borel's method in slightly modified form and the use of continued fractions have produced the same result.

If, in place of (8), conditions (10) hold (now $0<t_\nu^{(i)}<\infty$ in formulae (13,15)), conditions (12) still holding, the $\{F_i(u)\}$ converge to the bounded function $\hat{F}(u)$ for $Re(u)\le\xi-\delta$, and $\{C_i(\lambda)\}$ converges to the function $F(\lambda)$ of the form (6) for all finite λ not lying on the nonnegative real axis (this can be demonstrated by rotating the integration path in formulae (14,17); we remark that extraction of the function $F(\lambda)$ by application of the Watson-Nevanlinna theory (conditions (10,12) holding) to the function $F(-\lambda)$ leads to an integral of Borel type defining $F(\lambda)$ only in a sector with included angle slightly less than π).

Simultaneously and independently, F. Bernstein attacked the Stieltjes case (in which conditions (10) hold) directly, assuming that $f_\nu\le K(\chi\nu)!\xi^{-\nu}$ $(\nu=0,1,\ldots)$ $(0<K,\xi<\infty)$ where $0<\chi<2$ (a later remark [4] extended the work to the case in which $\chi=2$) and that $|\lambda|<\infty$, $\lambda\notin[0,\infty]$. He made use of Mittag-Leffler's function $E_\chi(u)=\Sigma u^\nu/(\chi\nu)!$ and the result that

$$(\lambda-t_\nu^{(i)})^{-1}=\chi^{-1}\lambda^{-1+1/\chi}\int_0^\infty\exp\{-(\lambda u)^{1/\chi}\}u^{(1/\chi)-1}E_\chi(ut_\nu^{(i)})du$$

for $0<t_\nu^{(i)}<\infty$, to derive, in place of formula (14),

$$C_i(\lambda)=\chi^{-1}\lambda^{-1+1/\chi}\int_0^\infty\exp\{-(\lambda u)^{1/\chi}\}u^{(1/\chi)-1}F_i(u)du$$

where

(18) $F_i(u)=\Sigma_1^i M_\nu^{(i)} E_\chi(t_\nu^{(i)} u)$.

The remainder of Bernstein's argument closely resembled that of Hamburger: again subject to the stated conditions the methods of Le Roy and Stieltjes are equivalent.

6 Ascending power series

The theory described above is easily adapted to the transformation of ascending power series of the form $\Sigma f_\nu z^\nu$. Subject to conditions (8) this series generates an associated continued fraction whose convergents have the form

$$(19)\quad \hat{C}_1(z)=\frac{v_1}{1-w_1 z},\ \hat{C}_i(z)=\frac{v_1}{1-w_1 z-}\ \frac{v_2}{1-w_2 z-}\ \cdots\ \frac{v_i}{1-w_i z}.\qquad (i=2,3,\ldots)$$

In the notation of formula (11), we have $C_i(\lambda)=\lambda^{-1}\hat{C}_i(\lambda^{-1})$ ($i=1,2,\ldots$) and the decomposition corresponding to (13) is

$$(20)\quad \hat{C}_i(z)=\Sigma_1^i M_\nu^{(i)}(1-t_\nu^{(i)} z)^{-1}.\qquad (0<M_\nu^{(i)}<\infty, -\infty<t_\nu^{(i)}<\infty)$$

Subject to conditions easily transliterated from the above theory, $\{\hat{C}_i(z)\}$ converges to $f(z)=\int_{-\infty}^{\infty}(1-zt)^{-1}d\sigma(t)$. In the Stieltjes case in which conditions (10) hold, convergence is to $f(z)=\int_0^{\infty}(1-zt)^{-1}d\sigma(t)$.

The convergents (19) are Padé quotients: denoting the Padé quotients derived from the series $\Sigma f_\nu z^\nu$ with ((8) holding) by $\{P_{i,j}(z)\}$, $\hat{C}_i(z)=P_{i,i-1}(z)$ ($i=1,2,\ldots$). The more general Padé quotients $P_{i,i+m-1}(z)$ with $m>0$ may be expressed as the sum of a polynomial and an irreducible fraction ([23] Ch. 5) and the remaining Padé quotients $P_{i,j}(z)$ with $i>j+1$ as the reciprocal of such a sum. Using these decompositions, convergence of general forward diagonal sequences of Padé quotients may be demonstrated. Indeed, Wall [32] has shown that under certain conditions which do not now concern us, such sequences of Padé quotients converge to the Borel sum of their generating function.

7 General approximants

As described above, Hamburger's work provides a theory of the approximation of a function $\hat{F}(u)$ defined by the series (16) by linear combinations of exponential functions of the form (15); Bernstein's work provides a similar theory of approximation by linear combinations of Mittag-Leffler functions. The work of these authors may be viewed as the first steps towards a generalisation of the Padé quotient.

Greater strides in this direction were made possible after the appearance of a fundamental and far reaching memoir of M. Riesz ([26], see [1] Ch. 2 and [28] Ch. 2). Formulae such as (13,20) are special cases of quadrature formulae

$$(21) \quad R_i=\Sigma_1^i M_\nu^{(i)} g(t_\nu^{(i)})$$

for the approximation of integrals of the form $R=\int_{-\infty}^{\infty} g(t)d\sigma(t)$ (we have $g(t)=(\lambda-t)^{-1}$ in the case of formula (13) and $g(t)=(1-zt)^{-1}$ in the case of formula (20)). The methods of proof deriving from the theory of quadrature which suffice to demonstrate convergence of the convergents $\{C_i(\lambda)\}$ or $\{\hat{C}_i(z)\}$ to a function $F(\lambda)$ or $f(z)$ under appropriate conditions also serve to demonstrate convergence of approximants of the form (21) derived from functions $g(t)$ of for more general structure than $(\lambda-t)^{-1}$ or $(1-zt)^{-1}$. In particular, the function $g(t)$ is not restricted to being analytic in the complex variable t for all $t\in[-\infty,\infty]$; it is principally from this simple fact that the power and generality of M. Riesz' theory derives.

In his investigations of the conditions under which Parseval's formula holds for the Fourier expansions in terms of an orthonormal polynomial system generated by the function $\sigma\in\mathbb{B}(-\infty,\infty)$ for all $\mathbb{L}_\sigma^2$ functions ($\mathbb{L}_\sigma^2$ is the class of functions g for which $\int_{-\infty}^{\infty}|g(t)|^2 d\sigma(t)<\infty$) M. Riesz showed that $\{R_i\}$ converges to R for all $\mathbb{L}_\sigma^2$ functions if and only if σ is either the unique normalised solution of the Hamburger moment problem (7) or is, in the sense of R. Nevanlinna [19] an extremal solution of this moment problem; the theory associated with the Stieltjes moment problem (9) is slightly simpler. Functions g(t) that are continuous over every bounded closed interval in $(-\infty,\infty)$ and tend to finite limits as t tends to $\pm\infty$ belong to the class $\mathbb{L}_\sigma^2$. As is pointed out in the classic memoir of Shohat and Tamarkin ([28] Ch. 4), the convergence theory of continued fractions associated with a power series of the form $\Sigma f_\nu \lambda^{-\nu-1}$ (the $\{f_\nu\}$ being a moment sequence), in particular, can be picked up from the above results by setting $g(t)=(\lambda-t)^{-1}$ (although this method of derivation is circular; M. Riesz made use of the convergence theory of such continued fractions and the result that the linear manifold determined by the functions $(\lambda-t)^{-1}$ for nonreal λ is dense in $\mathbb{L}_\sigma^2$). But convergence results for the $\{R_i\}$ are not confined to functions g(t) (of possibly other variables besides t) which tend to finite limits as t tends to $\pm\infty$. Continuing the work of M. Riesz, Shohat [27] showed that similar convergence results hold for certain functions of exponential growth in t (special cases of these results lead back to the Hamburger approximants (14)).

The above theory is trivially extended to functions of other variables besides t, and to approximants derived from more general Padé quotients: assuming that

(22) $G(z,t)\sim\Sigma G_\nu(z)t^\nu$

for small real t in the Hamburger case in which conditions (8) hold, for all z in $\mathbb{M}$, setting

$$G(z,t)=\Sigma_0^{m-1}(G_\nu(z)t^\nu+t^m G_m(z,t)$$

and letting $\{t_\nu^{(i,i+m-1)}\},\{M_\nu^{(i,i+m-1)}\}$ be the zeros and weight factors of order i generated by the measure $t^m d\sigma(t)$, the Riesz-Shohat approximant $R_{i,i+m-1}(z)$ is defined by

(23) $$R_{i,i+m-1}(z)=\Sigma_0^{m-1} f_\nu G_\nu(z)+\Sigma_1^i M_\nu^{(i,i+m-1)} G_m(z,t_\nu^{(i,i+m-1)})$$

for m=0,2,4,... . In the Stieltjes case of conditions (10), (22) is assumed to hold for small positive real t, and the Riesz-Shohat approximants (21) are defined for m=0,1,... .

We now step ahead from the classical theory. The generalisations of the Padé quotient dealt with so far may be unified and extended [45]: we decompose the Padé quotient derived from a general series $\Sigma f_\nu z^\nu$ $(f_o \neq 0)$ in the irreducible form

$$\lambda^{-1} P_{i,j}(\lambda^{-1}) = \sum_{\nu=o}^{r(i,j)} \sum_{\tau=o}^{s(i,j;\nu)} M_{\nu,\tau}^{(i,j)} \tau! (\lambda - t_\nu^{(i,j)})^{-\tau-1}$$

(one of the $\{t_\nu^{(i,j)}\}$ is always zero) and, for the purposes of motivation, assume that $G(z,\lambda)$ is analytic in λ within and upon a contour C enclosing all $\{t_\nu^{(i,j)}\}$ for all z belonging to a point set $\mathbb{M}$; we set

$$W_{i,j}(z) = (2\pi i)^{-1} \int_C \lambda^{-1} P_{i,j}(\lambda^{-1}) G(z,\lambda) d\lambda$$

and obtain

$$(24) \quad W_{i,j}(z) = \sum_{\nu=o}^{r(i,j)} \sum_{\tau=o}^{s(i,j;\nu)} M_{\nu,\tau}^{(i,j)} G^{(\tau)}(z, t_\nu^{(i,j)})$$

where

$$(25) \quad G^{(\tau)}(z,t) = (\partial/\partial t)^\tau G(z,t).$$

We now discard the assumptions concerning the analytic nature of G imposed in the above motivation, and simply define generalised approximants in terms of formulae (24,25). Hamburger and Bernstein approximants (with u replaced by z in formulae (15,18)) are obtained by replacing $G(z,t)$ by e^{zt} and $E_\chi(zt)$ respectively; Riesz-Shohat approximants are obtained by assuming $G(z,t)$ to satisfy a relationship of the form (22) and the coefficients of the series $\Sigma f_\nu z^\nu$ to have representations of the form (7) or (9). Use of formulae (24,25) permits us to construct general approximants of order (i,j) with i>j+1 from a Stieltjes or Hamburger series, when the $\{t_\nu^{(i,j)}\}$ become complex (we must now assume, of course, that $G(z,t)$ is analytic in t over certain domains). Indeed, the convergence behaviour of the $\{W_{i,j}(z)\}$ can easily be studied in any case in which the convergence behaviour of the $\{P_{i,j}(z)\}$ is known (for example, when $\Sigma f_\nu z^\nu$ is the exponential series, or the series expansion of more general series investigated by Edrei and others [2]).

For the sake of completeness, we mention that an equivalent but less perspicuous unification and extension of generalised Padé quotients may be obtained by considering linear functionals of the form $W_{i,j}=\int_{\mathbb{E}} P_{i,j} d\Xi$, where $d\Xi$ is a prescribed measure, $\mathbb{E}$ is an appropriate set, and the integral is suitably defined.

The behaviour of the functions $\{W_{i,j}(z)\}$ may differ markedly from that of Padé quotients. For example, as is easily verified, if $\Sigma f_\nu z^\nu$ is a normal series, then for all polynomials $G(z,t)=\Sigma_0^m G_\nu(z)t^\nu$, $W_{i,j}(z)=\Sigma_0^m f_\nu G_\nu(z)$ for all $i+j\geq m$. This result, which implies a simple convergence result, is independent of the convergence behaviour of the $\{P_{i,j}(z)\}$. A more recondite example of difference concerns the case in which the $\{f_\nu\}$ are moments of an indeterminate Stieltjes moment problem, and $G(z,t)=\Sigma G_\nu(z)t^\nu$ is an entire function such that $\Sigma f_\nu |G_\nu(z)|$ converges. It is then found that $W_{i,i+m-1}(z)$ converges to $\Sigma f_\nu G_\nu(z)$ as i tends to infinity for $m=0,1,\ldots$. In the corresponding case with the $\{f_\nu\}$ as described, as was shown by Wall in his doctoral dissertation [31], for finite $z\notin[0,\infty]$ each diagonal sequence of Padé quotients $P_{i,i+m-1}(z)$ derived from the series $\Sigma f_\nu z^\nu$ converges, as i tends to infinity, to a finite limit for $m=0,1,\ldots$, but no two neighbouring limits are equal.

An extensive structural and convergence theory of the approximants $\{W_{i,j}(z)\}$ has been given [45]. We mention some features of the theory of special interest: there exists a theory of asymptotic relationships between the $\{W_{i,j}(z)\}$ and the formal sum of the series $\Sigma f_\nu G_\nu(z)$ when $\{G_\nu(z)\}$ is an asymptotic sequence with gaps, so that with $\tau(\nu)$ $(\nu=0,1,\ldots)$ a prescribed increasing non-negative integer sequence, $G_{\tau(\nu+1)}(z)=o\{G_{\tau(\nu)}(z)\}$ as z tends to a fixed number z', $G_{\nu'}(z)$ being identically zero when ν' is unequal to any of the $\{\tau(\nu)\}$; with $z'=\infty$, $G_{\tau(\nu)}(z)$ may, for example, have the form $\lambda^z_{\tau(\nu)}(|\lambda_{\tau(\nu)}|<1)$, $\Sigma f_\nu G_\nu(z)$ now being a Dirichlet series, or $G_{\tau(\nu)}(z)$ may have the form $G_{\tau(\nu)}(z)=\prod_{k=1}^{\tau(\nu)}(z+\omega_k)^{-1}$, $\Sigma f_\nu G_\nu(z)$ now being a Newton series. (In the case of the Padé table, in which $G_\nu(z)=z^\nu$ $(\nu=0,1,\ldots)$ the above asymptotic theory deals with the extent of agreement between the series expansion of $P_{i,j}(z)$ and the generating series; in this case there are no gaps.) Extremal properties analogous to those existing for Padé quotients [36,41] have been derived, and the monotonic nature of forward and backward diagonal sequences of certain real approximants $\{W_{i,j}(z)\}$ has been established. The theory takes a specially simple form when $G_\nu(z)=g_\nu z^\nu$ $(\nu=0,1,\ldots)$, the $\{g_\nu\}$ being finite complex numbers. If $\Sigma g_\nu t^\nu$ defines an entire function, all $W_{i,j}(z)$ are entire in this case. With $\Sigma g_\nu t^\nu$ being the series expansion of a function of the form $\Sigma_1^n v_\nu(1-w_\nu t)^{-\lambda_\nu}$, the $\{v_\nu\}$, $\{w_\nu\neq 0\}$ and $\{\lambda_\nu\}$ being fixed finite complex numbers, and $\Sigma f_\nu z^\nu$ being the series expansion of a function of the form $f(z)=\int_0^1(1-zt)^{-1}d\sigma(t)$ with σ bounded and nondecreasing over $[0,1]$, the $\{W_{i,j}(z)\}$ become approximants to a function defined over a star shaped domain whose rays emanate from the points $z=w_\nu^{-1}$ $(\nu=1,2,\ldots,n)$ (with $n=v_1=w_1=\lambda_1=1$, the $\{W_{i,j}(z)\}$ reduce to Padé quotients generated by $f(z)$, which is of course defined over a primitive star shaped domain with single ray emanating from the point $z=1$). The $\{W_{i,j}(z)\}$ may be constructed recursively; $\{f_\nu^{(\tau)}\}$ $(\tau=0,1,\ldots)$ being a sequence of sequences, and $\{g_\nu^{(o)}\}$ also prescribed, $\{f_\nu\}$ $\{g_\nu\}$ are replaced by $\{f_\nu^{(\tau)}\},\{g_\nu^{(\tau)}\}$ in the above to derive approximations to $g^{(\tau)}(z)$, the sum of the series $\Sigma f_\nu^{(\tau)}g_\nu^{(\tau)}z^\nu=\Sigma g_\nu^{(\tau+1)}z^\nu$ $(\tau=0,1,\ldots)$. In summary, a coherent supporting theory of the approximants of formulae (24,25) has been established, just as has also been [34] for Padé quotients derived from series whose coefficients are elements of a ring over which an inverse is defined (for example, matrix valued Padé quotients) and, by Pincherle [24,25], Hermite [14], Padé [21,22] and Cordone [8] (see the memoir of Van Vleck [30] for further references to this classical subject) for Padé quotients derived simultaneously from a number of power series.

Although the theory outlined above has considerable intrinsic interest, the importance of the approximants $\{W_{i,j}(z)\}$ derives from the scope of their application. These functions offer approximations to the sum or formal sum of a series of the form $\Sigma f_\nu G_\nu(z)$. Given a series $\Sigma f'_\nu G'_\nu(z)$ to be transformed, where the $\{G'_\nu(z)\}$ are, for example, higher transcendental functions, one selects a decomposition of the form $f'_\nu G'_\nu(z)=f_\nu G_\nu(z)$ $(\nu=0,1,\ldots)$, where $\Sigma f_\nu z^\nu$ is a series for which a great deal is known about the convergence behaviour of its associated Padé quotients (it may, for example, be a Hamburger or Stieltjes series, or the exponential series) and values of a function $G(z,t)$ connected with the $\{G_\nu(z)\}$ by relationships similar to (22) or having some other appropriate form may easily be computed, and derives approximants of the form (24,25) for the sum of the given series. The choice of functions $G_\nu(z)=z^\nu$ $(\nu=0,1,\ldots)$ leads to the construction of Padé quotients. But this choice is in many cases highly unnatural, and not surprisingly many series decline, politely but resolutely, to be transformed to any good effect by the use of Padé quotients. Use of approximants of the form (24,25) allows a given series to be transformed by the use of approximating functions whose behaviour is far more in accord with that of its sum or formal sum.

8 The Euler Maclaurin series

Imposing suitable restrictions upon Ψ, and integrating by parts the expression

$$\hat{R}_j=\{h^{2j+2}/(2j+2)!\}\int_0^{nh}\beta_{2j+2}(\mu/h)\mathcal{D}^{2j+2}\Psi(\mu)d\mu,$$

where $0\le j<\infty$, $0<n<\infty$ (j,n integers), $0<h<\infty$, and β_{2j+2} is a periodic Bernoulli polynomial defined by the formulae

$$t(e^{yt}-1)/(e^t-1)=\Sigma\hat{\beta}_\nu(y)t^\nu/\nu!,$$

$$\beta_\nu(y)=\hat{\beta}_\nu(y)\ (0\le y<1),\ \beta_\nu(y+1)=\beta_\nu(y)\ (y>0),$$

the identity

$$\int_0^{nh}\Psi(\mu)d\mu=T(0,n;h)+h^2\{C_j(0)-C_j(nh)\}+\hat{R}_j,$$

where $T(0,n;h)$ is the trapezium term

$$T(0,n;h)=h[\tfrac{1}{2}\{\Psi(0)+\Psi(nh)\}+\Sigma_1^{n-1}\Psi(\nu h)],$$

$$(26)\quad C_j(\mu)=\Sigma_0^{j-1}b_\nu\mathcal{D}^{2\nu+1}\Psi(\mu)h^{2\nu}$$

and $b_\nu=B_{2\nu+2}/(2\nu+2)!$ $(\nu=0,1,\ldots)$ (the $\{B_{2\nu+2}\}$ being Bernoulli numbers, so that $b_0=1/12, b_1=-1/720,\ldots$) may be obtained. Assuming the two integrals over semi-infinite intervals to exist, $\hat{R}_j$ may be expressed as the difference of two such integrals, and we obtain

$$\int_0^{nh}\Psi(\mu)d\mu=T(0,n;h)+h^2\{C_j(0)+R_j(0)-C_j(nh)-R_j(nh)\}$$

where

$$(27)\quad R_j(\hat{\mu})=\{h^{2j}/(2j+2)!\int_{\hat{\mu}}^{\infty}\beta_{2j+2}\{(\hat{\mu}-\mu)/h\}\mathcal{D}^{2j+2}\Psi(\mu)d\mu.$$

The series whose successive partial sums and remainder terms are given by formulae (26,27) respectively for $j=0,1,\ldots$ is the Euler Maclaurin series.

Hardy [12], using properties of the periodic Bernoulli polynomials, has shown that if $\Psi(\mu)$ is real for $0\le\mu<\infty$ with $(-1)^\nu\mathcal{D}^\nu\Psi(\mu)\ge 0$ for $0\le\mu<\infty$ $(\nu=2J+1,2J+2,\ldots)$, then the terms of the series $\Sigma_J b_\nu\mathcal{D}^{2\nu+1}\Psi(\mu)h^{2\nu}$ alternate in sign and this series is semi-convergent, i.e. $|R_j(\mu)|<|b_j\mathcal{D}^{2j+1}\Psi(\mu)h^{2j}|$ $(j=J,J+1,\ldots)$. Hardy has also shown that without the above restriction upon Ψ, but with Ψ assumed regular for $\mathrm{Re}(\mu)\ge-\delta$ $(0<\delta<\infty)$ and $\Psi(\mu)=0(|\mu|^s)$ in this half-plane, the intercalated series $b_J\mathcal{D}^{2J+1}\Psi(\mu)+0+b_{J+1}\mathcal{D}^{2J+3}\Psi(\mu)h^2+0+\ldots$ with $2J+1>s$ is summable (B*).

The theory outlined in §5 strongly suggests that the Euler Maclaurin series should be amenable to continued fraction transformation. The following result has been derived: let $(-\mathcal{D})^\nu\Psi(\mu)\ge 0$ for all $\mu\in[-\delta,\infty)$ $(\delta\in(0,\infty)$ being fixed) for $\nu=2J+1,2J+2,\ldots,J\ge 0$

being a fixed integer; then for fixed $\mu \in [0,\infty)$ (i)

$$(28)\quad b_{J+\nu} \mathcal{D}^{2J+2\nu+1} \Psi(\mu) = (-1)^{\nu} \int_{0}^{\infty} t^{\nu} d\sigma(t) \qquad (\sigma \in \mathbb{B}(0,\infty); \nu=0,1,\ldots)$$

and (ii) for real h

$$(29)\quad \Sigma b_{J+\nu} \mathcal{D}^{2J+2\nu+1} \Psi(\mu) h^{2\nu}$$

is a semi convergent asymptotic series in ascending powers of h^2; (iii) for fixed $h \in (0,\infty)$ the series (29) is summable (B,2) and (iv) all forward diagonal sequences of Padé quotients $P_{i,i+m}(h^2)$, $P_{i+m,i}(h^2)$ (m=0,1,...) derived from (29) converge to its (B,2) sum (the proof follows by exhibiting the numbers $\{b_\nu\}$ in the form $b_\nu = (-1)^{\nu} \int_{0}^{(2\pi)^{-2}} t^{\nu} d\hat{\sigma}(t)$ $(\nu=0,1,\ldots)$, by exhibiting the derivatives of $\Psi(\mu)$ in the form $\mathcal{D}^{2J+2\nu+1}\Psi(\mu) = \int_{0}^{\infty} t^{\nu} d\tilde{\sigma}(t)$ $(\nu=0,1,\ldots)$, and hence deriving formula (28), and by examining the orders of magnitude of the terms $\{b_{J+\nu} \mathcal{D}^{2J+2\nu+1}\Psi(\mu)\}$; details are given in [44]; for a further continued fraction transformation of the Euler-Maclaurin series, see [40] and supporting material in [37]).

The Euler-Maclaurin series may also be transformed by the use of approximants of the form (24,25). Firstly, $b_\nu \mathcal{D}^{2\nu+1}\Psi(\mu) h^{2\nu} = f_\nu G_\nu(\mu)$ $(\nu=0,1,\ldots)$ where

$$f_\nu = B_{2\nu+2}/(2\nu+2) = (-1)^{\nu} \int_{0}^{\infty} t^{\nu} d\sigma(t), \qquad (\nu=0,1,\ldots)$$

$$\sigma(t) = \int_{0}^{t} \{\exp(2\pi\sqrt{t'})-1\}^{-1} dt',$$

the Stieltjes moment problem associated with this distribution being determinate, and $G_\nu(\mu) = \mathcal{D}^{2\nu+1}\Psi(\mu) h^{2\nu}/(2\nu+1)!$ $(\nu=0,1,\ldots)$. It is assumed that the function

$$G(\mu,t) = \{\Psi(\mu+h\sqrt{t}) - \Psi(\mu-h\sqrt{t})\}/\{2h\sqrt{t}\}$$

is, for fixed real μ and $h(\neq 0)$ continuous for $t \in (-\infty,\infty)$, that $\Psi(\mu)$ tends to finite limits as μ tends to $\pm\infty$, and that

$$G(\mu,t) \sim \Sigma \mathcal{D}^{2\nu+1}\Psi(\mu) h^{2\nu} t^{\nu}/(2\nu+1)!$$

as t tends to zero through positive real values. The diagonal sequences $W_{i,i+m-1}(\mu)$ (m=0,1,...) then tend, as i tends to infinity, to

$$\int_0^\infty \frac{\Psi(\mu+h\sqrt{t})-\Psi(\mu-h\sqrt{t})}{2h\sqrt{t}\{\exp(2\pi\sqrt{t})-1\}}\,dt.$$

9 The anti-derivative

Recently [35,38,42,43] four processes for approximating a (possibly divergent) integral over a semi-infinite interval in terms of the values of the derivatives of the integrand at the finite end point have been introduced. Assuming all derivatives of the function ψ to exist at the point μ, the processes involve the construction of a sequence of functions whose limit $I(\psi,\mu)$ satisfies the relationship

$$\text{(30)} \quad \mathcal{D}I(\psi,\mu)=-\psi(\mu)$$

and whose members approximate $I(\psi,\mu)$.

The theories underlying the methods are, upon this occasion, sufficiently illustrated by considering one of them. The functions of the sequence involved are

$$\text{(31)} \quad \rho_{2r}(\psi,\mu)=H[\mathcal{D}^{\tau-1}\psi(\mu)/\tau!]_r/H[\mathcal{D}^{\tau+1}\psi(\mu)/(\tau+2)!]_{r-1}$$

for r=1,2,..., where the first element in the first row of the numerator determinant is taken to be zero. If

$$\text{(32)} \quad \psi(\mu)=\sum_{\nu=1}^{N}\sum_{\tau=1}^{n(\nu)} A_{\nu,\tau}(\mu-\alpha_\nu)^{-\tau-1} \quad (A_{\nu,n(\nu)}\neq 0,\nu=1,2,\ldots,N<\infty)$$

and

$$\rho(\psi,\mu)=\sum_{\nu=1}^{N}\sum_{\tau=1}^{n(\nu)} A_{\nu,\tau}\tau^{-1}(\mu-\alpha_\nu)^{-\tau}$$

then $\mathcal{D}\rho(\psi,\mu)=-\psi(\mu)$ for all finite $\mu\neq\alpha_\nu$ ($\nu=1,2,\ldots,N$), and we have the algebraic result that for these values of μ, $\rho_{2n}(\psi,\mu)=\rho(\psi,\mu)$, where $n=\Sigma_1^N n(\nu)$. The functions $\{\rho_{2r}(\psi,\mu)\}$ may be constructed by means of a simple recursive algorithm: setting $\omega_{-1}^{(m)\prime}=0$ (m=1,2,...), $\omega_0^{(o)\prime}=0$, $\omega_0^{(m)\prime}=\mathcal{D}^{m-1}\psi(\mu)/m!$ (m=1,2,...) and determining further numbers $\{\omega_r^{(m)\prime}\}$ by use of the relationships

$$\omega_{2r+1}^{(m)'}=\omega_{2r-1}^{(m+1)'}+\omega_{2r}^{(m)'}/\omega_{2r}^{(m+1)'},\quad \omega_{2r+2}^{(m)'}=\omega_{2r}^{(m+1)'}(\omega_{2r+1}^{(m)}-\omega_{2r+1}^{(m+1)'})$$

for $r,m=0,1,\ldots$ we have, in the absence of breakdown, $\omega_{2r}^{(o)'}=\rho_{2r}(\psi,\mu)$. A substantial convergence theory of the functions $\{\rho_{2r}(\psi,\mu)\}$ has been given.

The improper Riemann integral $R(\psi,\mu)=\int_\mu^\infty\psi(\mu')d\mu'$, should it exist, satisfies a relationship similar to (30), and in this case the limit of the sequence $\{\rho_{2r}(\psi,\mu)\}$ is, subject to convergence and consistency, a Riemann integral. However, convergence to a limit function satisfying a relationship analogous to (30) may hold even when the Riemann integral fails to exist. In these circumstances, the sequence $\{\rho_{2r}(\psi,\mu)\}$ serves both to define and to approximate a divergent integral. By way of illustrating the above principle, we consider the function $A(\mu-t)^{-2}$ $(-\infty<A,t<\infty)$, the simplest possible example of the general functions given by formula (32). In this case the sequence $\{\rho_{2r}(\psi,\mu)\}$ terminates abruptly with

$$\rho_2(\psi,\mu)=-2\psi(\mu)^2/\mathcal{D}\psi(\mu)=A/(\mu-t).$$

We have $\rho_2(\psi,\mu)=R(\psi,\mu)$ when $\mu>t$; $\rho_2(\psi,\mu)$ defines a divergent integral when $\mu<t$.

To give a less primitive example, we remark that it has been shown that when $\psi(\mu)=\int_\alpha^\beta(\mu-t)^{-2}d\sigma(t)$, $-\infty<\alpha<\beta<\infty$ and σ is bounded and nondecreasing over $[\alpha,\beta]$, $\{\rho_{2r}(\psi,\mu)\}$ tends to $\int_\alpha^\beta(\mu-t)^{-1}d\sigma(t)$ both for real $\mu>\beta$, when $R(\psi,\mu)$ exists, and real $\mu<\alpha$, when it does not. Taking $\alpha=0,\beta=1,d\sigma(t)/dt=(\nu!)^{-1}$ for $(2\nu)^{-1}<t<(2\nu-1)^{-1}$ and $d\sigma(t)/dt=0$ for $(2\nu+1)^{-1}<t<(2\nu)^{-1}$ $(\nu=1,2,\ldots)$, σ being continuous, we have

$$\psi(\mu)=\Sigma_1(\nu!)^{-1}\{\frac{1}{\mu-(2\nu-1)^{-1}}-\frac{1}{\mu-(2\nu)^{-1}}\}$$

and the limit of the sequence $\{\rho_{2r}(\psi,\mu)\}$ in this case is

$$\rho(\psi,\mu)=\Sigma_1(\nu!)^{-1}\ell n[\{\mu-(2\nu)^{-1}\}/\{\mu-(2\nu-1)^{-1}\}].$$

$\psi(\mu)$ is composed of dipole functions over $[(2\nu)^{-1},(2\nu-1)^{-1}]$ ($\nu=1,2,\ldots$) and has a limit point of dipoles at the origin $\mu=0$. Taking $\mu=-1$, we find that $\rho_{10}(\psi,-1)=-0.32539\ 72125$, whereas $\rho(\psi,-1)=-0.32539\ 72135$.

The above example points to an important feature of the use of the anti-derivative in the computation of singular integrals, namely that precise knowledge of the location of the singularities is not required in the implementation of the method. In this way the difficult topological problems associated with integrands having algebraic singularities (as occur, for example, in Feynman path integrals [9,15,16]) may be avoided.

10 A transformation of series

The Euler-Maclaurin formula for integration over a semi-infinite interval may be rearranged as

$$(33)\quad \Sigma\psi(\mu+\nu h)=h^{-1}\int_{\mu}^{\infty}\psi(\mu')d\mu'+\tfrac{1}{2}\psi(\mu)-\Sigma b_{\nu}\mathcal{D}^{2\nu+1}\psi(\mu)h^{2\nu+1}$$

and in this form has been proposed [5] (see also [13] §11.12) as a method for approximating the sum of an infinite series $\Sigma\psi(\mu+\nu h)$ in terms of an integral and a linear combination of derivatives of its first term (i.e. a truncated version of the last sum in formula (33)). In trivial cases in which the integral involved may be evaluated in closed form (e.g. with $\psi(\mu)=(\mu+1)^{-2}$, when $\int_{\mu}^{\infty}\psi(\mu')d\mu'=(\mu+1)^{-1}$) this method functions effectively; in many nontrivial cases it turns out that it is more difficult to evaluate the integral in formula (33) than to sum the infinite series on the left hand side. Nevertheless, as we have seen in §9, effective methods now exist for approximating an integral over a semi-infinite interval: the method of transforming a series described above can be resuscitated.

We consider in detail the function

$$(34)\quad \psi(\mu)=\ell n(1-\mu^{-1})+(\mu-1)^{-1}=\int_{0}^{1}(\mu-t)^{-2}t\,dt.$$

The functions $\{\rho_{2r}(\psi,\mu)\}$ in this case (see §9) converge to $\int_0^1(\mu-t)^{-1}tdt$ when $\mu>1$. Furthermore $(-\mathcal{D})^{\nu}\psi(\mu)\geq 0$ for $1<\mu<\infty$ $(\nu=0,1,\ldots)$ so that the series $\Sigma b_\nu \mathcal{D}^{2\nu+1}\psi(\mu)h^{2\nu}$ is (see §8) a Stieltjes series in ascending powers of $-h^2$ when $\mu>1$.

When $\mu=2,h=1$ the sum of the first series in formula (33) for the above function ψ becomes

$$\Sigma[\ell n\{1-(2+\nu)^{-1}\}+(\nu+1)^{-1}]=\gamma=0.57721\ 56649\ 02.$$

For these values of μ and h the series $\Sigma b_\nu \mathcal{D}^{2\nu+1}\psi(\mu)$ in the case being considered, although semi-convergent, diverges sharply. For this reason, we delay application of the transformation being considered, rewriting formula (33) as

$$(35)\quad \Sigma\psi(\mu+\nu h)=\Sigma_0^{j-1}\psi(\mu+\nu h)+h^{-1}\int_{\mu+jh}^{\infty}\psi(\mu')d\mu'$$
$$+\tfrac{1}{2}\psi(\mu+jh)-\Sigma b_\nu \mathcal{D}^{2\nu+1}\psi(\mu+jh)h^{2\nu+1}.$$

We have taken $\mu=2,h=1,j=5$, have approximated $\int_7^{\infty}\psi(\mu')d\mu'$ by $\rho_8(\psi,7)$, and the sum of the last series in formula (35) by $\Sigma_0^4 b_\nu \mathcal{D}^{2\nu+1}\psi(\mu+5)$ in this case. Replacing the constituents of the right hand side of formula (35) by their numerical values, we obtain the sum

$$\begin{aligned}&0.49157\ 38641\ 05+0.07905\ 47587\ 90\\&+0.00625\ 79934\ 20+0.00032\ 90485\ 91\\&=0.57721\ 56649\ 06.\end{aligned}$$

Both the sequence $\{\rho_{2r}(\psi,\mu)\}$ used to approximate the integral in formula (33) and the sequence of partial sums of the last series to occur in that formula may be transformed by corresponding repeated application of the ε-algorithm (see, for example, [39]): given a sequence $\{S_m\}$ to be transformed, arrays of numbers $\{{}_\tau\varepsilon_r^{(m)}\}$ are determined from the sets of initial values ${}_\tau\varepsilon_{-1}^{(m)}=0$ $(\tau=0,1,\ldots;$ $m=1,2,\ldots)$, ${}_0\varepsilon_0^{(m)}=S_m$ $(m=0,1,\ldots)$, ${}_{\tau+1}\varepsilon_0^{(2m)}={}_\tau\varepsilon_{2m}^{(o)}$, ${}_{\tau+1}\varepsilon_0^{(2m+1)}={}_\tau\varepsilon_{2m}^{(1)}$ $(\tau=0,1,\ldots;m=0,1,\ldots)$; it often occurs that the numbers $\{{}_\tau\varepsilon_0^{(m)}\}$ $(\tau>0)$ offer far better approximations to the limit or formal limit of the sequence $\{S_m\}$ than do any of the members of this sequence from which the numbers are derived. (For example, taking $S_m = \sum_o^{m-1} \binom{-\frac{1}{2}}{\nu} (\nu+1)^{-1}$ (m=0,1,...), $S_6 = {}_o\varepsilon_o^{(6)} = 0.81\ldots$, ${}_1\varepsilon_o^{(6)} = 0.82840\ldots$, ${}_2\varepsilon_o^{(6)} = 0.82842\ 71247\ 43$, whereas $S_\infty = 0.82842\ 71247\ 49$: we have extracted from six terms of the series being transformed information concerning its sum alternatively to be obtained by direct summation of more than ten thousand million terms.) Taking $\mu=2$, h=1, and $S_o=0, S_m=\rho_{2m}(\psi,2)$ (m=1,2,...) (ψ being the function of formula (34)) we find that ${}_2\varepsilon_o^{(5)} = 0.38629\ 43611$; the value of the isolated term $\frac{1}{2}\psi(\mu)$ in formula (33) is 0.15342 64097; in the transformation of the last series (taking S_m in the above to be the sum of the first m terms) we have ${}_3\varepsilon_o^{(9)} = 0.03749\ 48940$; the sum of these three numbers is 0.57721 56648; we have approximated the value of a sum directly (without delay) in terms of the derivatives of its first term.

References

1 Akhiezer, N.I., The classical moment problem, Moscow (1961), Oliver and Boyd (1965).

2 Arms, R.J. and A. Edrei, The Padé tables and continued fractions generated by totally positive sequences, in Mathematical essays dedicated to A.J. Macintyre, Ohio (1970) 1-21.

3 Bernstein, F., Die Übereinstimmung derjenigen beiden Summationsverfahren einer divergenten Reihe, welche von T.J. Stieltjes und E. Borel herrühren, Jahresbericht der Deutschen Math. Vereinigung, 28 (1919) 50-63; 29 (1920) 94.

4 Bernstein, F., Bemerkung zu der bevorstehenden Abhandlung: Über die Konvergenz eines mit einer Potenzreihen assoziierten Kettenbruches, von H. Hamburger in Berlin, Math. Ann., 81 (1920) 46-47.

5 Bickley, W.G. and J.C.P. Miller, The numerical summation of slowly convergent series, unpublished memoir (available from National Physical Laboratory Mathematics Division library, Teddington, U.K.).

6 Borel, E., Leçons sur les séries divergentes, Gauthier-Villars, Paris (1928).

7 Carleman, T., Les fonctions quasi-analytiques, Gauthier-Villars, Paris (1926).

8 Cordone, S., Sopra un problema fundamentale delle teoria delle frazioni continue algebriche generalizzate, Rend. Circ. Mat. Palermo, 12 (1898) 240-257.

9 Feynman, R.P. and A.R. Hibbs, Quantum mechanics and path integrals, McGraw-Hill, New York (1965).

10 Hamburger, H., Über die Konvergenz eines mit einer Potenzreihe assoziierten Kettenbruchs, Math. Ann., 81 (1920) 31-45.

11 Hamburger, H., Über eine Erweiterung des Stieltjes'schen Momentenproblems, Math. Ann., 81 (1920) 235-319; 82 (1921) 120-164; 168-187.

12 Hardy, G.H., Divergent series, Oxford (1949).

13 Hartree, D.R., Numerical analysis, Oxford (1955).

14 Hermite, C., Sur la généralisation des fractions continues algébriques, Annali di Mat., ser. 2, 21 (1893) 289-308.

15 Hwa, C.R. and V.L. Teplitz, Homology and Feynman integrals, Benjamin, New York (1966).

16 Lefschetz, S., Applications of algebraic topology: graphs and networks: the Picard-Lefschetz theory and Feynman integrals, Springer, Berlin (1975).

17 Le Roy, E., Sur les séries divergentes et les fonctions définies par un développement de Taylor, Ann. Fac. Sci. Toulouse, 2 (1900) 317-430.

18 Nevanlinna, F., Zur Theorie der asymptotischen Potenzreihen, Ann. Acad. Sci. Fenn., A, 12 (1916) 1-81.

19 Nevanlinna, R., Asymptotische Entwickelungen beschränkter Funktionen und das Stieltjes'schen Momentenproblem, Ann. Acad. Sci. Fenn., A, 18 (5) (1922).

20 Ostrowski, A., Über quasianalytische Funktionen und Bestimmtheit asymptotischer Entwickelungen, Acta Math., 53 (1929) 180-266.

21 Padé, H., Sur la généralisation des fractions continues algébriques, Jour. de Math., sér. 4, 10 (1894) 291-329.

22 Padé, H., Sur la généralisation des fractions continues algébriques, Comptes Rendus de l'Acad. Sci. (Paris) 118 (1894) 848.

23 Perron, O., Die Lehre von den Kettenbrüchen, 2, Teubner, Stuttgart (1957).

24 Pincherle, S., Saggio di una generalizzazione delle frazioni continue algebriche, Mem. della R. Acad. delle Scienze dell'Istituto di Bologna, ser. 4, 10 (1890) 513-538.

25 Pincherle, S., Sulla generalizzazione delle frazioni continue, Annali di Mat., ser. 2, 19 (1891) 75-95.

26 Riesz, M., Sur le problème des moments, Arkiv. för matematik, astronomi och fysik, 16 (1921) (12); 16 (1922) (19); 17 (1923) (16).

27 Shohat, J.A., Sur les quadratures mécaniques et sur les zéros de polynômes de Tchebycheff dans un intervalle infini, Comptes Rendus Acad. Sci. (Paris), 185 (1927) 597-598.

28 Shohat, J.A. and J.D. Tamarkin, The problem of moments, Amer. Math. Soc. Math. Surveys, 1 (1943).

29 Stieltjes, T.J., Recherches sur les fractions continues, Ann. Fac. Sci. Toulouse, 8 (1894) 1-122; 9 (1895) 1-47 (also: Oeuvres complètes de Thomas Jan Stieltjes, Noordhoff, Groningen (1918) 402-566).

30 Van Vleck, E.B., Selected topics in the theory of divergent series and of continued fractions, Amer. Math. Soc. Coll. Pubns. 1, Boston Colloquium 1903, New York (1905) 75-187.

31 Wall, H.S., On the Padé approximants associated with the continued fraction and series of Stieltjes, Trans. Amer. Math. Soc., 31 (1929) 91-116.

32 Wall, H.S., General theorems on the convergence of sequences of Padé approximants, Trans. Amer. Math. Soc., 34 (1932) 409-416.

33 Watson, G.N., A theory of asymptotic series, Phil. Trans. Roy. Soc., A, 211 (1911) 279-313.

34 Wynn, P., Continued fractions whose coefficients obey a non-commutative law of multiplication, Arch. Rat. Mech. Anal., 12 (1963) 273-312.

35 Wynn, P., Upon the definition of an integral as the limit of a continued fraction, Arch. Rat. Mech. Anal., 28 (1968) 83-148.

36 Wynn, P., Upon the Padé table derived from a Stieltjes series, Jour. SIAM Numer. Anal., 5 (1968) 805-834.

37 Wynn, P., On an extension of a result due to Pólya, Jour. f. d. reine u. ang. Math. (Crelle's Journal), 248 (1971) 127-132.

38 Wynn, P., Upon some continuous prediction algorithms, Calcolo, 9 (1972) 197-278.

39 Wynn, P., Transformation de séries à l'aide de l'ε-algorithme, Comptes Rendus de l'Acad. Sci. (Paris), 275 A (1972) 1351-1353.

40 Wynn, P., Accélération de la convergence de séries d'opérateurs en analyse numérique, Comptes Rendus de l'Acad. Sci. (Paris), 276 (1973) 803-806.

41 Wynn, P., Extremal properties of Padé quotients, Acta Math. Acad. Sci. Hungaricae, 25 (1974) 291-298.

42 Wynn, P., A convergence theory of some methods of integration, Jour. f. d. reine u. ang. Math. (Crelle's Journal) 285 (1976) 181-208.

43 Wynn, P., The evaluation of singular and highly oscillatory integrals by use of the anti-derivative, School of Computer Science, McGill University, Report (1976).

44 Wynn, P., A continued fraction transformation of the Euler-Maclaurin series, School of Computer Science, McGill University, Report (1976).

45 Wynn, P., An array of functions, School of Computer Science, McGill University, Report (1976).

P. Wynn*
School of Computer Science
McGill University
Montreal, Quebec, Canada

* Research supported by the Canadian Department of National Defence.

SPECIAL FUNCTIONS AND CONTINUED FRACTIONS

A RELATIONSHIP BETWEEN LIE THEORY AND CONTINUED FRACTION EXPANSIONS FOR SPECIAL FUNCTIONS

C.P. Boyer and W. Miller, Jr.

We adapt and extend a technique due to D. Widder to show that the symmetry groups associated with the special functions of hypergeometric type can be used to derive and analyse the structure of continued fraction expansions for these functions.

1 Expansions for Hermite Functions

The special functions of hypergeometric type arise as separable solutions for certain partial differential equations of mathematical physics in appropriate coordinate systems. The differential recurrence relations obeyed by these functions are consequences of the Lie symmetries enjoyed by the associated differential equations, [4] . As an example we consider the relation between Hermite functions and the heat equation

(1.1) $Q\Phi(t,x) = 0 \ , \ Q = \partial_t - \partial_{xx} \ , \ (t,x) \in \mathbb{C}^2 \ .$

In the coordinates $s = t^{\frac{1}{2}}/2i$, $z = xi/2t^{\frac{1}{2}}$, (1.1) admits solutions

(1.2) $\Phi_n = H_n(z)s^n \ , \ n \in \mathbb{C}$

where $H_n(z) = 2^{n/2}e^{z^2/2}D_n(2^{\frac{1}{2}}z)$ is a Hermite function, a polynomial for $n = 0,1,2,\ldots$, [1]. Furthermore, $Q = \partial_{zz} - 2z\partial_z + 2s\partial_s$. The symmetry algebra of (1.1), i.e., the Lie algebra of all first order differential operators L such that $Q(L\Phi) = 0$ whenever $Q\Phi = 0$, is six-dimensional with basis

(1.3) $L_2 = -2s^2(z\partial_z + s\partial_s + 1 - 2z^2) \ , \ L_1 = s(-\partial_z + 2z), \ L_0 = 1$

$L_{-1} = \frac{1}{4}s^{-1}\partial_z, \ L_{-2} = \frac{1}{8}s^{-2}(z\partial_z - s\partial_s), \ L^0 = s\partial_s + \frac{1}{2}$

and commutation relations

$$(1.4)\quad [L^0,L_j] = jL_j,\ j = \pm 2,\pm 1,0\ ,\ [L_{\pm 1},L_{\pm 2}] = 0$$

$$[L_{\pm 1},L_{\mp 2}] = \mp L_{\mp 1}\ ,\ [L_{-1},L_1] = \frac{1}{2}L_0,\ [L_{-2},L_2] = L^0\ .$$

It is straightforward to show that

$$(1.5)\quad \begin{aligned} &L_1\Phi_n = \Phi_{n+1}\ ,\ L^0\Phi_n = (n+\tfrac{1}{2})\Phi_n\ ,\ L_{-1}\Phi_n = \frac{n}{2}\Phi_{n-1} \\ &L_2\Phi_n = \Phi_{n+2}\ ,\ L_{-2}\Phi_n = \frac{1}{4}n(n-1)\Phi_{n-2} \end{aligned}$$

and that these are the only first order recurrence relations obeyed by Hermite functions. Such relations can be viewed as a consequence of (1.4) and the fact that the L_j map solutions of (1.1) to solutions.

Now consider the set $\{\Phi_n | n = n_0+k\ ,\ k = 0,\pm 1,\pm 2,\ldots\}$ where n_0 is a fixed complex number. We see from (1.5) that the action of the symmetry operators on this basis set defines a representation of the symmetry algebra. This representation is algebraically irreducible for n_0 nonintegral but for $n_0 = 0$ it is merely indecomposable. Indeed in the latter case the space spanned by the positive weight functions $\{\Phi_k | k = 0,1,2,\ldots\}$ is irreducible but the representation space cannot be decomposed into a direct sum of irreducible subspaces. (It is not possible to go down the weight ladder past Φ_0, but one can go up the ladder from Φ_{-1} to Φ_0.)

The functions

$$(1.6)\quad \hat{\Psi}_n = e^{z^2}H_{-n-1}(iz)(4is)^n$$

form another set of solutions for (1.1) which are eigenfunctions of L^0 and linearly independent of (1.2). In fact, all such solutions are linear combinations of these two. The action of the symmetry operators on this set is

$$(1.7)\quad \begin{aligned} &L_1\hat{\Psi}_n = \frac{1}{2}(n+1)\hat{\Psi}_{n+1},\ L^0\hat{\Psi}_n = (n+\tfrac{1}{2})\hat{\Psi}_n,\ L_{-1}\hat{\Psi}_n = \hat{\Psi}_{n-1} \\ &L_2\hat{\Psi}_n = \frac{1}{4}(n+1)(n+2)\hat{\Psi}_{n+2}\ ,\ L_{-2}\hat{\Psi}_n = \hat{\Psi}_{n-2}\ . \end{aligned}$$

and for $n = n_0 + k$ defines a Lie algebra representation. For n_0 not an integer this representation is equivalent to that defined by (1.5). Indeed the action of the L-operators on the basis $\Psi_n = \Gamma(n+1)2^{-n}\hat{\Psi}_n$ agrees exactly with (1.5). For $n_0 = 0$ however, the indecomposable representation defined by (1.7) is inequivalent to (1.5) since now one cannot reach Φ_0 from below. On the other hand, setting $\Psi_k = k!2^{-k}\hat{\Psi}_k$, $k = 0,1,2,\ldots$, we can make (1.5) and (1.7) agree as long as all subscripts in (1.7) are nonnegative.

We next relate these remarks to continued fraction expansions via an idea contained in [7, pp. 223-25]. We use the following notation for the continued fraction

$$(1.8)\quad \frac{a_1/b_1}{+}\ \frac{a_2/b_2}{+}\ \frac{a_3/b_3}{+}\ \cdots\ .$$

where a_j, b_j are complex numbers. Let $p_0 = 0$, $p_1 = a_1$, $q_0 = 1$, $q_1 = b_1$ and define the numbers $p_m, q_m, m > 1$, recursively by

$$(1.9)\quad p_m = p_{m-1}b_m + p_{m-2}a_m\ ,\quad q_m = q_{m-1}b_m + q_{m-2}a_m\ .$$

Then the mth convergent c_m of (1.8) is $c_m = p_m/q_m$ and the value of the continued fraction is $c = \lim_{m\to\infty} c_m$, provided this limit exists.

To develop a continued fraction expansion for Hermite functions we will use relations (1.5) and (1.7) to find nondifferential recurrence relations for functions on the weight ladders. One of the simplest is provided by the pair of operators $L_{\pm 1}$. Eliminating the term in ∂_z we find

$$(1.10)\quad \Phi_n = 2zs\Phi_{n-1} - 2(n-1)s^2\Phi_{n-2}$$

with the identical relation for Ψ_n (provided none of the subscripts is a negative integer). Now consider the continued fraction with $p_m = \Phi_{m-1}$, $q_m = \Psi_{m-1}$, $m = 1,2\ldots$. A simple computation in the asymptotics of Hermite functions shows $c = \lim_{m\to\infty} \Phi_m/\Psi_m = \sqrt{\pi}$ for $z = iy, y > 0$, e.g., [7, p. 223]. From relations (1.10) we have $b_m = 2zs$, $a_m = -2(m-2)s^2$, $m \geq 3$. Furthermore, $b_1 = \Psi_0 = \frac{1}{2}\pi^{\frac{1}{2}}\mathrm{Erfc}(iz)$, $a_1 = \Phi_0 = 1$, $b_2 = \Phi_1 = 2zs$, $a_2 = \Psi_1 - \Psi_0\Phi_1 = ise^{z^2}$. Thus,

(1.11) $\pi^{-\frac{1}{2}} = [\Psi_0 + ise^{z^2}d]^{-1}$, $d = 1/2zs\,\_\; 2s^2/2zs\,\_\; 4s^2/2zs\,\_\ldots$

or, with $y = -iz > 0$, $s = -i$,

(1.12) $\frac{1}{2}\pi^{\frac{1}{2}}\mathrm{Erfc}(y)e^{y^2} = 1/2y\,_+\; 2/2y\,_+\; 4/2y\,_+\; 6/2y\,_+\ldots$.

Next we consider the weight ladders $\{\Phi_n\}$, $\{\Psi_n\}$ for non-integral n. Our technique is to construct the continued fraction with $p_m = \tilde{\Phi}_{\nu+m}$, $q_m = \Psi_{\nu+m}$, $m = 1,2,\ldots, \nu \in ¢$. However, this time it is more convenient to use instead of the basis (1.2) the functions $\tilde{\Phi}_{\nu+m} = \Gamma(\nu+m+1)2^{-\nu-m}e^{z^2}H_{-\nu-m-1}(-iz)(4is)^{\nu+m}$ which also satisfy (1.5). Then $c = \lim_{m\to\infty} \tilde{\Phi}_{\nu+m}/\Psi_{\nu+m} = \infty$ for $\mathrm{Im}z > 0$ as follows from a simple asymptotic formula for Hermite functions [1, p.123]. From relations (1.10), which hold for both $\tilde{\Phi}_n$ and Ψ_n, it follows that $b_m = 2zs$, $a_m = -2(\nu+m-1)s^2$, $m \geq 3$. Furthermore, $b_1 = \Psi_{\nu+1}$, $a_1 = \tilde{\Phi}_{\nu+1}$, $b_2 = \tilde{\Phi}_{\nu+2}/\tilde{\Phi}_{\nu+1}$, $a_2 = (\tilde{\Phi}_{\nu+1}\Psi_{\nu+2} - \tilde{\Phi}_{\nu+2}\Psi_{\nu+1})/\tilde{\Phi}_{\nu+1}$. Thus

$$0 = b_1 + a_2/(b_2+d)\;,\quad d = -2(\nu+2)s^2/2zs\,\_\; 2(\nu+3)s^2/2zs\,\_\ldots$$

or

(1.13) $-\Psi_{\nu+2}/\Psi_{\nu+1} = d$,

a simple expansion of the ratio of two successive functions on the weight ladder.

The above examples can be greatly multiplied. In each case one eliminates the derivative terms from some combination of the L operators to obtain a recurrence relation which can be used to define the expansion. Using the operators $L_{\pm 2}$, for instance, one can develop a recurrence relation involving every other function on the weight ladder while with $L_{\pm 1}, L_{\pm 2}$ one can develop a relation which skips every third function. It is, however, typical of all these examples that the irreducible representations always lead to expansions for ratios of functions on the weight ladder while the indecomposable but reducible representations lead to expansions for the second kind function $\hat{\Psi}_0$. (This is related to the fact that $L_1\hat{\Psi}_{-1} = L_2\hat{\Psi}_{-2} = 0$, i.e., one cannot reach $\hat{\Psi}_0$ from below.)

2 Expansions for Hypergeometric Functions

Next we consider continued fraction expansions of the hypergeometric functions ${}_2F_1$. These functions arise via separation of variables in the complex wave equation

$$(2.1)\quad (\partial_{tt} - \partial_{x_1x_1} - \partial_{x_2x_2} - \partial_{x_3x_3})F = 0 \quad .$$

Indeed, in the coordinates $z = (x_1^2 + x_2^2)/(t^2 - x_3^2)$, $s = 2/(x_3+t)$, $u = 2/(x_3-t)$, $v = \frac{1}{2}(ix_1+x_2)$, (2.1) has the separable solutions $F = v^{-1}\Phi_{abc}$ where

$$(2.2)\quad \Phi_{abc}(z,s,u,v) = {}_2F_1(a,b;c;z)s^a u^b v^c \ , \ a,b,c \in ¢ \ ,$$

[2] , [5] . The symmetry algebra of (2.1) is $sl(4,¢) \cong o(6,¢)$ and a basis for this algebra (given in terms of its action on the functions $\Phi = vF$) is

$$(2.3)\quad \begin{aligned}
&E^{\alpha} = s(z\partial_z+s\partial_s) \ , \ E^{\alpha\beta\gamma} = suv\partial_z \ , \ E^{\beta} = u(z\partial_z+u\partial_u) \\
&E_{\gamma} = v^{-1}(z\partial_z+v\partial_v-1), \ E^{\alpha\gamma} = sv((1-z)\partial_z-s\partial_s), \\
&E^{\gamma} = v((1-z)\partial_z+v\partial_v-s\partial_s-u\partial_u), \ E_{\alpha} = s^{-1}(z(1-z)\partial_z + \\
&v\partial_v-s\partial_s-zu\partial_u), \ E_{\alpha\gamma} = s^{-1}v^{-1}(z(1-z)\partial_z-zu\partial_u+v\partial_v-1) \\
&E_{\alpha\beta\gamma} = s^{-1}u^{-1}v^{-1}(z(z-1)\partial_z-v\partial_v+zs\partial_s+zu\partial_u-z+1), \ E^{\beta\gamma}, E_{\beta\gamma}, E_{\beta} \\
&J(\alpha) = s\partial_s - \frac{1}{2}v\partial_v \ , \ J(\beta) = u\partial_u - \frac{1}{2}v\partial_v \ , \\
&J(\gamma) = v\partial_v - \frac{1}{2}(s\partial_s+u\partial_u+1)
\end{aligned}$$

In particular, $E^{\alpha}\Phi_{abc} = \mathcal{E}^{\alpha}(abc)\Phi_{a+1,bc}$, $E_{\alpha}\Phi_{abc} = \mathcal{E}_{\alpha}(abc) \times \Phi_{a-1,bc}$, $E^{\alpha\gamma}\Phi_{abc} = \mathcal{E}^{\alpha\gamma}(abc)\Phi_{a+1,b,c+1}$, etc., where

(2.4) $E^{\alpha} = a \ , \ E^{\alpha\beta\gamma} = ab/c \ , \ E^{\beta} = b \ , \ E^{\alpha\gamma} = a(b-c)/c$

$$E^{\gamma} = (c-b)(c-a)/c \ , \ E_{\alpha} = c-a \ , \ E_{\alpha\gamma} = c-1 = E_{\beta\gamma} = E_{\gamma}$$

$$E_{\alpha\beta\gamma} = 1-c \ , \ E^{\beta\gamma} = b(a-c)/c \ , \ E_{\beta} = c-b \quad .$$

The action of $J(\alpha)$, $J(\beta)$, $J(\gamma)$ on the eigenfunctions Φ_{abc} is obvious. For $a,b,c,b-c,a-c$ nonintegers, a linearly independent set of basis functions also satisfying relations (2.4) is

(2.5) $$\Psi_{abc} = \frac{\Gamma(c)\Gamma(b-c+1)\Gamma(a-c+1)}{\Gamma(2-c)\Gamma(a)\Gamma(b)} z^{1-c} {}_2F_1(a-c+1,b-c+1;2-c;z) \times s^a u^b v^c \quad .$$

For $a-\mu$, $b-\nu$, $c-\omega$ integers and $\mu,\nu,\omega,\omega-\mu,\omega-\nu$ nonintegers the sets $\{\Phi_{abc}\}$ and $\{\Psi_{abc}\}$ separately define a model of an algebraically irreducible representation of $sl(4,\mathbb{C})$. If any of μ, ν, $\omega-\mu$, $\omega-\nu$ is integral then the set $\{\Phi_{abc}\}$ defines a model of a reducible but indecomposable representation of $sl(4,\mathbb{C})$ whereas the $\{\Psi_{abc}\}$ are no longer defined for all permissable a,b,c. (For ω an integer the $\{\Phi_{abc}\}$ are no longer defined in the case $c = 0,-1,-2,\ldots$ but this can be remedied by consideration of the alternate basis ${}_2F_1(a,b;c;z)s^a u^b v^c/\Gamma(c)$.) Note that the operators (2.3) yield all of the differential recurrence relations obeyed by the ${}_2F_1$.

Just as in the preceding section, we can use a subset of the operators (2.3), eliminate the terms in ∂_z to obtain non-differential recurrence relations for the Φ_{abc} and the Ψ_{abc}, and then construct an associated continued fraction expansion. Again the irreducible representations lead to expansions for ratios of contiguous functions whereas the reducible indecomposable representations yield expansions for a second kind function at the extreme end of a weight ladder.

The simplest types of continued fractions could be constructed by considering a pair of operators such as $E^{\alpha\gamma}$, $E_{\alpha\gamma}$.

To derive the famous Gaussian continued fraction, however, it is necessary to consider a weight ladder of rising steps obtained by applying $E^{\alpha\gamma}$ and $E^{\beta\gamma}$ successively to a given $\Phi_{\mu\nu\omega}$. To go back down the ladder one applies $E_{\alpha\gamma}$ and $E_{\beta\gamma}$ successively. Thus, the mth rung on the ladder is occupied by $\Phi_m = \Phi_{\mu+[(m+1)/2], \nu+[m/2], \omega+m}$ where $[n]$ is the greatest integer in n. The recurrence formulas relating successive rungs on the ladder are obtained by eliminating the ∂_z terms from the differential relations for $E^{\alpha\gamma}$ and $E_{\beta\gamma}$ (as well as $E_{\alpha\gamma}$ and $E^{\beta\gamma}$):

$$(2.6)\quad \Phi_m = \frac{-(\omega+m-1)(\omega+m-2)}{z(\mu+m/2-\frac{1}{2})(\mu-\omega-m/2+\frac{1}{2})}(sv\Phi_{m-1}-suv^2\Phi_{m-2}),\quad m \text{ odd}$$

$$= \frac{-(\omega+m-1)(\omega+m-2)}{z(\nu+m/2-1)(\mu-\omega-m/2+1)}(uv\Phi_{m-1}-suv^2\Phi_{m-2})\quad m \text{ even}.$$

The independent solutions $\Psi_m = \Psi_{\mu+[(m+1)/2],\nu+[m/2],\omega+m}$ satisfy the identical relations. We construct the continued fraction for which $p_m = \Psi_m$, $q_m = \Phi_m$. From well-known asymptotic formulas for the ${}_2F_1$, [3, pp. 235-239], [6] we find $c = \lim_{m\to\infty} p_m/q_m = \infty$. Expression (2.6) implies

$$(2.7)\quad b_m = -a_m = \alpha_k = \frac{(\omega+2k)(\omega+2k-1)}{z(\mu+k)(\omega-\nu+k)},\quad m = 2k+1 \quad \text{odd}$$

$$b_m = -a_m = \beta_k = \frac{(\omega+2k-1)(\omega+2k-2)}{z(\nu+k-1)(\omega-\mu+k-1)},\quad m = 2k \quad \text{even}$$

for $m \geq 3$ and $u = v = s = 1$; furthermore $b_1 = \Phi_1$, $a_1 = \Psi_1$, $b_2 = \Psi_2/\Psi_1$, $a_2 = (\Psi_1\Phi_2-\Psi_2\Phi_1)/\Psi_1$. Thus,

$$0 = \Phi_1 + \frac{\Psi_1\Phi_2-\Psi_2\Phi_1}{\Psi_2+\Psi_1 d},\quad d = -1/1 - \alpha_1^{-1}/1 - \beta_1^{-1}/1 - \alpha_2^{-1}/1 - \beta_2^{-1}/1 - \cdots$$

or $d = -\Phi_2/\Phi_1$. Written out explicitly and simplified this identity reads

(2.8) $\quad {}_2F_1(\mu+1,\nu+1;\omega+2;-z)/{}_2F_1(\mu+1,\nu;\omega+1;-z) = 1/1 + \alpha_1^{-1}/1 + \beta_2^{-1}/1 + \alpha_2^{-1}/1 + \beta_3^{-1}/1 + \cdots .$

For $\nu = 0$ the representation of $sl(4,\mathbb{C})$ becomes reducible but indecomposable and the related continued fraction yields a simple expansion for ${}_2F_1(\mu+1,1;\ \omega+2;-z)$.

Similar techniques yield continued fraction expansions for other special functions of hypergeometric type. In particular the Gegenbauer functions (the hypergeometric functions which satisfy quadratic transformation formulas) are obtained as separable solutions of the wave equation $(\partial_{tt}-\partial_{x_1x_1}-\partial_{x_2x_2})\Phi = 0$ in an appropriate coordinate system. The symmetry algebra of this equation is $o(5,\mathbb{C})$. The confluent hypergeometric functions ${}_1F_1$ are obtained as separable solutions of $(\partial_t-\partial_{x_1x_1}-\partial_{x_2x_2})\Phi = 0$ with the 9-dimensional Schrödinger algebra as symmetry algebra. Finally the Bessel functions are obtainable as separated solutions of the Helmholtz equation $(\partial_{x_1x_1}+\partial_{x_2x_2}+\omega^2)\Phi = 0$ with 3-dimensional symmetry algebra $E(2)$.

In conclusion, the traditional method for establishing continued fraction expansions for functions of hypergeometric type employs the contiguous function relations to obtain formal expansions and deeper theorems from continued fraction theory to establish convergence of the expansions. The method presented here shows the explicit Lie algebraic significance of the contiguous function relations and uses asymptotic formulas for special functions to obtain the continued fractions. The contrast between irreducible and reducible but indecomposable representations is very significant in this regard.

References

1 Erdelyi, A. et al., Higher Transcendental Functions, Vol. 2, McGraw-Hill, New York, 1951.

2 Kalnins, E.G. and W. Miller, Jr., Lie theory and the wave equation in space-time. 3, J. Math. Phys. (to appear).

3 Luke, Y., The Special Functions and their Approximations, Vol. 1, Academic Press, New York, 1969.

4 Miller, W. Jr., Lie algebras and generalizations of the hypergeometric function, Proceedings of Symposia in Pure Mathematics, Vol. 26, American Mathematical Society, Providence, R.I., 1973.

5 Miller, W. Jr., Lie theory and generalizations of the hypergeometric functions, SIAM J. Appl. Math., 25 (1973), 226-235.

6 Watson, G.N., Asymptotic expansions of hypergeometric functions, Trans. Cambridge Philos. Soc., 22 (1918), 277-308.

7 Widder, D.V., The Heat Equation, Academic Press, New York, 1975.

Charles Boyer
IIMAS
Universidad Nacional Autonoma de Mexico
Mexico 20, D.F.

Willard Miller, Jr.*
School of Mathematics
University of Minnesota
Minneapolis, Minnesota 55455

*Research supported in part by NSF Grant MCS 76-04838.

NOTE ON A THEOREM OF SAFF AND VARGA

P. Henrici

A recent result by E.B. Saff and R.S. Varga on zero-free parabolic regions for partial sums of certain power series is generalized so as to apply to sequences of polynomials interpolating the sum of the power series. The proof is by continued fraction methods.

1 Introduction

In a recent paper, E.B. Saff and R.S. Varga [3] showed that certain polynomials generated by three-term recurrence relations have no zeros in parabolic regions P_α in the complex plane defined by $y^2 < 4\alpha(x + \alpha)$, $x > -\alpha$. Their result applies, in particular, to the partial sums of the series

$$(1) \quad f(z) = \sum_{k=0}^{\infty} \alpha_k z^k$$

where $\alpha_k > 0$, $k = 0,1,2,\ldots$, and

$$(2) \quad \alpha: = \inf_{k>0} \left\{ \frac{\alpha_{k-1}}{\alpha_k} - \frac{\alpha_{k-2}}{\alpha_{k-1}} \right\} > 0,$$

and thus to the exponential series, where $\alpha = 1$.

In our Theorem below we give a generalization of the result of Saff and Varga which permits to reach the same conclusion for arbitrary sequences of interpolating polynomials, provided that the divided differences formed with the interpolated values satisfy a condition similar to (2). The proof uses some basic facts of continued fraction theory.

2 The Theorem

THEOREM. Let $\{\beta_n\}_1^\infty$, $\{\varepsilon_n\}_1^\infty$ be sequences of positive real numbers such that

(3) $\alpha := \inf_{n>0}(\beta_n - \varepsilon_n) > 0,$

<u>and let</u> $\{z_n\}_1^\infty$ <u>be a sequence of complex numbers belonging to the parabolic region</u>

(4) $P_\alpha : |z| < \text{Re } z + 2\alpha.$

<u>Then the numbers</u> q_n <u>defined by</u>

(5) $q_{-1} := 0,\ q_0 := 1,\ q_n := (\beta_n + z_{n+1})q_{n-1} - \varepsilon_n z_n q_{n-2},$
$n = 1,2,\ldots,$

<u>are different from zero for</u> $n \geqq 0$.

<u>Proof</u>. The q_n are the denominators of the continued fraction

$$C := \frac{z_1\varepsilon_1|}{|\beta_1 + z_2} - \frac{z_2\varepsilon_2|}{|\beta_2 + z_3} - \frac{z_3\varepsilon_3|}{|\beta_3 + z_4} - \cdots .$$

We first assume that all $z_n \neq 0$. Then the numerators p_n and the denominators of C cannot vanish simultaneously, and in order to show that $q_n \neq 0$ it suffices that all approximants

$$w_n = \frac{p_n}{q_n},\ n = 0,1,\ldots,$$

of C are finite. Now for $n \geqq 1$ [2]

$$w_n = t_1 \circ t_2 \circ \cdots \circ t_n(0),$$

where t_n denotes the Moebius transformation

$$t_n(u) := -\frac{z_n\varepsilon_n}{\beta_n + z_{n+1} + u} .$$

We have $t_n = t_n^* \circ t_n^{**}$, where

$$t_n^{**}(u) := z_{n+1} + u,\ t_n^*(u) := -\frac{z_n\varepsilon_n}{\beta_n + u} .$$

Letting $s_n := t_{n-1}^{**} \circ t_n^*$,

$$s_n(u) = z_n(1 - \frac{\varepsilon_n}{\beta_n + u}),$$

there follows

$$w_n = w_n^* - z_1,$$

where

$$w_n^* = s_1 \circ s_2 \circ \dots \circ s_n(z_{n+1}).$$

The w_n are finite if and only if the w_n^* are finite. The w_n^* are finite if there exists a region H in the extended complex plane such that

$$\text{(i)}\ P_\alpha \subset H, \quad \text{(ii)}\ \infty \notin H, \quad \text{(iii)}\ s_k(H) \subset H,$$

$k = 1,2,\dots$. It is asserted that the open half-plane H: Re $u > -\alpha$ has the required properties. It is clear that (i) and (ii) hold. To verify (iii) we first note that because $s_k^{-1}(\infty) = -\beta_k$ does not belong to the closure of H, $s_k(H)$ is a disk. The center c_k of that disk is $s_k(u_0)$, where u_0 is the point symmetric to $-\beta_k$ with respect to Re $u = -\alpha$, the boundary of H. Hence $u_0 = \beta_k - 2\alpha$, and

$$c_k = s_k(\beta_k - 2\alpha) = z_k(1 - \frac{\varepsilon_k}{2\beta_k - 2\alpha}).$$

The boundary of the disk $s_k(H)$ passes through $s_k(\infty) = z_k$. Hence the disk has the radius

$$\rho_k := \frac{|z_k|\varepsilon_k}{2\beta_k - 2\alpha},$$

and its leftmost point has the real part

$$\xi := \text{Re } c_k - \rho_k = \text{Re } z_k - \frac{\varepsilon_k(\text{Re } z_k + |z_k|)}{2\beta_k - 2\alpha}.$$

It remains to be shown that $\xi > -\alpha$. Using (4),

$$\xi > \mathrm{Re}\, z_k - \frac{\varepsilon_k(2\mathrm{Re}\, z_k + 2\alpha)}{2\beta_k - 2\alpha}$$

$$= (1 - \frac{\varepsilon_k}{\beta_k - \alpha})\mathrm{Re}\, z_k + \frac{\varepsilon_k}{\beta_k - \alpha}(-\alpha).$$

By (3), $0 \leq \varepsilon_k(\beta_k - \alpha)^{-1} \leq 1$, and because (4) implies $\mathrm{Re}\, z_k > -\alpha$, $\xi > -\alpha$ follows.

If some z_n are zero, let z_{m+1} be the first such. It follows as above that $q_k \neq 0$, $k = 0,1,\ldots,m$. The recurrence relation (5) shows that $q_{m+n} = q_m q_n^*$, $n = 0,1,2,\ldots$, where the q_n^* are the denominators of the fraction

$$C^* := -\frac{z_{m+1}\varepsilon_{m+1}|}{|\beta_{m+1} + z_{m+2}} - \frac{z_{m+2}\varepsilon_{m+2}|}{|\beta_{m+2} + z_{m+3}} - \cdots$$

and thus are different from zero at least until another z_n vanishes. Proceeding in the same manner, we find as above that all $q_n \neq 0$.

3 An Application

Let $\{x_n\}_0^\infty$ be a sequence of real numbers, and let ϕ be a real function defined at least on the points x_n. Let

$$\alpha_k := \phi[x_0,x_1,\ldots,x_k],$$

the k-th divided difference of ϕ (see [1], p. 277). The unique polynomial p_n of degree $\leq n$ interpolating ϕ at the points $x_0,\ldots,x_n$ then is given by

$$p_n(z) := \alpha_0 + \sum_{j=1}^{n} \alpha_j(z-x_0)\ldots(z-x_{j-1}).$$

Suppose all $\alpha_k \neq 0$. The normalized polynomials $q_n(z) := \alpha_n^{-1}p_n(z)$ are then easily seen to satisfy the recurrence relation $q_{-1} = 0$, $q_0 = 1$,

$$q_n = (\frac{\alpha_{n-1}}{\alpha_n} + z - x_{n-1})q_{n-1} - (z - x_{n-1})\frac{\alpha_{n-2}}{\alpha_{n-1}}q_{n-2} ,$$

$n = 1,2,\ldots$. This is of the form (5) where

$$z_n = z - x_{n-1}, \ \beta_n = \frac{\alpha_{n-1}}{\alpha_n} + x_n - x_{n-1}, \ \varepsilon_n = \frac{\alpha_{n-2}}{\alpha_{n-1}},$$

$n = 1,2,\ldots$. If the sequence $\{x_n\}$ is nondecreasing, there follows $p_n(z) \neq 0$ for $z - x_{n-1} \in P_\alpha$ where α is defined by (3). If the sequence $\{x_n\}$ is nonincreasing, we have $p_n(z) \neq 0$ for $z \in P_\alpha$, where

$$\alpha: = \inf_{n>0} \left\{ \frac{\alpha_{n-1}}{\alpha_n} - \frac{\alpha_{n-2}}{\alpha_{n-1}} + x_n - x_{n-1} \right\}.$$

For example, if $\phi(x): = e^x$ and $x_n = -nh$ $(h > 0)$, we have

$$\alpha_k = \frac{1}{k!} \left(\frac{1 - e^{-h}}{h}\right)^k, \quad k = 0,1,\ldots,$$

and thus may conclude that the polynomials

$$p_n(z): = \sum_{k=0}^{n} \frac{1}{k!} (1 - e^{-h})^k \left(\frac{z}{h}\right)_k$$

are $\neq 0$ for $z \in P_\alpha$ where

$$\alpha: = \frac{h}{e^h - 1}.$$

References

1 G. Dahlquist and A. Björck, Numerical Methods, Prentice-Hall, 1974.

2 P. Henrici and P. Pfluger, Truncation error estimates for Stieltjes fractions. Numer. Math. 9, 120-138 (1966).

3 E.B. Saff and R.S. Varga, Zero-free parabolic regions for sequences of polynomials. SIAM J. Math. Anal. 7, 344-357, (1975).

P. Henrici
Seminar für angewandte Mathematik
ETH - Zentrum
8092 Zürich, Switzerland

MULTIPLE-POINT PADÉ TABLES

William B. Jones

The concept of the Padé table has been generalized recently to give rational approximants for formal Newton series (called Newton-Padé approximants) and for approximation alternately at 0 and ∞ (called two-point Padé approximants). In the case of the Newton-Padé table the approximation is at a sequence of (not necessarily distinct) interpolation points in the finite complex plane. Recent results on the Newton-Padé table are reviewed briefly. Some new results are given on continued fractions (T-fractions) which lie in the two-point Padé table. These include necessary and sufficient conditions for the continued fraction (with non-zero coefficients) to form a diagonal in the table and explicit formulas for the coefficients of the continued fraction.

1 The Newton-Padé table

A series of the form

$$(1)\qquad f(z) = c_0 + \sum_{k=1}^{\infty} c_k \prod_{i=1}^{k} (z-\beta_i)$$

is called a formal Newton series (fNs) with sequence of (not necessarily distinct) interpolation points $\{\beta_i\}$ in the finite complex plane. A rational function

$$(2)\qquad \frac{a_0 + a_1 z + \dots + a_m z^m}{b_0 + b_1 z + \dots + b_n z^n}$$

is said to be of type $[m,n]$ if its denominator is not identically zero. The (m,n) Newton-Padé approximant (denoted by $R_{m,n}(f,z)$) of a fNs $f(z)$ is a rational function of type $[m,n]$ defined in the manner completely analogous to that of the (m,n) Padé approximant (see, [7], [21]). Existence and uniqueness proofs can be found in the preceding references. If all $\beta_i = 0$ then (1) reduces to a formal power series and $R_{m,n}(f,z)$ becomes its (m,n) Padé approximant.

Many of the classical properties of the Padé table have been extended recently to the more general Newton-Padé table. Although space does not permit a detailed statement of these results here, we give a brief review, citing principal references where the statements and proofs can be found.

A large number of identities and algorithms that can be used to construct (both algebraically and numerically) Newton-Padé approximants are contained in the following: [5], [6], [13], [18], [19], [23] and [25]. The most complete sets are contained in the Ph.D. theses by Warner [21] and Claessens [4]. A Newton-Padé approximant

(3) $R_{m,n}(f,z) = P_{m,n}(f,z) / Q_{m,n}(f,z)$

for which $P_{m,n}$ and $Q_{m,n}$ are relatively prime polynomials, is said to be normal if the degrees of $P_{m,n}$ and $Q_{m,n}$ are exactly m and n, respectively, and $R_{m,n}$ occurs only once in the table. Necessary and sufficient conditions for normality of $R_{m,n}$ in terms of determinants (analogous to Hankel determinants) are given by [7]. Necessary and sufficient conditions for $R_{m,n}(f,z)$ to be a solution of the Hermite interpolation problem for the sequence $\{\beta_1, \beta_2, \ldots, \beta_{m+n+1}\}$ can be found in [21]. These conditions are implied by normality. It has also been shown that certain "staircase" sequences of (normal) Newton-Padé approximants are the approximants of Thiele-type continued fractions [6].

The convergence theorem of de Montessus de Ballore was extended by Saff [15] in 1972 to the Newton-Padé table. A further extension of this theorem has been given recently by Warner [22]. Karlsson [12] has extended to the Newton-Padé table the theorems on convergence in measure of Nuttall, and convergence in capacity of Pommerenke. The author, in a joint paper with M.A. Gallucci [7], has shown that, under suitable conditions, uniform convergence of a sequence of Newton-Padé approximants is equivalent to uniform boundedness. In the same paper it is also shown that, with suitable restrictions (implied by normality), $R_{m,n}(f,z)$ behaves continuously as a function of the c_k and β_i in (1). Bounds for Newton-Padé approximants of series of Stieltjes have been given by Baker [2] and Barnsley [3].

2 The two-point Padé table

Corresponding to a pair of formal Laurent series (fLs)

(4a) $L = c_0 + c_1 z + c_2 z^2 + \ldots ,$ (increasing powers)

(4b) $L^* = c^*_\mu z^\mu + c^*_{\mu-1} z^{\mu-1} + c^*_{\mu-2} z^{\mu-2} + \ldots ,$ (decreasing powers)

rational function approximants can be formed in a manner analogous to that for Padé and Newton-Padé approximants. In the present case the sequence of interpolation points is given by $\{0,\infty,0,\infty,0,\infty,\ldots\}$. Following Baker [1], we call these <u>two-point Padé approximants</u>. The (m,n) two-point Padé approximant of L and L^* will be denoted by $R_{m,n}(L,L^*,z)$, or, more simply, $R_{m,n}(z)$ or $R_{m,n}$. Applications of two-point Padé approximants in theoretical physics have been made in a number of papers (see, for example, [1], [8], [16], [17]).

In this section we discuss some new results, obtained jointly with W.J. Thron, regarding a special class of continued fractions and its relation to two-point Padé tables. Proofs of existence and uniqueness of two-point Padé approximants and proofs of all theorems stated below will be given in a subsequent joint paper by Thron and the author.

In 1948 Thron [20] introduced a class of continued fractions of the form

(5) $$c_0 + d_0 z + \frac{z}{1 + d_1 z} + \frac{z}{1 + d_2 z} + \cdots$$

in which the d_n are arbitrary complex constants. He showed that corresponding to an arbitrary formal power series (fps)

(6) $L = c_0 + c_1 z + c_2 z^2 + \ldots ,$

there exists a unique continued fraction (5) such that its

nth approximant $w_n(z)$ has a Taylor series expansion at $z = 0$ of the form

$$(7) \quad w_n(z) = c_0 + c_1 z + \ldots + c_n z^n + \gamma_{n+1}^{(n)} z^{n+1} + \ldots, \quad n \geq 0 .$$

Conversely, he showed that for each continued fraction (5) there exists a unique fps (6) such that (7) holds. Perron in 1957 [14, Section 31] considered the more general class of continued fractions of the form

$$(8) \quad e_0 + d_0 z + \frac{z}{e_1 + d_1 z} + \frac{z}{e_2 + d_2 z} + \cdots, \quad e_n \neq 0, \quad n \geq 0,$$

and observed that, if $d_n \neq 0$ for $n \geq 1$, then (8) corresponds to a fLs of the form

$$(9) \quad L^* = c_1^* z + c_0^* + c_{-1}^* z^{-1} + c_{-2}^* z^{-2} + \ldots$$

in the sense that the nth approximant $w_n(z)$ of (8) has a Laurent expansion at $z = \infty$ of the form

$$(10) \quad w_n(z) = c_1^* z + c_0^* + \ldots + c_{-(n-1)}^* z^{-(n-1)} + \gamma_n^{*(n)} z^{-n} + \ldots .$$

Perron referred to the continued fraction (8) as the Thronschen Kettenbrüche. As an abbreviation the continued fractions (5) have been called the T - fractions [10]. Hereafter we shall refer to (8), or its equivalent form

$$(11) \quad F_0 + G_0 z + \frac{F_1 z}{1 + G_1 z} + \frac{F_2 z}{1 + G_2 z} + \cdots, \quad F_n \neq 0, \quad n \geq 0,$$

as a <u>general</u> T - <u>fraction</u>. The following theorem relates the general T - fractions, with all $G_n \neq 0$, to two-point Padé tables.

THEOREM 1. (A) <u>If for a given pair of</u> fLs

$$(12) \quad L = \sum_{k=-\nu}^{\infty} c_k z^k \quad \underline{\text{and}} \quad L^* = \sum_{k=-\infty}^{\mu} c_k^* z^k, \quad (\nu \geq 0, \ \mu \geq 0)$$

<u>there exists a general</u> T - <u>fraction</u>

(13a) $\sum_{k=1}^{\mu} c_k^* z^k + \sum_{k=-\nu}^{-1} c_k z^k + c_0 + \frac{F_1 z}{1+G_1 z} + \frac{F_2 z}{1+G_2 z} + \cdots ,$

with

(13b) $F_n \neq 0$ and $G_n \neq 0$, $n \geq 1$

corresponding to L and L^* in the sense that the nth approximant $w_n(z)$ of (13) has Laurent expansions of the forms

(14a) $w_n(z) = c_{-\nu} z^{-\nu} + c_{-(\nu-1)} z^{-(\nu-1)} + \ldots + c_n z^n + \gamma_{n+1}^{(n)} z^{n+1} + \ldots$ (at $z = 0$)

and

(14b) $w_n(z) = c_\mu^* z^\mu + c_{\mu-1}^* z^{\mu-1} + \ldots + c_{-(n-1)}^* z^{-(n-1)} + \gamma_{-n}^{*(n)} z^{-n} + \ldots$ (at $z = \infty$),

then (letting $\delta_k = c_k^* - c_k$; $c_k = 0$ for $k < -\nu$; $c_k^* = 0$ for $k > \mu$)

(15a) $\Delta_n \neq 0$ and $\Phi_n \neq 0$, $n \geq 1$,

where

(15b) $\Delta_n = \begin{vmatrix} \delta_{-(n-1)} & \cdots & \delta_0 \\ \vdots & & \vdots \\ \delta_0 & \cdots & \delta_{n-1} \end{vmatrix}, \quad n \geq 1; \qquad \Delta_{-1} = \Delta_0 = 1, \quad \Delta_1 = \delta_0, \quad \Delta_2 = \begin{vmatrix} \delta_{-1} & \delta_0 \\ \delta_0 & \delta_1 \end{vmatrix},$

and

(15c) $\Phi_{n+1} = \begin{vmatrix} \delta_{-(n-1)} & \cdots & \delta_1 \\ \vdots & & \vdots \\ \delta_1 & \cdots & \delta_{n+1} \end{vmatrix}, \quad n \geq 1; \qquad \Phi_{-1} = -1, \quad \Phi_0 = 1, \quad \Phi_1 = \delta_1, \quad \Phi_2 = \begin{vmatrix} \delta_0 & \delta_1 \\ \delta_1 & \delta_2 \end{vmatrix}.$

Moreover,

(16a) $F_1 = -\Phi_1$; $F_n = -\Delta_{n-1}\Phi_{n-1} / \Delta_{n-2}\Phi_n$, $n \geq 2$,

(16b) $G_n = -\Delta_{n-1}\Phi_n / \Delta_n\Phi_{n-1}$, $n \geq 1$.

(B) Conversely, if for given fLs (12), conditions (15) hold, then the general T-fraction (13a), with coefficients defined by (16), satisfies (13b) and corresponds to L and L^* in the sense of (14).

(C) If $\nu = 0$ and (15) and (16) hold, then the nth numerator $A_n(z)$ and denominator $B_n(z)$ of (13) are polynomials of exact degrees $n+\mu$ and n, respectively, and for $n \geq 1$,

(17) $w_n(z) = A_n(z)/B_n(z) = R_{n+\mu,n}(L, L^*, z)$.

The following a posteriori truncation error bounds for general T-fractions is a slight extension of results previously given by [9] and [11].

THEOREM 2. Let $w_n(z)$ denote the nth approximant of a general T-fraction

(18) $$F_0 + G_0 z + \frac{F_1 z}{1+G_1 z} + \frac{F_2 z}{1+G_2 z} + \cdots, \quad F_n \neq 0, \quad n \geq 1.$$

If

(19) $F_n > 0$ and $G_n \geq 0$, $n \geq 1$,

then for $|\arg z| < \pi$,

(20) $$|w_{n+m}(z) - w_n(z)| \leq K(z)|w_n(z) - w_{n-1}(z)|, \quad n \geq 2, \quad m \geq 0,$$

where

(21) $$K(z) = \begin{cases} 1, & \text{if } |\arg z| \leq \frac{\pi}{2} \\ \sec(|\arg z| - \frac{\pi}{2}), & \text{if } \frac{\pi}{2} < |\arg z| < \pi. \end{cases}$$

We conclude this paper with a theorem stating sufficient conditions for the existence of a solution to a moment problem for powers of t with both positive and negative exponents.

THEOREM 3. Let $\{c_n\}_{n=0}^{\infty}$ and $\{c_n^*\}_{n=0}^{-\infty}$ be sequences of numbers such that the determinants (15) satisfy

$$\Phi_{n+1} = (-1)^{n+1} \Delta_n ,$$

$$0 < m' \le \Delta_n \le M', \quad n \ge 0 ,$$

Let

$$M = \left(\frac{M'}{m'}\right)^2 , \quad m_1 = \left(\sqrt{\frac{1}{M} + 1} - 1\right)^{-2} ,$$

$$M_1 = (\sqrt{M+1} + 1)^{-2} .$$

Then there exists a bounded non-decreasing function $\varphi(t)$ such that

$$c_1 = G_0 + \int_{M_1}^{m_1} \frac{d\,\varphi(t)}{t} ; \quad c_k = (-1)^{k-1} \int_{M_1}^{m_1} \frac{d\,\varphi(t)}{t^k} , \quad k \ge 2 ,$$

$$c_0^* = c_0 + \int_{M_1}^{m_1} d\,\varphi(t) ; \quad c_k^* = (-1)^k \int_{M_1}^{m_1} t^k \, d\,\varphi(t) , \quad k \ge 1 .$$

References

1 Baker, G.A., Jr., G.S. Rushbrooke, and H.E. Gilbert, High-temperature series expansions for the spin -1/2 Heisenberg model by the method of irreducible representations of the symmetry group, Physical Review, 135, No. 5A (August 31, 1964), A1272-A1277.

2 Baker, G.A., Jr., Best error bounds for Padé approximants to convergent series of Stieltjes, J. Mathematical Phys., 10 (1964), 814-820.

3 Barnsley, M., The bounding properties of the multiple-point Padé approximant to a series of Stieltjes, Rocky Mountain J. of Mathematics, 4, No. 2 (Spring 1974), 331-333.

4 Claessens, G., Some aspects of the rational Hermite interpolation table and its applications, Ph.D. thesis, Universitaire Instelling Antwerpen, Wilrijk (1976).

5 Claessens, G., The rational Hermite interpolation problem and some related recurrence relations, Comp. and Maths. with Appls., 2 (1976), 117-123.

6 Claessens, G., A new algorithm for oscillatory rational interpolation, to appear in Numerische Mathematik.

7 Gallucci, Michael A. and William B. Jones, Rational approximation corresponding to Newton series (Newton-Padé approximants), J. Approximation Theory, 17, No. 4, (August 1976), 366-392.

8 Isihara, A. and E.W. Montroll, A note on the ground state energy of an assembly of interacting electrons, Proc. Nat. Acad. Sci. USA, 68, No. 12, (December 1971), 3111-3115.

9 Jefferson, T.H., Truncation error estimates for T-fractions, SIAM J. Numer. Anal., 6 (1969), 359-364.

10 Jones, William B. and W.J. Thron, Further properties of T-fractions, Math. Annalen, 166 (1966), 106-118.

11 Jones, William B. and W.J. Thron, A posteriori bounds for the truncation error of continued fractions, SIAM J. Numer. Anal., 8 (December 1971), 693-705.

12 Karlsson, J., Rational interpolation and best rational approximation, J. Mathematical Analysis and Applications, 53, No. 1, (January 1976), 38-51.

13 Larkin, F.M., Some techniques for rational interpolation, Computer J., 10 (1967), 178-187.

14 Perron, O., Die Lehre von den Kettenbrüchen, Band II, Stuttgart, Teubner, (1957).

15 Saff, E.B., An extension of Montessus de Ballore's theorem on the convergence of interpolating rational functions, J. Approximation Theory, 6 (1972), 63-67.

16 Shing, P., Application of two-point Padé approximants to some solid state problems, Rocky Mountain Journal of Mathematics, 4, No. 2 (Spring 1974), 385-386.

17 Shing, P. and J.D. Dow, Intermediate coupling theory: Padé approximants for polarons, Phys. Rev. B4 (1974), 1343-1359.

18 Stoer, J., Über zwei Algorithmen zur Interpolation mit rationalen Funktionen, Numer. Math., 3 (1961), 285-304.

19 Thacher, H.C. and J. Tukey, Recursive algorithm for interpolation by rational functions, Unpublished manuscript, (1960).

20 Thron, W.J., Some properties of continued fractions $1 + d_0 z + K\left(\frac{z}{1 + d_n z}\right)$, Bull. Am. Math. Soc., 54 (1948), 206-218.

21 Warner, D.D., Hermite interpolation with rational functions, Ph.D. Thesis, University of California, San Diego, (1974).

22 Warner, D.D., An extension of Saff's theorem on the convergence of interpolating rational functions, J. Approximation Theory, to appear.

23 Wuytack, L., An algorithm for rational interpolation similar to the qd-algorithm, Numer. Math., 20 (1973), 418-424.

24 Wuytack, L., On some aspects of the rational interpolation problem, SIAM J. Numer. Anal., 11, No. 1 (March 1974), 52-60.

25 Wuytack, L., On the osculatory rational interpolation problem, Math. of Comp., 29, No. 131 (July 1975), 837-843.

William B. Jones
Department of Mathematics
University of Colorado
Boulder, Colorado 80309

Research supported in part by the National Science Foundation under Grant No. MPS 74-22111.

APPLICATION OF STIELTJES FRACTIONS TO BIRTH-DEATH PROCESSES

William B. Jones and Arne Magnus

Murphy and O'Donohoe [3] consider a birth-death process with birth rates λ_r and death rates μ_r for population size r. They treat the differential difference equations involved by means of continued fractions. We continue this investigation by establishing convergence of the continued fractions involved and by employing an estimate of Henrici and Pfluger for the rate of convergence of related Stieltjes fractions. Under relatively mild conditions on the λ_r's and μ_r's we establish convergence of approximations to the probabilities $p_r(t)$ of having population size r at the time t.

In the paper referred to above J. A. Murphy and O'Donohoe consider a birth-death process in which a population of size m at time t = 0 is changing due to birth or immigration at a rate of λ_r and death or emigration at a rate μ_r when the population has size r. That is, each individual has a probability of $\lambda_r\Delta t + 0((\Delta t)^2)$ of producing a new individual and a probability of $\mu_r\Delta t+0((\Delta t)^2)$ of dying during a short time interval $(t, t + \Delta t)$.

If $p_r(t)$ is the probability that the population has size r at time t then the differential difference equations that govern the growth of the population are

$$p_0'(t) = -\lambda_0 p_0(t) + \mu_1 p_1(t)$$
$$p_r'(t)=\lambda_{r-1}p_{r-1}(t)-(\lambda_r+\mu_r)p_r(t)+\mu_{r+1}p_{r+1}(t),\ r=1,2,\dots .$$

To solve these equations Murphy and O'Donohoe introduce the Laplace transform $P_r(s) = \int_0^\infty e^{-st}p_r(t)dt$ of $p_r(t)$, and obtain

$$\mu_1P_1(s) = -\delta_{0,m} + (\lambda_0 + s)P_0(s)$$
$$\mu_{r+1}P_{r+1}(s) = -\lambda_{r-1}P_{r-1}(s) + (\lambda_r+\mu_r+s)P_r(s)-\delta_{r,m},\ r=1,2,\dots ,$$

where $\delta_{r,m}$ is the Kronecker delta. The rates λ_r and μ_r will be assumed positive for all r and we define L_r and M_r by

$$L_r=\lambda_0\lambda_1\dots\lambda_r,\ r=0,1,\dots;\ M_r=\mu_1\mu_2\dots\mu_r,\ r=1,2,\dots .$$

We normalize the above equations by

(1) $f_0^{(m)}(s) = P_0(s),\ f_r^{(m)}(s) = (-1)^r M_r P_r(s),\ r = 1, 2, \dots$

and obtain

$$(2)\begin{cases} f_1^{(m)}(s) = \delta_{0,m} - (\lambda_0 + s) f_0^{(m)}(s) \\ f_{r+1}^{(m)}(s) = -\lambda_{r-1}\mu_r f_{r-1}^{(m)}(s) - (\lambda_r + \mu_r + s) f_r^{(m)}(s) + (-1)^r M_r \delta_{r,m},\ r = 1, 2, \dots . \end{cases}$$

If m = 0 we may deduce that: setting $f_r^{(0)}(s) = f_r(s)$

$$f_0(s) = \frac{1}{\lambda_0 + s + f_1(s)/f_0(s)}$$
$$= \frac{1}{\lambda_0+s} - \frac{\lambda_0\mu_1}{\lambda_1+\mu_1+s} - \dots - \frac{\lambda_{r-1}\mu_r}{\lambda_r+\mu_r+s+f_{r+1}(s)/f_r(s)},\quad r=1, 2, \dots,$$

which leads us to consider the J-fraction

(3) $$\frac{1}{\lambda_0 + s} - \frac{\lambda_0\mu_1}{\lambda_1 + \mu_1 + s} - \frac{\lambda_1\mu_2}{\lambda_2 + \mu_2 + s} - \dots$$

whose nth approximant we denote by A_n/B_n. This J-fraction is the even part of the Stieltjes fraction

(4) $$\frac{1}{s} + \frac{\lambda_0}{1} + \frac{\mu_1}{s} + \frac{\lambda_1}{1} + \frac{\mu_2}{s} + \frac{\lambda_2}{1} + \dots$$

and (4) converges uniformly to an analytic function of s in any compact set whose distance from the negative half of the real axis is positive, whenever

$$\sum \mu_1 \cdots \mu_r/\lambda_0 \cdots \lambda_r = \sum M_r/L_r \text{ or } \sum L_r/M_{r+1} = \sum \lambda_0 \cdots \lambda_n/\mu_1 \cdots \mu_{r+1}$$

diverges. Henceforth we will assume that the λ_r's and μ_r's are so restricted. We denote the value of (3) by f_0 and define $f_1, f_2, \dots$ recursively by (2) with m = 0 thus obtaining one of the solutions of that system (when m = 0). Two consecutive f_r's cannot equal zero so $f_{r+1}/f_r \in \bar{\mathbb{C}}$, the extended complex plane.

With s restricted as above the following two continued fractions are seen to converge to meromorphic functions, namely

$$\frac{f_{r+1}}{f_r} = -\frac{\lambda_r\mu_{r+1}}{\lambda_{r+1} + \mu_{r+1} + s} - \frac{\lambda_{r+1}\mu_{r+2}}{\lambda_{r+2} + \mu_{r+2} + s} - \dots$$

$$(5)\quad f_r = \frac{(-1)^r L_{r-1} M_r / B_r}{B_{r+1}/B_r} - \frac{\lambda_r \mu_{r+1}}{\lambda_{r+1} + \mu_{r+1} + s} - \frac{\lambda_{r+1}\mu_{r+2}}{\lambda_{r+2} + \mu_{r+2} + s} - \dots .$$

The latter fraction is equivalent to

$$\frac{(-1)^r L_{r-1} M_r}{B_{r+1}} - \frac{\lambda_r \mu_{r+1} B_r}{\lambda_{r+1}+\mu_{r+1}+s} - \frac{\lambda_{r+1}\mu_{r+2}}{\lambda_{r+2}+\mu_{r+2}+s} - \dots$$

which is seen to have approximants with denominators

$$(6)\quad B_n^{(r)} = B_{r+n}.$$

If $m > 0$ and $r = 1, 2, \dots, m$ induction on r shows that

$$B_r f_{r-1}^{(m)} + B_{r-1} f_r^{(m)} = 0$$

for all solutions of (2). In particular, for $r = m$

$$(7)\quad B_m f_{m-1}^{(m)} = -B_{m-1} f_m^{(m)},$$

which, when inserted in (2) with $r = m$, gives

$$(8)\quad f_{m+1}^{(m)} = (-1)^m M_m - \frac{B_{m+1}}{B_m} f_m^{(m)}.$$

Equation (8), together with (2), where $r = m + 1, m + 2, \dots,$ forms a system of equations analogous to (2) with $m=0$ and leads us to consider a convergent continued fraction similar to (3) whose possibly infinite value we denote by $f_m^{(m)}$, namely

$$(9)\quad f_m^{(m)} = \frac{(-1)^m M_m}{B_{m+1}/B_m} - \frac{\lambda_m \mu_{m+1}}{\lambda_{m+1}+\mu_{m+1}+s} - \frac{\lambda_{m+1}\mu_{m+2}}{\lambda_{m+2}+\mu_{m+2}+s} - \dots .$$

By the theory of Stieltjes fractions the zeros of the B_r's all lie on the negative real axis. Thus (7) and (2) determine the other $f_r^{(m)}$'s in terms of $f_m^{(m)}$, giving one solution to the system (2). Comparing (5) and (9), applying (7) repeatedly and using (2) we find

$$(10)\begin{cases} f_r^{(m)} = (-1)^{m-r} \dfrac{B_r}{L_{m-1}} f_m & r = 0, 1, \dots, m \\[2ex] f_r^{(m)} = \dfrac{B_m}{L_{m-1}} f_r & r = m, m+1, \dots . \end{cases}$$

Equations (1), (10) and (5) show that each P_r, $r = 0, 1, \dots$ has a continued fraction expansion

$$P_r = \frac{M_m}{M_r}\frac{B_r}{B_m}\left(\frac{1}{B_{m+1}/B_m} - \frac{\lambda_m \mu_{m+1}}{\lambda_{m+1}+\mu_{m+1}+s} - \dots\right),\quad r = 0, 1, \dots, m,$$

$$P_r = \frac{L_{r-1}}{L_{m-1}}\frac{B_m}{B_r}\left(\frac{1}{B_{r+1}/B_r} - \frac{\lambda_r\mu_{r+1}}{\lambda_{r+1}+\mu_{r+1}+s} - \cdots\right), \quad r = m, m+1, \ldots,$$

with approximants denoted by $P_{r,n}$.

We now define the inverse Laplace transform

$$p_{r,n}(t) = \mathcal{L}^{-1}(P_{r,n}) = \frac{1}{2\pi}\int_{-\infty}^{\infty} e^{st}P_{r,n}(s)\,dw,$$

where $s = c + iw$, $c > 0$ and investigate the convergence of $p_{r,n}(t)$ to $p_r(t) = \mathcal{L}^{-1}(P_r)$. Since

$$|p_r - p_{r,n}| \leq \frac{e^{ct}}{2\pi}\int_{-\infty}^{\infty} |P_r(c + iw) - P_{r,n}(c + iw)|\,dw,$$

we seek an estimate for $|P_r - P_{r,n}|$.

We denote the nth approximants of $g = f_{r+1}/f_r$ and of the convergent J-fraction

$$h = 1/(\lambda_r + s + g) = \frac{1}{\lambda_r+s} - \frac{\lambda_r\mu_{r+1}}{\lambda_{r+1}+\mu_{r+1}+s} - \frac{\lambda_{r+1}\mu_{r+2}}{\lambda_{r+2}+\mu_{r+2}+s} - \cdots$$

by g_n and h_n, respectively, and find for $r = m, m+1, \ldots$

$$|P_r - P_{r,n}| = \frac{L_{r-1}}{L_{m-1}}\left|\frac{B_m}{B_{m+1}}\right|\left|\frac{B_{m+1}}{B_{m+2}}\right|\cdots\left|\frac{B_{r-1}}{B_r}\right|\cdot$$

$$\frac{1}{|B_{r+1}/B_r+g|\,|B_{r+1}/B_r+g_{n-1}|\,|h|\,|h_n|}\,|h - h_n|.$$

From $B_{r+1} = (\lambda_r + \mu_r + s)B_r - \lambda_{r-1}\mu_r B_{r-1}$ one can prove by induction that $\mathrm{Im}(B_{r+1}/B_r) \geq \mathrm{Im}(s) = w$ when $w > 0$ and $\mathrm{Im}(B_{r+1}/B_r) \leq w$ when $w < 0$.

Since $-\lambda_{t-1}\mu_t/(\lambda_t + \mu_t + s)$ maps the half plane $w > \Omega > 0$ onto the interior of a circle in the upper half plane touching the real axis at the origin, we see that by repeated application of such mappings that g_{n-1}, and therefore g, lies in the half plane $\mathrm{Im}\, g_{n-1} > 0$, $\mathrm{Im}\, g > 0$. A similar argument shows that h_n and h lie in the lower half plane, in fact, that they are bounded away from zero when $|w| \geq \Omega > 0$. Thus for $|w| \geq \Omega > 0$, $|B_{r+1}/B_r| \geq |w|$, for all r and $|B_{r+1}/B_r + g|$ and $|B_{r+1}/B_r+g_n|$ are all greater than $|w|$. This shows that $B_{r+1}/B_r + g$ and B_{r+1}/B_r+g_n are not identically zero, nor are h and h_n.

When s is positive so that $w = 0$ then B_{r+1}/B_r, g_n, g, h_n and h are all real. Thus $B_{r+1}/B_r + g$, $B_{r+1}/B_r + g_n$, h or h_n could equal zero so that P_r or $P_{r,n}$ could be infinite. Since none of these functions are identically zero they have at most a countable number of isolated zeros (all of which are real). Therefore, for any fixed $\Omega > 0$ and all n there exists $c > 0$ and $\delta > 0$ such that for $s = c + iw$, $-\Omega \le w \le \Omega$,

$$|B_{m+1}/B_m|, |B_{m+2}/B_{m+1}|, \dots, |B_r/B_{r-1}|, |B_{r+1}/B_r+g| \text{ and}$$

$$|B_{r+1}/B_r + g_n|$$

are all greater than δ and for $s = c + iw$, $-\infty < w < \infty$, $|h| > \delta$ and $|h_n| > \delta$.

We fix such a value of c and observe first that $f_m^{(m)}$ given by (9) is then finite for all w. Further, we arrive at the following estimates for $|P_r - P_{r,n}|$,

$$|P_r - P_{r,n}| \le \begin{cases} \dfrac{L_{r-1}}{L_{m-1}} \cdot \dfrac{1}{\delta^{r-m}} \cdot \dfrac{1}{\delta^4} \cdot |h - h_n|, & |w| \le \Omega \\[2ex] \dfrac{L_{r-1}}{L_{m-1}} \cdot \dfrac{1}{|w|^{r-m}} \cdot \dfrac{1}{|w|^2} \cdot \dfrac{1}{\delta^2} \cdot |h - h_n|, & |w| \ge \Omega. \end{cases}$$

The quantity $|h - h_n|$ is estimated by Henrici and Pfluger [2],

$$|h - h_n| \le K \prod_{k=2}^{2n} (1 + 2\xi/\sqrt{a_k})^{-1/2},$$

where h_n is the 2n-th approximant of a Stieltjes fraction analogous to (4), a_k is the k-th numerator of this Stieltjes fraction, $a_1 = 1$, $a_2 = \lambda_r$, $a_3 = \mu_{r+1}$, $a_4 = \lambda_{r+1}$, $\dots$,

$$\xi = \operatorname{Re}\sqrt{s} = \sqrt{\frac{c + \sqrt{c^2 + w^2}}{2}} > 0 \text{ and } K = a_1\sqrt{(1+\sqrt{5})/2}\Big/|s|\cos(\arg s)/2$$

$$\le \sqrt{1 + \sqrt{5}}\,/\sqrt{c^2 + w^2}.$$

When $|w| \leq \Omega$ we replace w by 0 in the integrand and find

$$\frac{e^{ct}}{2\pi}\int_{-\Omega}^{\Omega}|P_r-P_{r,n}|dw \leq \frac{e^{ct}}{2\pi}\cdot\frac{L_{r-1}}{L_{m-1}}\cdot\frac{1}{\delta^{r-m+4}}\cdot\frac{\sqrt{1+\sqrt{5}}}{c}\cdot 2\Omega\Big/\sqrt{\prod_{k=2}^{2n}(1+2\sqrt{c}/\sqrt{a_k})}$$

while for $|w| \geq \Omega$ we first replace c by 0 in the integrand and then $1 + 2\sqrt{|w|}/\sqrt{a_k}$ by $2\sqrt{|w|}/\sqrt{a_k}$ and obtain

$$\frac{e^{ct}}{2\pi}\int_{\Omega}^{\infty}|P_r-P_{r,n}|dw \leq \frac{e^{ct}}{2\pi}\cdot\frac{L_{r-1}}{L_{m-1}}\cdot\frac{\sqrt{1+\sqrt{5}}}{\delta^2}\prod_{k=2}^{2n}\sqrt[4]{\frac{a_k}{2}}\,\frac{1}{r-m+3/4+n/2}\cdot$$

$$\frac{1}{\Omega^{r-m+3/4+n/2}}\,.$$

The same estimate applies to $\int_{-\infty}^{-\Omega}$.

If we now minimize the resulting upper estimate for $|p_r-p_{r,n}|$ with respect to Ω we find

$$(11)\quad |p_r - p_{r,n}| < \text{constant}\cdot\left(\prod_{k=2}^{2n} a_k\right)^{1/4x}\left(\prod_{k=2}^{2n}(1 + 2\sqrt{c}/\sqrt{a_k})\right)^{-1/2+1/2x},$$

where $x = r - m + 7/4 + n/2$. A similar estimate applies for $r = 0, 1, \ldots, m$. If therefore the λ_r's and μ_r's are further restricted to insure that the quantity on the right hand side of (11) tends to zero with n then $p_{r,n} \to p_r$ as $n \to \infty$. This will, for example, be the case if the λ_r's and μ_r's are bounded above, or if they don't tend to infinity too fast with r.

To calculate the inverse Laplace transform $p_{r,n} = \mathcal{L}^{-1}(P_{r,n})$ it is very helpful to know the location of the poles of $P_{r,n}$ or, according to (6), the zeros of B_{r+n}, which all lie on the negative axis. The qd algorithm of Rutishauser - see, for example [1], in particular page 614, and Chapter 12 in the forthcoming volume II - can be used to locate these zeros. Preliminary computations indicate that this algorithm is quite effective for our purposes.

References

1 Henrici, P. Applied and computational complex analysis, Vol. 1, Wiley-Interscience series, New York.

2 Henrici, P. and P. Pfluger, Truncation error estimates for Stieltjes fractions, Numer. Math. 9 (1966), 120-138.

3 Murphy, J. A. and M. R. O'Donohoe, Some properties of continued fractions with applications in Markov processes, J. Inst. Maths. Appl. 16 (1975), 57-71.

William B. Jones*
Department of Mathematics
University of Colorado
Boulder, Colorado 80302

Arne Magnus
Department of Mathematics
Colorado State University
Fort Collins, Colorado 80523

*Research supported in part by the National Science Foundation under Grant MPS 74-22111.

ON GEOMETRIC CHARACTERIZATIONS OF AN INDETERMINATE STIELTJES MOMENT SEQUENCE

E. P. Merkes and Marion Wetzel

Necessary and sufficient conditions for a Stieltjes moment sequence to be indeterminate are obtained in terms of two-dimensional cross sections of certain cones in Euclidean space R^{n+1} $(n \geq 0)$ and backward extensions of Stieltjes sequences. Earlier results for indeterminacy of both Hamburger and Stieltjes moment sequences are reformulated as limits of certain projections of points in these cones.

1 Introduction

A sequence of real numbers $\{c_j\}_{j=0}^{\infty}$ is an H (Hamburger moment) sequence if there is a bounded nondecreasing left-continuous function γ on $(-\infty,\infty)$ such that $\gamma(0) = 0$ and

$$(1) \quad c_j = \int_{-\infty}^{\infty} t^j \, d\gamma(t) \qquad (j = 0,1,2,\ldots).$$

An S (Stieltjes moment) sequence is an H sequence such that there is a mass function γ that is constant on $(-\infty,0)$ and (1) holds. If the mass function γ for an H sequence is unique, then the sequence is said to be determinate. A determinate S sequence is one that has a unique mass function γ that is constant on $(-\infty,0)$.

Let M_{n+1} be the subset of Euclidean R^{n+1} space defined by $(c_0,c_1,\ldots,c_n) \in M_{n+1}$ whenever there is a bounded nondecreasing mass function γ such that (1) holds for $j = 0,1,2,\ldots,n$. The set M_{n+1} is a convex cone with vertex at the origin. In a similar way, we define the convex cone $N_{n+1} \subset M_{n+1}$ where $(c_0,c_1,\ldots,c_n) \in N_{n+1}$ whenever (1) holds for $j = 0,1,2,\ldots,n$ and for a γ that is constant on $(-\infty,0)$. When $\{c_j\}_{j=0}^{\infty}$ is an S or H sequence, the point $(c_0, c_1, \ldots, c_n)$ is in N_{n+1} or M_{n+1} respectively for all $n \geq 0$. Many properties of these cones are included in the book [2] by Karlin and Studden. Knowledge of the position of certain points in these cones can be utilized to characterize indeterminate moment sequences.

2 Preliminaries

Let p be an integer and let $\{c_{n+p}\}_{n=0}^{\infty}$ be a sequence of real numbers. The Hankel determinants $\Delta_{n,p}$ for this sequence are defined by $\Delta_{-1,p} = 1$,

$$(2)\quad \Delta_{n,p} = \begin{vmatrix} c_p & c_{p+1} & \cdots & c_{p+n} \\ c_{p+1} & c_{p+2} & \cdots & c_{p+n+1} \\ \vdots & \vdots & & \vdots \\ c_{p+n} & c_{p+n+1} & \cdots & c_{p+2n} \end{vmatrix} \quad (n = 0,1,2,\ldots).$$

These determinants satisfy Jacobi's identity [1, p. 38]

$$(3)\quad \Delta_{n,p+1}\Delta_{n,p-1} - \Delta_{n-1,p+1}\Delta_{n+1,p-1} = \Delta_{n,p}^2 \quad (n = 0,1,2,\ldots).$$

If $\Delta_{n-1,p+2} \neq 0$ for some integer $n \geq 0$, then there is a unique real number $\underline{c}_{p,n}$ such that the determinate (2) is zero when the element c_p is replaced by $\underline{c}_{p,n}$. For convenience we set $\underline{c}_{p,-1} = 0$.

A sequence $\{c_{n+p}\}_{n=0}^{\infty}$ is positive definite if $\Delta_{n,p} > 0$ $(n = 0,1,2,\ldots)$. A necessary and sufficient condition for $\{c_n\}_{n=0}^{\infty}$ to be an H sequence corresponding to a mass function γ whose spectrum is not finite is that the sequence be positive definite [4, p.5]. When the latter holds, $\{c_{n+2p}\}_{n=0}^{\infty}$ is positive definite for each integer $p \geq 0$. To be an S sequence corresponding to a mass function γ with infinitely many points of increase on $[0,\infty)$, it is necessary and sufficient that both $\{c_n\}_{n=0}^{\infty}$ and $\{c_{n+1}\}_{n=0}^{\infty}$ be positive definite [4, p.6]. In this case $\{c_{n+p}\}_{n=0}^{\infty}$ is positive definite for $p = 0,1,2,\ldots$.

LEMMA 2.1. *If* $\{c_{n+p}\}_{n=0}^{\infty}$ *is positive definite, then* $\{\underline{c}_{p,n}\}_{n=0}^{\infty}$ *is strictly increasing and* $\underline{c}_{p,n} \to \underline{c}_p \le c_p$ *as* $n \to \infty$.

Proof. By (2), we have

$$c_p - \underline{c}_{p,n} = \Delta_{n,p}/\Delta_{n-1,p+2} > 0.$$

Hence, by (3)

$$\underline{c}_{p,n+1} - \underline{c}_{p,n} = \frac{\Delta_{n,p}}{\Delta_{n-1,p+2}} - \frac{\Delta_{n+1,p}}{\Delta_{n,p+2}}$$

$$= \frac{\Delta_{n,p+1}^2}{\Delta_{n-1,p+2}\Delta_{n,p+2}} > 0,$$

and the proof is complete.

The quantities $c_p - \underline{c}_{p,n}$ measure the distance of the point $(c_p, c_{p+1}, \ldots, c_{p+n})$ to the boundary $(\underline{c}_{p,n}, c_{p+1}, \ldots, c_{p+n})$ of M_{n+1} in the direction of the first coordinate whenever $\{c_{p+j}\}_{j=0}^{\infty}$ is positive definite.

3 *Indeterminate sequences*

For $(c_0, c_1, \ldots, c_{2n})$ in M_{2n+1}, let

$$D_n = \{(x,y) \in R^2 : (x, y, c_2, c_3, \ldots, c_{2n}) \in M_{2n+1}\}.$$

When $\{c_n\}_{n=0}^{\infty}$ is positive definite, the region D_n is a two-dimensional section of M_{2n+1} bounded by a proper parabola with horizontal axis and vertex in the right half-plane [3]. The length of the latus rectum of this parabola is $c_2 - \underline{c}_{2,n-1} = \Delta_{n-1,2}/\Delta_{n-2,4}$. A positive definite H sequence $\{c_n\}_{n=0}^{\infty}$ is indeterminate if and only if $D = \bigcap_{n=1}^{\infty} D_n$ is a region bounded by a nondegenerate parabola containing (c_0, c_1) in its interior [3]. Now D is bounded by a nondegenerate parabola if and only if $c_2 - \underline{c}_2 > 0$. Indeed, $c_2 - \underline{c}_2$ is the length of the latus rectum of the parabola. Furthermore, (c_0, c_1) is an interior point of D if and only if $c_0 - \underline{c}_0 > 0$. These observations enable us to paraphrase a previous result as follows.

THEOREM 3.1. *Let $\{c_n\}_{n=0}^{\infty}$ be a positive definite* H *sequence. A necessary and sufficient condition for this sequence to be indeterminate is that $c_0 > \underline{c}_0$ and $c_2 > \underline{c}_2$.*

An S sequence $\{c_n\}_{n=0}^{\infty}$ is indeterminate if and only if $\{d_n\}_{n=0}^{\infty}$, where $d_{2n} = c_n$ and $d_{2n+1} = 0$ $(n = 0,1,2,\ldots)$ is an indeterminate H sequence. By Theorem 3.1 we immediately obtain the following characterization.

THEOREM 3.2. *Let $\{c_n\}_{n=0}^{\infty}$ be a positive definite* S *sequence. A necessary and sufficient condition for this sequence to be indeterminate is that $c_0 > \underline{c}_0$ and $c_1 > \underline{c}_1$.*

It is known [3] that a positive definite S sequence is indeterminate if and only if the point (c_0, c_1) is an interior point of D and $c_1 > \underline{c}_1$. The last condition implies D is bounded by a nondegenerate parabola with axis horizontal and the condition $c_0 > \underline{c}_0$ insures that (c_0, c_1) is an interior point of D. The two dimensional section

$$S_n = \{(x,y) \in R^2 : (x,y,c_2,c_3,\ldots,c_{2n}) \in N_{2n+1}\}$$

is the intersection of D_n and the half plane $y \geq \underline{c}_{1,n}$.

4 S-fractions

A real sequence $\{c_j\}_{j=0}^{\infty}$ is a positive definite S sequence if and only if the formal power series $\sum_{j=0}^{\infty} c_n/z^{n+1}$ corresponds to an S-fraction

$$(4) \quad \frac{1}{k_1 z} - \frac{1}{k_2} - \frac{1}{k_3 z} - \cdots - \frac{1}{k_{2n-1} z} - \frac{1}{k_{2n}} - \cdots,$$

where $k_j > 0$ $(j = 1,2,\ldots)$. The S sequence is indeterminate if and only if $\sum_{j=1}^{\infty} k_j$ diverges [4, viii].

An S sequence $\{c_n\}_{n=0}^{\infty}$ is said to be extended backwards if there is a real number c_{-1} such that $\{c_n\}_{n=-1}^{\infty}$ is an S sequence. A positive definite S sequence when extended backwards is positive definite. Backward extensions of S sequences was first considered by Wall [5]. We review his results below and prove some additional characterizations of determinacy.

THEOREM 4.1. <u>An S sequence $\{c_n\}_{n=0}^{\infty}$ is indeterminate if and only if $c_0 > \underline{c}_0$ and there exists a backward extension of the sequence. In this case, the extended sequence $\{c_n\}_{n=-1}^{\infty}$ is indeterminate if and only if $c_{-1} > \underline{c}_{-1}$.</u>

<u>Proof</u>. If (4) is the S-fraction corresponding to $\sum_{n=0}^{\infty} c_n / z^{n+1}$, then [4, p. 78]

$$(5)\quad k_{2n+1} = \frac{\Delta_{n-1,1}^{2}}{\Delta_{n-1,0}\Delta_{n,0}}, \quad k_{2n} = \frac{\Delta_{n,0}^{2}}{\Delta_{n-1,1}\Delta_{n,1}} \quad (n = 0,1,2,\ldots).$$

We have by (3) that

$$(6)\quad \begin{aligned} k_{2n+1} &= \frac{\Delta_{n-1,2}}{\Delta_{n,0}} - \frac{\Delta_{n-2,2}}{\Delta_{n-1,0}} = \frac{1}{c_0 - \underline{c}_{0,n}} - \frac{1}{c_0 - \underline{c}_{0,n-1}}, \\ k_{2n} &= \frac{\Delta_{n,-1}}{\Delta_{n-1,1}} - \frac{\Delta_{n+1,-1}}{\Delta_{n,1}} = \underline{c}_{-1,n} - \underline{c}_{-1,n-1}, \end{aligned}$$

$(n = 0,1,2,\ldots)$. Thus, for $m \geq 1$

$$\sum_{n=0}^{m} k_{2n+1} = \frac{1}{c_0 - \underline{c}_{0,m}}, \quad \sum_{n=1}^{m} k_{2n} = \underline{c}_{-1,m}.$$

The series $\sum_{n=1}^{\infty} k_n$ converges if and only if $c_0 > \underline{c}_0$ and $\underline{c}_{-1} < \infty$. For any real number $c_{-1} \geq \underline{c}_{-1}$, the sequence $\{c_n\}_{n=-1}^{\infty}$ is a backward extension of the original S sequence [5]. When $c_{-1} \geq \underline{c}_{-1}$, moreover, we conclude by Theorem 3.2 that $\{c_n\}_{n=-1}^{\infty}$ is itself an indeterminate S sequence. If $c_{-1} = \underline{c}_{-1}$, the backward extended sequence is determinate.

Since $\{c_n\}_{n=-1}^{\infty}$ is indeterminate when $c_{-1} > \underline{c}_{-1}$ and the original S sequence is indeterminate, it can also be extended backwards. By repeated application of this process, there exist indeterminate S sequences $\{c_n\}_{n=-m}^{\infty}$ for each positive integer m. Such extensions are not unique.

Let $\{c_n\}_{n=0}^{\infty}$ be a positive definite S sequence. From (5), (6), and the Jacobi identity, we have for $n \geq 1$ that

$$\underline{c}_{-1,n} - \underline{c}_{-1,n-1} = \frac{\Delta_{n,0}^2}{\Delta_{n-1,1}\Delta_{n,1}} = (c_0 - \underline{c}_{0,n})^2 \frac{\Delta_{n-1,2}^2}{\Delta_{n-1,1}\Delta_{n,1}}$$

$$= (c_0 - \underline{c}_{0,n})^2 \frac{\Delta_{n-1,1}\Delta_{n-1,3} - \Delta_{n-2,3}\Delta_{n-1}}{\Delta_{n-1,1}\Delta_{n,1}}$$

$$= (c_0 - \underline{c}_{0,n})^2 \left[\frac{1}{c_1 - \underline{c}_{1,n}} - \frac{1}{c_1 - \underline{c}_{1,n-1}}\right] .$$

If $0 \leq \ell \leq c_0 - \underline{c}_{0,n} \leq L$ for $n \geq 1$, we conclude

$$\ell^2 \frac{\underline{c}_{1,n}}{c_1(c_1 - \underline{c}_{1,n})} \leq \underline{c}_{-1,n} \leq L^2 \frac{\underline{c}_{1,n}}{c_1(c_1 - \underline{c}_{1,n})} \quad (n = 1,2,\ldots).$$

Thus, $\underline{c}_{-1,n}$ is bounded when $\underline{c}_{1,n} \to \underline{c}_1 < c_1$ and $\underline{c}_{-1,n} \to \infty$ as $n \to \infty$ when $\underline{c}_1 = c_1$, $c_0 > \underline{c}_0$.

Suppose now the S sequence $\{c_n\}_{n=0}^{\infty}$ is determinate. By Theorem 3.2, either $c_0 = \underline{c}_0$ or $c_1 = \underline{c}_1$. If $c_0 > \underline{c}_0$, then $\underline{c}_{-1,n} \to \infty$ as $n \to \infty$ and no backward extension is possible. If $c_1 > \underline{c}_1$, then $\underline{c}_{-1} < \infty$ and backward extensions $\{c_n\}_{n=-1}^{\infty}$, where $c_{-1} \geq \underline{c}_{-1}$, exist. Furthermore, the limit parabola in the

(c_{-1},c_0) plane is nondegenerate, the latus rectum being $c_1 - \underline{c}_1 > 0$. If $c_0 = \underline{c}_0$ and $c_1 = \underline{c}_1$, backward extensions may or may not occur [5]. When such exist, however, the limit parabola in the (c_{-1},c_0) plane is degenerate. In particular, these observations prove the following characterization.

THEOREM 4.2. <u>An</u> S <u>sequence</u> $\{c_n\}_{n=0}^{\infty}$ <u>is indeterminate if and only if there exist backward extensions</u> $\{c_n\}_{n=-1}^{\infty}$, $\{c_n\}_{n=-2}^{\infty}$ <u>such that the limit parabola in the</u> (c_{-2},c_{-1}) <u>plane is nondegenerate</u>.

When the S sequence is determinate, we showed above that two backward extensions are possible only if $c_{-1} = \underline{c}_{-1}$. If such extensions exist, then $c_0 = \underline{c}_0$ so the limit parabola in the (c_{-2},c_{-1}) plane is degenerate. When the S sequence is indeterminate, the limit parabola in the (c_{-2},c_{-1}) plane is nondegenerate. Since $c_0 > \underline{c}_0$, this is the case even when the S sequence has been extended backwards to a determinate sequence ($c_{-1} = \underline{c}_{-1}$ or $c_{-2} = \underline{c}_{-2}$).

References

1 Gragg, W. B., The Padé table and its relations to certain algorithms of numerical analysis,, SIAM Review, 14 (1972), 1-62.

2 Karlin, S. and W. J. Studden, Tchebycheff Systems: with Applications in Analysis and Statistics, Interscience, New York, 1966.

3 Merkes, E. P. and Marion Wetzel, A geometric characterization of indeterminate moment sequences, Pac. J. Math., to appear.

4 Shohat, J. A. and J. D. Tamarkin, The Problem of Moments, Math. Surveys #1, Amer. Math. Soc., R. I., 1943.

5 Wall, H. S., On the Padé approximants associated with the continued fraction and series of Stieltjes, Trans. Amer. Math. Soc., 31 (1929), 91-116.

Marion Wetzel
Department of Mathematics
Denison University
Granville, Ohio 43023

E. P. Merkes
Department of Mathematical Sciences
University of Cincinnati
Cincinnati, Ohio 45221

RATIONAL APPROXIMATION TO THE EXPONENTIAL FUNCTION

Q.I. Rahman and G. Schmeisser

In this paper we mention certain recent results concerning rational approximation to e^{-x}. In particular, the Chebyshev approximation to e^{-x} on $[0,\infty)$ by rational functions of degree n is compared with the Chebyshev approximation to e^{-x} on $[0,\infty)$ by reciprocals of polynomials of degree n.

1 Introduction

In 1969 Cody, Meinardus, and Varga [1] raised and studied the problem of approximating e^{-x} by rational functions on the positive real axis. More precisely, if Π_n denotes the class of all polynomials of degree at most n and

$$\lambda_{m,n} := \inf_{\substack{p(x)\in\Pi_m \\ q(x)\in\Pi_n}} \sup_{x\geq 0} \left| e^{-x} - \frac{p(x)}{q(x)} \right| \quad (m\leq n)$$

they wanted to know the asymptotic behaviour of $\lambda_{m,n}$ for large n, in particular they asked for

$$\sigma_1 := \overline{\lim_{n\to\infty}} (\lambda_{0,n})^{\frac{1}{n}} \text{ and } \sigma_2 := \overline{\lim_{n\to\infty}} (\lambda_{n,n})^{\frac{1}{n}}.$$

As to the first quantity they obtained the bounds

$$1/6 \leq \sigma_1 \leq 0.43501\ldots .$$

A few years later, Schönhage [6] succeeded in determining the precise value of σ_1. He showed that $\lim_{n\to\infty} (\lambda_{0,n})^{1/n}$ exists and that

$$\sigma_1 = \lim_{n\to\infty} (\lambda_{0,n})^{1/n} = \frac{1}{3} .$$

Schönhage's proof is based on the interesting observation that the Chebyshev approximation to e^{-x} by reciprocals of polynomials of degree at most n has asymptotically the same speed of convergence as a certain weighted least square approximation to the exponential function by "ordinary" polynomials of degree at most n. More precisely, he showed that

$$\lim_{n\to\infty} (\lambda_{0,n})^{1/n} = \lim_{n\to\infty} \left\{ \inf_{q(x)\in\Pi_n} \int_0^\infty e^{-4x}(e^x-q(x))^2\, dx \right\}^{\frac{1}{2n}} .$$

Thus the originally nonlinear problem could be attacked via a linear one which is easily solvable.

The problem of determining σ_2 seems to be considerably more difficult, and at the present time we only know a lower and an upper bound for it.

2 Lower bound for σ_2

A first step towards the determination of σ_2 is due to D.J. Newman [2] who showed that $\underline{\lim}_{n\to\infty} (\lambda_{n,n})^{1/n} \geq \frac{1}{1280}$. Hence $\sigma_2 \geq (1280)^{-1}$ and approximating by rational functions instead of reciprocals of polynomials does not give a faster than geometric speed of convergence.

Although this fact needed a very clever proof, it is somehow not surprising: the function e^{-x} tends to zero very rapidly as $x \to \infty$, whereas a polynomial tends to infinity. Therefore, it is hard to believe that a polynomial in the numerator can give a remarkable improvement.

In order to see how much better an approximation we could possibly get, let τ be a large positive number and suppose that there exist polynomials $p(x)$, $q(x) \in \Pi_n$ such that

$$\left| e^{-x} - \frac{p(x)}{q(x)} \right| < e^{-\tau n}$$

for all $x \in [0,\infty)$. Putting $x=nt$, $P(t):=p(nt)$, and $Q(t):=q(nt)$ we obtain

(1) $$\left|e^{-nt} - \frac{P(t)}{Q(t)}\right| < e^{-\tau n} \quad \text{for all } t \geq 0.$$

Let the polynomial $Q(t)$ be normalized such that

(2) $$\max_{x\varepsilon[0,1]} |Q(t)| = 1.$$

If ξ is a point in $I_1:=[0,1]$ where $\max_{x\varepsilon[0,1]} |Q(x)|$ is attained, then by the triangle inequality

(3) $$|P(\xi)| \geq e^{-n}(1-e^{-(\tau-1)n}).$$

If I_2 denotes the interval $[\tau,\tau+1]$, then using a well known inequality of Chebyshev we deduce from (2) that

(4) $$\max_{t\varepsilon I_2} |Q(t)| \leq (c\tau)^n$$

where c is a constant. Next, from (1) we get

(5) $$|P(t)| \leq 2e^{-n\tau} \max_{t\varepsilon I_2} |Q(t)| \leq 2(c\tau e^{-\tau})^n$$

for all $t\varepsilon I_2$. Now, going back to the point ξ by applying the Chebyshev's inequality again (this time to $P(t)$) we obtain

(6) $$|P(\xi)| \leq 2(c^2\tau^2 e^{-\tau})^n.$$

This contradicts (3) for large τ, if we simply use the property that the exponential function grows faster than a polynomial of degree 2.

Of course, every τ leading to a contradiction furnishes the lower bound $e^{-\tau}$ for σ_2. We worked out this idea carefully, constructing curved majorants for $Q(t)$, $P(t)$ on I_2 instead of (4) and (5), and using a refined version of Chebyshev's inequality for polynomials with curved majorants we obtain

$$\sigma_2 \geq (308)^{-1}.$$

This improves Newman's result quite a bit. Moreover, our method is flexible enough to extend to certain classes of entire functions. (For details see [3]).

3 Upper bound for σ_2

Trivially, Schönhage's result yields $\sigma_2 \leq \frac{1}{3}$. However, the question -raised in particular by P. Erdös and D.J. Newman- "whether $\sigma_2 < \frac{1}{3}$" remained open for some time. A natural way to attack this question is to see if the elements of the Padé table for e^{-x} already lead us to an affirmative answer. It was shown by Saff, Varga, and Ni [5] and independently in [4] that for $r\varepsilon(0,1)$, $m=[rn]$ if $P_{m,n}(x)/Q_{m,n}(x)$ is the element of the Padé table for e^{-x} with degree $P_{m,n}=m$, degree $Q_{m,n}=n$, then

$$\lim_{n\to\infty}\left\{\sup_{x\geq 0}\left|e^{-x}-\frac{P_{m,n}(x)}{Q_{m,n}(x)}\right|\right\}^{1/n} = c(r) \geq \frac{1}{3},$$

where $c(r) = r^r(\frac{1-r}{2})^{1-r}$. Here equality is attained if and only if $r = 1/3$. Thus, we do not get more than what we know already from Schönhage's result.

Indeed the elements of the Padé table cannot furnish a good approximation because

$$E(x) := e^{-x} - \frac{P_{m,n}(x)}{Q_{m,n}(x)}$$

attains its maximum modulus at only one point in $[0,\infty)$, whereas for the best approximation the error-function must have $m+n+2$ extremal points. However, $E(0)=0$ and we can, therefore, introduce a second extremal point by considering $E(x)$ on an appropriate interval $[-\alpha(r),+\infty)$, where $\alpha(r)>0$. In fact, it is shown in [4] that for every $r\varepsilon(0,1)$ there exists an $\alpha(r)>0$ such that for given $\varepsilon>0$ and all sufficiently large n

$$\sup_{x\geq-\alpha(r)n}\left|e^{-x}-\frac{P_{m,n}(x)}{Q_{m,n}(x)}\right| \leq (c(r)+\varepsilon)^n.$$

Hence, writing simply α instead of $\alpha(r)$, we have

$$\sup_{x\geq 0} \left| e^{-x} - e^{-\alpha n} \frac{P_{m,n}(x-\alpha n)}{Q_{m,n}(x-\alpha n)} \right| \leq \{e^{-\alpha}(c(r) + \varepsilon)\}^n .$$

This shows that $\sigma_2 < 1/3$. Looking for the best constant furnished by this idea we obtained [4]

$$\sigma_2 \leq (4.098...)^{-1} .$$

4 Concluding remarks

In [1] the values of $\lambda_{n,n}$ were calculated numerically up to $n=14$. Extrapolating $(\lambda_{n,n})^{1/n}$ from there it appears that σ_2 may possibly be equal to $1/9$. This would be an interesting conjecture which is a bit surprising for the following reason: if we approximate by reciprocals of polynomials of degree $2n$, then again, by Schönhage's result, the maximal error behaves asymptotically like 9^{-n}. Now, a polynomial of degree $2n$ and a rational function of degree n are both determined by $2n+1$ coefficients. Thus, if the conjecture were true, it would mean that only the total number of free parameters plays an important role in the approximation problem. In contrast to the reflection mentioned before, a non-constant polynomial in the numerator could be as useful as additional higher order terms in the denominator. Let us hope that these problems will be resolved in the near future.

References

1. Cody, W.J., G. Meinardus, and R.S. Varga, Chebyshev rational approximation to e^{-x} in $[0,+\infty)$ and applications to heat-conduction problems, J. Approximation Theory, 2 (1969), 50-65.

2. Newman, D.J., Rational approximation to e^{-x}, J. Approximation Theory, 10 (1974), 301-303.

3. Rahman, Q.I. and G. Schmeisser, Rational Approximation to e^{-x}, J. Approximation Theory (to appear).

4. Rahman, Q.I. and G. Schmeisser, Rational Approximation to e^{-x} II, Trans. Amer. Math. Soc. (to appear).

5. Saff, E.B., R.S. Varga, and W.C. Ni, Geometric convergence of rational approximation to e^{-z} in infinite sectors, Numer. Math., 26 (1976), 211-225.

6. Schönhage, A., Zur rationalen Approximierbarkeit von e^{-x} über $[0,+\infty)$, J. Approximation Theory, 7 (1973), 395-398.

Q.I. Rahman
Department of Mathematics
University of Montreal
Montreal, Canada

G. Schmeisser
Mathematisches Institut
Universität Erlangen-Nürnberg
D-852 Erlangen
West Germany

ON THE ZEROS AND POLES OF PADÉ APPROXIMANTS TO e^z. II.

E.B. Saff and R.S. Varga

In this paper, we continue our study of the location of the zeros and poles of general Padé approximants to e^z. We state and prove here two new results on improved estimates for the zeros of general Padé approximants $R_{n,\nu}(z)$ to e^z, and state results on the asymptotic location of the normalized zeros and poles for certain sequences of Padé approximants to e^z.

1 Introduction

A number of recent papers (cf. [1, 4, 6, 9, 15]) have been concerned with Padé rational approximations of e^z because of applications to the numerical analysis of methods for solving certain systems of ordinary differential equations. The purpose of this present paper is to continue our study [9] on the zeros and poles of general Padé approximants to e^z. In particular, for <u>every</u> Padé approximant we determine a "close-to-sharp" annulus, having center at $z = 0$, containing all the zeros and poles of this approximant. These results will be described in §2, with their proofs being given in §3.

In this paper, we also state more precise information about the asymptotic distribution of the zeros and poles for specific sequences of Padé approximants to e^z. What has motivated this work is an article by Szegö [13], which considers the zeros of the partial sums $s_n(z) := \sum_{k=0}^{n} z^k/k!$ of the Maclaurin expansion of e^z. Szegö [13] showed that $s_n(nz)$ has all its zeros in $|z| \leq 1$ for every $n \geq 1$, and that $\hat{z}$ is a limit point of zeros of $\{s_n(nz)\}_{n=1}^{\infty}$ iff

$$(1.1) \quad |\hat{z}\, e^{1-\hat{z}}| = 1 \text{ and } |\hat{z}| \leq 1.$$

(This result was also obtained later independently by Dieudonné [3].)

The connection of Szegö's result with Padé approximations of e^z is evident in that $s_n(z)$ is the (n, 0)-th Padé approximant to e^z. Our new results, giving sharp generalizations of Szegö's result to the asymptotic distribution of zeros of more general sequences of Padé approximants to e^z, will be stated explicitly in §2, but their proofs, being lengthy, will appear elsewhere. For the remainder of this section, we introduce necessary notation and cite needed known results.

Let π_m denote the set of all polynomials in the variable z having degree at most m, and let $\pi_{n,\nu}$ be the set of all complex rational functions r(z) of the form

$$r(z) = \frac{p(z)}{q(z)}, \text{ where } p \in \pi_n, \ q \in \pi_\nu, \text{ and } q(0) = 1.$$

Then, the (n,ν)-th Padé approximation to e^z is defined as that element $R_{n,\nu}(z) \in \pi_{n,\nu}$ for which

$$e^z - R_{n,\nu}(z) = \mathcal{O}(|z|^{n+\nu+1}), \text{ as } |z| \to 0.$$

In explicit form, it is known [8, p. 245] that

$$R_{n,\nu}(z) = P_{n,\nu}(z)/Q_{n,\nu}(z);$$

where

$$(1.2) \quad P_{n,\nu}(z) := \sum_{j=0}^{n} \frac{(n+\nu-j)!\ n!\ z^j}{(n+\nu)!\ j!\ (n-j)!},$$

and

$$(1.3) \quad Q_{n,\nu}(z) := \sum_{j=0}^{\nu} \frac{(n+\nu-j)!\ \nu!\ (-z)^j}{(n+\nu)!\ j!\ (\nu-j)!}.$$

We shall refer to the polynomials $P_{n,\nu}(z)$ and $Q_{n,\nu}(z)$ respectively as the <u>Padé numerator</u> and <u>Padé denominator</u> of type (n, ν) for e^z.

Generally, one is interested in both the zeros and the poles of the Padé approximants $R_{n,\nu}(z)$. However, since the polynomials of (1.2) and (1.3) satisfy the obvious relation

$$(1.4) \quad Q_{n,\nu}(z) = P_{\nu,n}(-z),$$

it suffices then to investigate only the zeros of the Padé approximants $R_{n,\nu}(z)$, or equivalently, the zeros of the Padé numerator $P_{n,\nu}(z)$.

The approximants $R_{n,\nu}(z)$ are typically displayed in the following infinite array, known as the Padé table for e^z:

$$(1.5)\quad \begin{bmatrix} R_{0,0}(z) & R_{1,0}(z) & R_{2,0}(z) & \cdots \\ R_{0,1}(z) & R_{1,1}(z) & R_{2,1}(z) & \cdots \\ R_{0,2}(z) & R_{1,2}(z) & R_{2,2}(z) & \cdots \\ \vdots & \vdots & \vdots & \end{bmatrix} .$$

Note that the first row $\{R_{n,0}(z)\}_{n=0}^{\infty}$ of the Padé table for e^z is, from (1.2), simply the sequence of partial sums $\{s_n(z) = \sum_{k=0}^{n} z^k/k!\}_{n=0}^{\infty}$ of e^z.

Essential for the statements and proofs of our main results are the following recent results on zeros of Padé approximants for e^z.

THEOREM 1.1. (Saff and Varga [9], [11], [12]). For every $\nu \geq 0$, $n \geq 2$, the Padé approximant $R_{n,\nu}(z)$ for e^z has all its zeros in the infinite sector

$$(1.6)\quad \$_{n,\nu} := \{z: |\arg z| > \cos^{-1}(\frac{n-\nu-2}{n+\nu})\}.$$

Furthermore, on defining generically the infinite sector $\$_\lambda$, $\lambda \geq 0$, by

$$(1.7)\quad \$_\lambda := \{z: |\arg z| > \cos^{-1}(\frac{1-\lambda}{1+\lambda})\},$$

consider any sequence of Padé approximants $\{R_{n_j,\nu_j}(z)\}_{j=1}^{\infty}$ satisfying

$$(1.8)\quad \lim_{j\to\infty} n_j = +\infty, \text{ and } \lim_{j\to\infty} \frac{\nu_j}{n_j} = \sigma,$$

for any σ with $0 < \sigma < \infty$. Then, for any ϵ with $0 < \epsilon < \sigma$, $\{R_{n_j,\nu_j}(z)\}_{j=1}^{\infty}$ has infinitely many zeros in $S_{\sigma-\epsilon}$ and only finitely many zeros in the complement of $S_{\sigma-\epsilon}$, and S_{σ} is the smallest sector of the form $|\arg z| > \mu$, $\mu > 0$, with this property.

THEOREM 1.2. (Saff and Varga [9]). If $1 < n < 3\nu + 4$, then all the zeros of the Padé approximant $R_{n,\nu}(z)$ for e^z lie in the half-plane

(1.9) $\operatorname{Re} z < n - \nu - 2.$

2 Statements of New Results

We list and discuss our main results in this section. Our first results give estimates for the zeros of general Padé approximants $R_{n,\nu}(z)$, which extend the results of Theorem 1.2 of §1.

THEOREM 2.1. For any $n \geq 1$ and any $\nu \geq 0$, all the zeros of the Padé approximant $R_{n,\nu}(z)$ satisfy

(2.1) $\operatorname{Re} z < n - \nu.$

THEOREM 2.2. For any $n \geq 1$ and any $\nu \geq 0$, all the zeros of the Padé approximant $R_{n,\nu}(z)$ lie in the annulus

(2.2) $(n+\nu)\mu < |z| < n+\nu+4/3 \quad (\mu \doteq 0.278\,465),$

where μ is the unique positive root of $\mu e^{1+\mu} = 1$. Moreover, the constant μ in (2.2) is best possible in the sense that

$$\mu = \inf_{\substack{n\geq 1\\ \nu\geq 0}} \left\{\frac{|z|}{(n+\nu)} : R_{n,\nu}(z) = 0\right\}.$$

We remark that while the first inequality of (2.2) of Theorem 2.2 is best possible in the above sense, the upper bound of (2.2) may not be best possible. In any event, because $R_{1,\nu}(z)$ has its sole zero at $z = -(\nu+1)$, we have

$$\sup_{n\geq 1,\ \nu\geq 0} \{|z| - (n+\nu) : R_{n,\nu}(z) = 0\} \geq 0$$

and thus, the constant 4/3 in (2.2) can however be decreased at most to zero. In fact, the Kakeya-Eneström Theorem (cf. [5, p. 106, Ex. 2]) directly gives that all the zeros of $s_n(z)$ lie in $|z| \leq n$, which sharpens the last inequality of (2.2) of Theorem 2.2 for the case $\nu = 0$. However, applying the Kakeya-Eneström Theorem to the general Padé numerator $P_{n,\nu}(z)$ gives only that $P_{n,\nu}(z)$ has all its zeros in $|z| \leq n(\nu+1)$ which, except in essentially trivial cases, gives a worse upper bound than that of the last inequality of (2.2) of Theorem 2.2.

Note that because of the relation (1.4), the inequalities of (2.2) of Theorem 2.2 hold for the zeros as well as for the poles of $R_{n,\nu}(z)$. Thus, given any compact subset Ω of the complex plane $\mathbb{C}$, there is a constant $\gamma > 0$, depending only on the geometry of Ω, such that all zeros and poles of any Padé approximant $R_{n,\nu}(z)$ lie outside of Ω if $(n+\nu) \geq \gamma$.

To describe the remaining results, for any σ with $0<\sigma<+\infty$, define the points

(2.3) $\quad z_\sigma^{\pm} := \{(1-\sigma) \pm 2\sqrt{\sigma}\, i\}/(1+\sigma),$

which have modulus unity, and consider the complex plane $\mathbb{C}$ slit along the two rays

$$\mathcal{R}_\sigma := \{z : z = z_\sigma^{+} + i\tau \text{ or } z = z_\sigma^{-} - i\tau,\ \tau \geq 0\},$$

as shown in Figure 1. Now, the function

(2.4) $\quad g_\sigma(z) := \sqrt{1 + z^2 - 2z\left(\frac{1-\sigma}{1+\sigma}\right)}$

has z_σ^{+} and z_σ^{-} as branch points, which

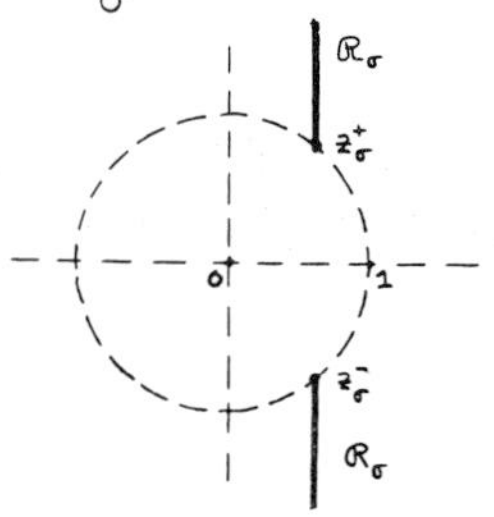

Figure 1

are the finite extremities of $\mathcal{R}_\sigma$. On taking the principal branch for the square root, i.e., on setting $g_\sigma(0) = 1$ and extending g_σ analytically on this doubly slit domain $\mathbb{C}\backslash\mathcal{R}_\sigma$, then g_σ is analytic and single-valued on $\mathbb{C}\backslash\mathcal{R}_\sigma$. Next, it can be verified that $1 \pm z + g_\sigma(z)$ does not vanish on $\mathbb{C}\backslash\mathcal{R}_\sigma$. Thus, we define, respectively, $(1 + z + g_\sigma(z))^{\frac{2}{1+\sigma}}$ and $(1 - z + g_\sigma(z))^{\frac{2\sigma}{1+\sigma}}$ by requiring that their values at $z = 0$ be $2^{\frac{2}{1+\sigma}}$ and $2^{\frac{2\sigma}{1+\sigma}}$, and by analytic continuation. These functions are also analytic and single-valued in $\mathbb{C}\backslash\mathcal{R}_\sigma$. With these conventions, we set

$$(2.5)\quad w_\sigma(z) := \frac{4\sigma^{\left(\frac{\sigma}{1+\sigma}\right)}\, z\, e^{g_\sigma(z)}}{(1+\sigma)(1+z+g_\sigma(z))^{\frac{2}{1+\sigma}}(1-z+g_\sigma(z))^{\frac{2\sigma}{1+\sigma}}},$$

$$0 < \sigma < +\infty,$$

and it follows that w_σ is analytic and single-valued on $\mathbb{C}\backslash\mathcal{R}_\sigma$. Next, on letting $\sigma \to 0$ in (2.5), we obtain that $w_0(z) := \lim_{\sigma\to 0} w_\sigma(z)$ satisfies

$$(2.5')\quad w_0(z) = z\, e^{1-z},$$

for Re $z \le 1$.

With these definitions, we state the following results, whose proofs will appear elsewhere.

THEOREM 2.3. <u>For any σ with $0 \le \sigma < \infty$, consider any sequence of Padé approximants $\{R_{n_j,\nu_j}(z)\}_{j=1}^{\infty}$ for e^z for which</u>

$$(2.6)\quad \lim_{j\to\infty} n_j = +\infty, \text{ and } \lim_{j\to\infty} \frac{\nu_j}{n_j} = \sigma.$$

<u>Then, $\hat{z}$ is a limit point of zeros of the normalized Padé approximants $\{R_{n_j,\nu_j}((n_j+\nu_j)z)\}_{j=1}^{\infty}$ iff $\hat{z}$ belongs to the curve</u>

$$(2.7)\quad D_\sigma := \{z \in \mathcal{S}_\sigma : |w_\sigma(z)| = 1 \text{ and } |z| \le 1\},$$

where $\overline{\$}_\sigma$ denotes the closure of $\$_\sigma$ (cf. (1.7)).

We first remark that the special case of Theorem 2.3 with $\sigma = 0$ and, in addition, with $\nu_j = 0$ and $n_j = j$ for all $j \geq 1$, reduces to Szegö's result (cf. (1.1) and (2.7)) since $\overline{\$}_0 = \mathbb{C}$, and since the normalized approximants $\{R_{j,0}(jz)\}_{j=1}^{\infty}$ are just $\{s_n(nz)\}_{n=1}^{\infty}$.

We illustrate this special case of $\sigma = 0$ of Theorem 2.3 by graphing respectively, in Figures 2 and 3, the twelve zeros of $s_{12}(12z)$ and the twenty-four zeros of $s_{24}(24z)$, along with D_0, indicated by the solid curve.

Similarly, for the choice $\nu_j = j = n_j$ for all $j \geq 1$, (2.6) is satisfied for $\sigma = 1$, and in this case, Figures 4 and 5 indicate the curve D_1, along with, respectively, the twelve zeros of the polynomial $P_{12,12}(24z)$, and the twenty-four zeros of the polynomial $P_{24,24}(48z)$. Next, we remark that the curve D_1, after rotations of $\pi/2$, form the boundaries of the eye-shaped domain consider by Olver [7, p. 336] in his asymptotic expansions of Bessel functions. That such a connection exists is not surprising, since the diagonal Padé numerators $P_{n,n}(z)$ satisfy $P_{n,n}(2iz) = n!\,(2z)^n e^{iz}(\pi z/2)^{1/2}\{(-1)^n J_{-(n+1/2)}(z) - iJ_{n+1/2}(z)\}/(2n)!$.

To further illustrate the result of Theorem 2.3, we note that the choice $\nu_j = j$, $n_j = 3j$ for all $j \geq 1$ satisfies (2.6) with $\sigma = 1/3$, and Figures 6 and 7 show $D_{1/3}$, along with the twelve zeros of $P_{12,4}(16z)$ and the twenty-four zeros of the polynomial $P_{24,8}(32z)$. Thus, we see from Figures 2-7 that Theorem 2.3 quite accurately predicts the zeros of $P_{n,\nu}((n+\nu)z)$, even for relatively small values of n and ν.

With the relationship of (1.4), Theorem 2.3 can directly be used to deduce the limit points of the poles of $\{R_{n_j,\nu_j}((n_j+\nu_j)z)\}_{j=1}^{\infty}$. Specifically, we state

COROLLARY 2.4. Let $\{R_{n_j,\nu_j}(z)\}_{j=1}^{\infty}$ be a sequence of Padé approximants to e^z satisfying (2.6) with $0 < \sigma < \infty$. Then $\hat{z}$ is a limit point of poles of $\{R_{n_j,\nu_j}((n_j+\nu_j)z)\}_{j=1}^{\infty}$ iff $\hat{z}$ belongs to the curve $-D_{1/\sigma}$ where (cf. (2.7))

$$(2.8) \quad -D_{1/\sigma} := \{z : -z \in D_{1/\sigma}\}.$$

Interestingly enough, the closed curve

$$(2.9) \quad J_\sigma := \{z: |w_\sigma(z)| = 1 \text{ and } |z| \le 1\}$$

can be represented as the union

$$(2.10) \quad J_\sigma = D_\sigma \cup \{-D_{1/\sigma}\}, \quad 0 < \sigma < \infty.$$

Thus, the limit points of zeros and poles of $\{R_{n_j,\nu_j}((n_j+\nu_j)z)\}_{j=1}^{\infty}$ play a complementary role to one another. To illustrate this, consider the sequence of Padé approximants $\{R_{3m,m}(4m\,z)\}_{m=1}^{\infty}$, for which $\sigma = 1/3$. In Figure 8, we have graphed $D_{1/3}$ and $(-D_3)$, along with the 24 zeros and 8 poles of $R_{24,8}(32z)$, denoted respectively by *'s and ◘'s.

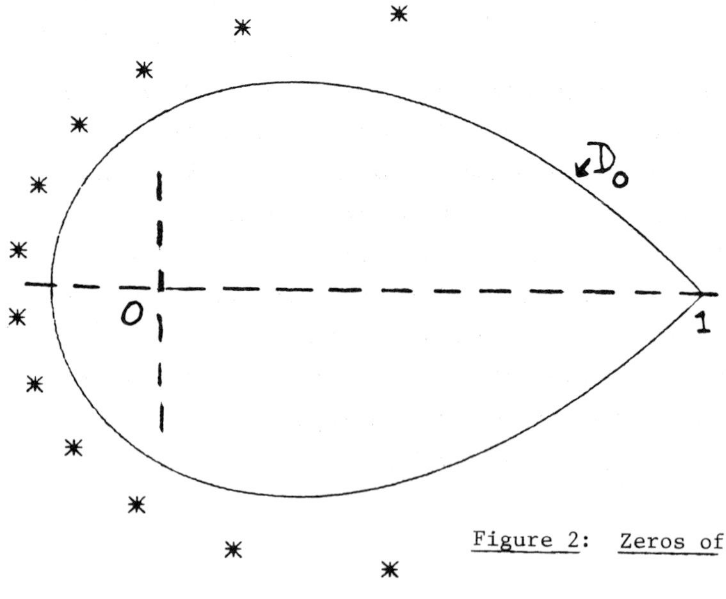

Figure 2: Zeros of $s_{12}(12z)$.

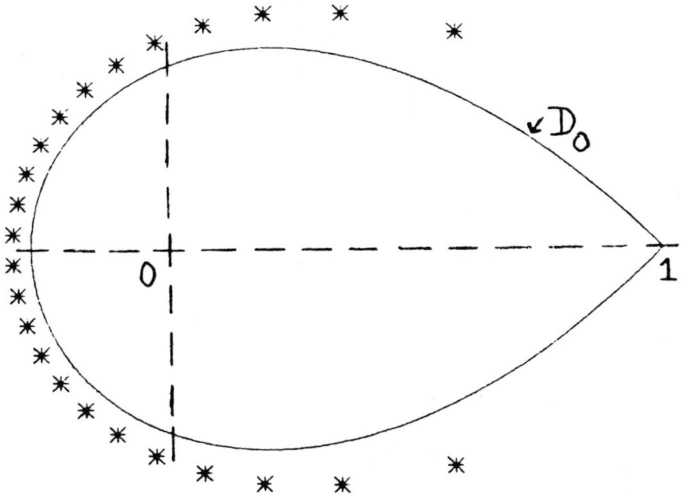

Figure 3: Zeros of $s_{24}(24z)$.

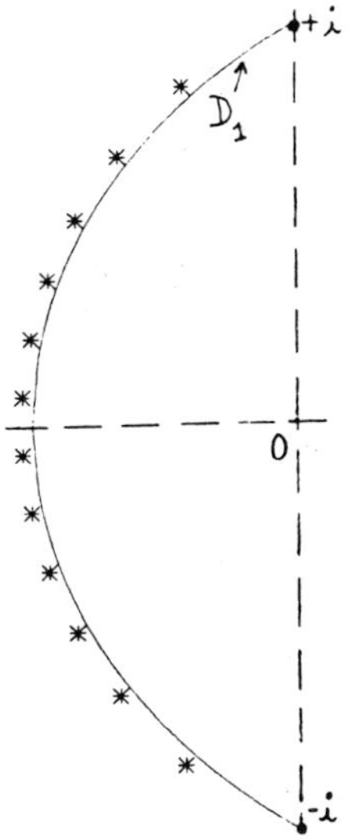

Figure 4: Zeros of $P_{12,12}(24z)$.

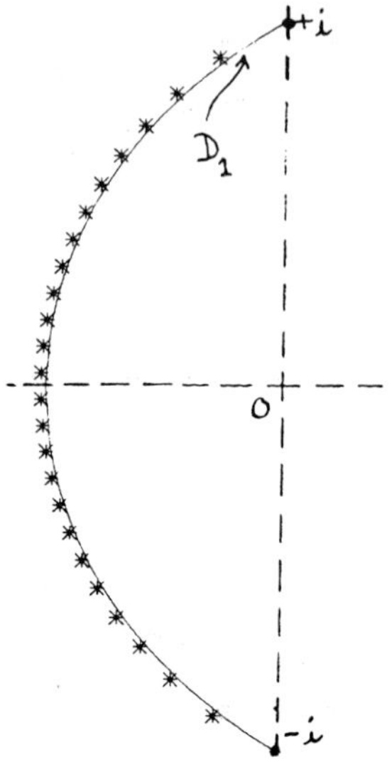

Figure 5: Zeros of $P_{24,24}(48z)$.

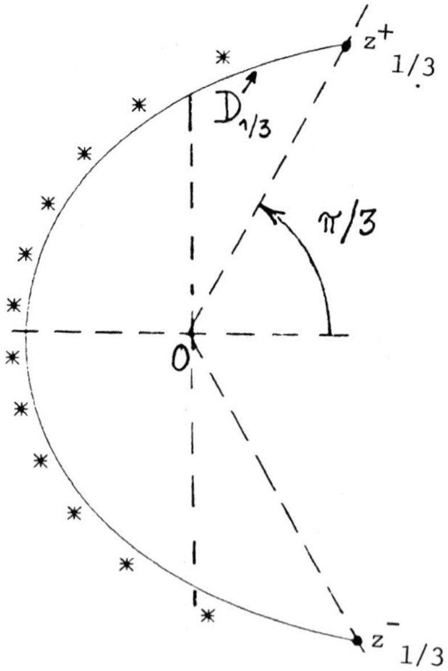

Figure 6: Zeros of $P_{12,4}(16z)$.

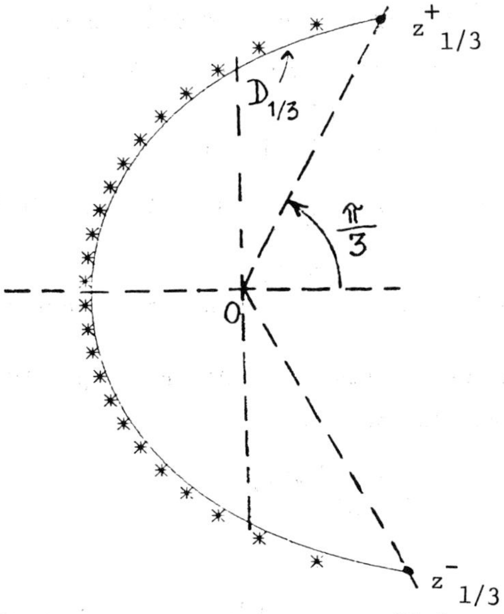

Figure 7: Zeros of $P_{24,8}(32z)$.

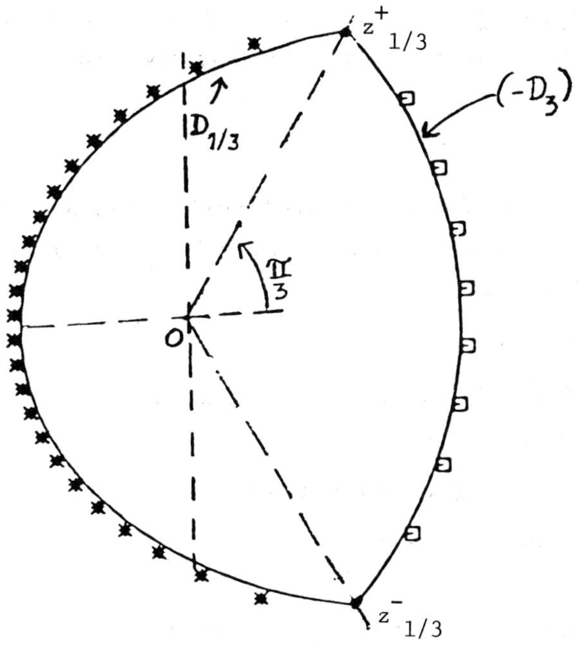

Figure 8: Zeros and Poles of $R_{24,8}(32z)$.

3 Proofs of Theorems 2.1 and 2.2

As in Saff-Varga [11], for any $n \geq 0$ and any $\nu \geq 0$, set

$$(3.1)\quad w_{n,\nu}(z) := e^{-z/2} z^{-(\frac{n+\nu}{2})} P_{n,\nu}(z),$$

where the Padé numerator $P_{n,\nu}(z)$ is defined in (1.2). In the case that $n+\nu$ is odd, $z^{-(\frac{n+\nu}{2})}$ denotes the principal branch of $z^{-(\frac{n+\nu}{2})}$. Then, as is known (cf. Olver [7, p. 260]), $w_{n,\nu}(z)$ satisfies Whittaker's equation

$$(3.2)\quad \frac{d^2w(z)}{dz^2} = [\frac{1}{4} - \frac{k}{z} + \frac{\lambda}{z^2}]\, w(z),$$

with

$$(3.3)\quad k := \frac{n-\nu}{2}, \text{ and } \lambda := (\frac{n+\nu+1}{2})^2 - \frac{1}{4} = \frac{(n+\nu)(n+\nu+2)}{4}.$$

Proof of Theorem 2.1. If $\nu \geq n > 1$, then $3\nu + 4 > n > 1$, so that from (1.9) of Theorem 1.2, we have that any zero z of $P_{n,\nu}(z)$ satisfies

$$\operatorname{Re} z < n - \nu - 2,$$

which is stronger than the desired result (2.1) of Theorem 2.1. Similarly, if $n = 1$, the single zero of $P_{1,\nu}(z)$ is $z = -(\nu+1)$, which implies that

$$\operatorname{Re} z = n - \nu - 2,$$

and again (2.1) of Theorem 2.1 is satisfied.

For the remaining case $0 \leq \nu < n$, define

$$(3.4)\quad y_\tau(x) := w_{n,\nu}(\tau x),\ \tau \neq 0,\ 0 \leq x < \infty,$$

which, using (3.2), satisfies

$$(3.5)\quad \frac{d^2y_\tau(x)}{dx^2} = \tau^2\{\frac{1}{4} - \frac{k}{\tau x} + \frac{\lambda}{\tau^2x^2}\}\, y_\tau(x) =: p_\tau(x)\cdot y_\tau(x).$$

Next, choose τ to be a zero of $P_{n,\nu}$, i.e., $w_{n,\nu}(\tau) = 0 = y_\tau(1)$. If Re $\tau \leq 0$, then by hypothesis Re $\tau \leq 0 < n - \nu$, and (2.1) of Theorem 2.1 is trivially satisfied. Hence, assume that Re $\tau > 0$, in which case $\bar{\tau}$ is also a distinct zero of $P_{n,\nu}$. Defining similarly

$$(3.6)\quad y_{\bar{\tau}}(x) := w_{n,\nu}(\bar{\tau}x),\quad 0 \leq x < \infty,$$

which satisfies

$$(3.7)\quad \frac{d^2 y_{\bar{\tau}}(x)}{dx^2} = \bar{\tau}^2\left\{\frac{1}{4} - \frac{k}{\bar{\tau}x} + \frac{\lambda}{\bar{\tau}^2 x^2}\right\} y_{\bar{\tau}}(x) =: p_{\bar{\tau}}(x) y_{\bar{\tau}}(x),$$

it follows from (3.5) and (3.7) that for real a and b,

$$(3.8)\quad \int_a^b (p_{\bar{\tau}}(x) - p_\tau(x))\, y_\tau(x) y_{\bar{\tau}}(x)\,dx$$

$$= \int_a^b \left\{y_\tau(x)\frac{d^2 y_{\bar{\tau}}(x)}{dx^2} - y_{\bar{\tau}}(x)\frac{d^2 y_\tau(x)}{dx^2}\right\}dx$$

$$= \left(y_\tau(x)\frac{dy_{\bar{\tau}}(x)}{dx} - y_{\bar{\tau}}(x)\frac{dy_\tau(x)}{dx}\right)\Bigg|_{x=a}^{x=b}.$$

Now, because Re $\tau > 0$, we see from (3.1) that $y_\tau(x)$, $y_{\bar{\tau}}(x)$, and their derivatives tend to zero as $x \to +\infty$, and as $y_\tau(1) = y_{\bar{\tau}}(1) = 0$, the choice $a = 1$, $b = +\infty$ in (3.8) gives

$$\int_1^\infty (p_{\bar{\tau}}(x) - p_\tau(x)) y_\tau(x) y_{\bar{\tau}}(x)\,dx = 0.$$

Equivalently, using the definitions of p_τ and $p_{\bar{\tau}}$, we have that

$$(\bar{\tau} - \tau)\int_1^\infty \left\{\frac{\tau + \bar{\tau}}{4} - \frac{k}{x}\right\} |y_\tau(x)|^2 dx = 0,$$

since $\overline{y_\tau(x)} = y_{\bar{\tau}}(x)$. But as $(\tau - \bar{\tau}) \neq 0$, this reduces to

$$(3.9)\qquad \int_1^{\infty} \left\{\frac{\operatorname{Re}\tau}{2} - \frac{k}{x}\right\} |y_{\tau}(x)|^2 \, dx = 0.$$

Clearly, the term $g(x) := \frac{\operatorname{Re}\tau}{2} - \frac{k}{x}$, which is monotone increasing on $[1, +\infty)$ and positive at infinity, cannot be positive for all $x \geq 1$, as this would contradict (3.9). Thus, $g(1) = \frac{\operatorname{Re}\tau}{2} - k < 0$ which implies from (3.3) that $\operatorname{Re}\tau < 2k = n - \nu$, the desired result of (2.1) of Theorem 2.1. ■

For the proof of Theorem 2.2, we need the following

LEMMA 3.1. *For any* $n \geq 1$, *and any* $\nu \geq 0$, *let* τ *be any maximal zero of* $P_{n,\nu}(z)$, *i.e.*,

$$(3.10)\qquad P_{n,\nu}(\tau) = 0 \text{ and } |\tau| = \max\{|z| : P_{n,\nu}(z) = 0\}.$$

Then,

$$(3.11)\qquad \operatorname{Re}\tau \geq -(\nu + 1).$$

Proof. First, it can be verified from (1.2) that $P_{n,\nu}(z)$ satisfies the differential equation

$$(3.12)\qquad n\,P_{n,\nu}(z) = (z + n + \nu)\,P'_{n,\nu}(z) - z\,P''_{n,\nu}(z).$$

Next, it is known (cf. Saff-Varga [10]) that all the zeros of $P_{n,\nu}(z)$ are simple. With τ a maximal zero of $P_{n,\nu}(z)$, define, for any $n > 1$,

$$(3.13)\qquad T := \tau - 2(n-1)\,\frac{P'_{n,\nu}(\tau)}{P''_{n,\nu}(\tau)}.$$

By definition, $P_{n,\nu}(z)$ has no zeros in $|z| > |\tau|$. Hence, using a result of Laguerre (cf. Szegö [14, p. 117]), T must lie in $|z| \leq |\tau|$, and, because $P_{n,\nu}(\tau) = 0$, equations (3.12) and (3.13) give that $T = \tau - \frac{2(n-1)\tau}{(\tau + n + \nu)}$. Then, a short calculation shows that $|T| \leq |\tau|$ implies (3.11) for any $n > 1$, any $\nu \geq 0$. If $n = 1$, the sole zero of $P_{1,\nu}(z)$ is $-(\nu + 1)$, which also satisfies (3.11). ■

Proof of Theorem 2.2. We first establish the second inequality of (2.2) of Theorem 2.2. Let τ be a maximal zero (cf. (3.10)) of $P_{n,\nu}(z)$. If τ is real, then τ is evidently negative since $P_{n,\nu}(z)$ has only positive coefficients, and thus, applying Lemma 3.1, we have $|\tau| \leq 1 + \nu \leq n + \nu$. But as τ is a maximal zero of $P_{n,\nu}(z)$, then $|z| \leq n + \nu$ for any zero of $P_{n,\nu}$, which satisfies the second inequality of (2.2) of Theorem 2.2.

Let τ then be any non-real maximal zero of $P_{n,\nu}(z)$ with $\operatorname{Im} \tau > 0$. With $w_{n,\nu}(z)$ defined in (3.1), set

(3.14) $$y(x) := w_{n,\nu}(\tau(1 + \gamma x)), \quad 0 \leq x < \infty,$$

where γ is a constant, to be selected later, such that

(3.15) $$\operatorname{Re}(\tau\gamma) > 0.$$

From (3.2), we see that y satisfies

(3.16) $$\frac{d^2 y(x)}{dx^2} = (\tau\gamma)^2 \left\{\frac{1}{4} - \frac{k}{\tau(1+\gamma x)} + \frac{\lambda}{\tau^2(1+\gamma x)^2}\right\} y(x) =: p(x) y(x).$$

Since $\bar{\tau}$ is also a maximal zero of $P_{n,\nu}(z)$, we also consider

(3.17) $$\overline{y(x)} = w_{n,\nu}(\bar{\tau}(1+\bar{\gamma}x)), \quad 0 \leq x < \infty,$$

which satisfies

(3.18) $$\frac{d^2 \overline{y(x)}}{dx^2} = \overline{p(x)}\;\overline{y(x)}.$$

As before (cf. (3.8)), we similarly have that

$$\int_a^b (\overline{p(x)} - p(x)) |y(x)|^2 dx = \left(y(x)\frac{d\overline{y(x)}}{dx} - \overline{y(x)}\frac{dy(x)}{dx}\right)\Bigg|_{x=a}^{x=b}.$$

Because of (3.15) and (3.1), $y(x)$, $\overline{y(x)}$, and their derivatives tend to zero as $x \to +\infty$, and as $y(0) = \overline{y(0)} = 0$, the choice $a = 0$ and $b = +\infty$ in the above expression then gives

(3.19) $$\int_0^\infty (\overline{p(x)} - p(x)) \, |y(x)|^2 dx = 0.$$

Now, $p(x) - \overline{p(x)}$ is, purely imaginary, and using (3.16),

$$\frac{p(x) - \overline{p(x)}}{2i} = \frac{\operatorname{Im}\{(\tau\gamma)^2|1+\gamma x|^4 - |1+\gamma x|^2(1+\bar{\gamma}x)4k\tau\gamma^2 + 4\lambda\gamma^2(1+\bar{\gamma}x)^2\}}{4|1+\gamma x|^4}. \tag{3.20}$$

Because of (3.19), note that the numerator of the right side of (3.20) cannot be of one sign for all $0 \le x < \infty$. Writing $\tau = \rho e^{i\theta}$ with $0 < \theta < \pi$, we choose $\gamma := \frac{1}{\rho} e^{-i\theta/2}$, so that (3.15) is satisfied. With this choice of γ, the numerator of (3.20) can be written, after some algebraic manipulations, as the product $2\rho^{-4} \sin \frac{\theta}{2} \cdot \sigma_4(x)$, where $\sigma_4(x)$, a quartic polynomial, is defined by

$$(3.21)\quad \begin{cases} \sigma_4(x) := x^4[\cos \frac{\theta}{2}] + x^3[4\rho \cos^2 \frac{\theta}{2} - 2k] \\ + x^2[(2\rho^2 - 4k\rho)\cos \frac{\theta}{2} + 4\rho^2 \cos^3 \frac{\theta}{2}] + x[4\rho^3 \cos^2 \frac{\theta}{2} - 2k\rho^2 - 4\lambda\rho] \\ + [\rho^4 - 4\lambda\rho^2]\cos \frac{\theta}{2} . \end{cases}$$

As before, σ_4 cannot be of one sign on $[0, +\infty)$ because of (3.19).

To complete the proof of the second inequality of (2.2) of Theorem 2.2, assume on the contrary that $\rho = |\tau| \ge n + \nu + 4/3$. Using the definitions of (3.3) and in particular the result of Lemma 3.1, a lengthy calculation (which we omit) shows that $\sigma_4(x) := \sum_{i=0}^{4} \beta_i x^i$ has only positive coefficients β_i, $0 \le i \le 4$, a contradiction. Thus, $|\tau| < n + \nu + 4/3$, and as τ is a maximal zero of $P_{n,\nu}(z)$, all zeros of $P_{n,\nu}(z)$ necessarily satisfy $|z| < n + \nu + 4/3$, the desired second inequality of (2.2) of Theorem 2.2.

We now establish the first inequality of (2.2). For the partial sums $s_n(z) := \sum_{k=0}^{n} z^k/k!$ of the Maclaurin expansion for e^z, it is known (cf. Buckholtz [2]) that, for any $n \ge 1$, any zero $\hat{z}$ of the normalized polynomial $s_n(nz)$ satisfies

(3.22) $|\hat{z}e^{1-\hat{z}}| > 1$ and $|\hat{z}| < 1$,

i.e., as Figures 2 and 3 show, $\hat{z}$ lies outside D_0 but in the unit disk. It is easy to verify (cf. Figures 2 and 3) that the point ζ in the unit disk satisfying $|\zeta e^{1-\zeta}| = 1$ which is closest to the origin is the real point $\zeta = -\mu, \mu > 0$, where $\mu e^{1+\mu} = 1$, and μ is given approximately by $\mu \doteq 0.278\,465$. Hence, for any $n \geq 1$, any zero z of $s_n(z)$ from (3.22) satisfies

(3.23) $|z| > n\mu$.

Since $s_n(z) = P_{n,0}(z)$, then (3.23) gives the desired first inequality of (2.2) of Theorem 2.2 for the case $\nu = 0$.

Next, on defining the "reciprocal" Padé numerators:

(3.24) $P^*_{n,\nu}(z) := z^n P_{n,\nu}(1/z)$ for all $n \geq 0$, $\nu \geq 0$,

it follows directly from the definition of $P_{n,\nu}(z)$ in (1.2) that

(3.25) $\frac{d^\nu}{dz^\nu} P^*_{n+\nu,0}(z) = \frac{(n+\nu)!}{n!} P^*_{n,\nu}(z)$ for all $n \geq 0$, $\nu \geq 0$.

Now, since $P^*_{n+\nu,0}(z)$ has, from (3.23), all its zeros in $|z| < ((n+\nu)\mu)^{-1}$, then, by the Gauss-Lucas Theorem (cf. Marden [5, p. 14]), the same is true for all derivatives of $P^*_{n+\nu,0}(z)$. In particular, from (3.25), $P^*_{n,\nu}(z)$ has all its zeros in $|z| < ((n+\nu)\mu)^{-1}$, which implies that $P_{n,\nu}(z)$ has all its zeros in $|z| > (n+\nu)\mu$, which gives the first inequality of (2.2) of Theorem 2.2.

Finally, the first inequality of (2.2) implies that

(3.26) $\mu \leq \inf\limits_{\substack{n\geq 1\\ \nu\geq 0}} \{\frac{|z|}{(n+\nu)} : P_{n,\nu}(z) = 0\}$.

To show that equality holds in (3.26), Szegö [13] has established that $\hat{z}$ is a limit point of zeros of $\{s_n(nz)\}_{n=1}^{\infty}$ iff $\hat{z}$ satisfies (1.1). As $\hat{z} = -\mu$ satisfies (1.1), then evidently equality holds in (3.26), completing the proof of Theorem 2.2. ∎

Acknowledgement. We are indebted to A. Ruttan of Kent State University for his computations on a Burroughs B-5700, which were used in preparing Figures 2-8.

References

1 Birkhoff, G. and R.S. Varga, Discretization errors for well-set Cauchy problems. I., J. Math. and Physics 44(1965), 1-23.

2 Buckholtz, J.D., A characterization of the exponential series, Amer. Math. Monthly 73(1966), 121-123.

3 Dieudonné, J., Sur les zéros des polynomes-sections de e^x, Bull. Sci. Math. 70(1935), 333-351.

4 Ehle, B.L., A-stable methods and Padé approximation to the exponential, SIAM J. Math. Anal. 4(1973), 671-680.

5 Marden, M., The Geometry of the Zeros, Mathematical Surveys III, American Mathematical Society, Providence, Rhode Island, 1949.

6 Nørsett, S.P., C-polynomials for rational approximation to the exponential function, Numer. Math. 25(1975), 39-56.

7 Olver, F.W.J., The asymptotic expansion of Bessel functions of large order, Phil. Trans. Roy. Soc. London Ser. A 247(1954), 328-368.

8 Perron, O., Die Lehre von den Kettenbrüchen, 3rd ed., B. G. Teubner, Stuttgart, 1957.

9 Saff, E.B., and R.S. Varga, On the zeros and poles of Padé approximants to e^z, Numer. Math. 25(1975), 1-14.

10 Saff, E.B., and R.S. Varga, Zero-free parabolic regions for sequences of polynomials, SIAM J. Math. Anal. 7(1976), 344-357.

11 Saff, E.B., and R.S. Varga, On the sharpness of theorems concerning zero-free regions for certain sequences of polynomials, Numer. Math. 26(1976), 345-354.

12 Saff, E.B., and R.S. Varga, The behavior of the Padé table for the exponential, Approximation Theory (G. G. Lorentz, C. K. Chui, and L. L. Schumacher, eds.), pp. 519-531. Academic Press, New York, 1976.

13 Szegö, G., Über eine Eigenschaft der Exponentialreihe, Sitzungsberichte der Berliner Mathematischen Gesselschaft 23(1924), 50-64.

14 Szegö, G., Orthogonal Polynomials, American Mathematical Society Colloquium Publications Volume XXIII, fourth ed., American Mathematical Society, Providence, Rhode Island, 1939.

15 Underhill, C., and A. Wragg, Convergence properties of Padé approximants to exp(z) and their derivatives, J. Inst. Maths. Applics. 11(1973), 361-367.

E. B. Saff*
Department of Mathematics
University of South Florida
Tampa, Florida 33620

R. S. Varga**
Department of Mathematics
Kent State University
Kent, Ohio 44242

*Research supported in part by the Air Force Office of Scientific Research under Grant AFOSR-74-2688.

**Research supported in part by the Air Force Office of Scientific Research under Grant AFOSR-74-2729, and by the Energy Research and Development Administration (ERDA) under Grant E(11-1)-2075.

TWO-POINT PADÉ TABLES, T-FRACTIONS AND SEQUENCES OF SCHUR

W.J. Thron

T-fractions and sequences of Schur correspond to a formal power series P at $z = 0$. Since additional conditions are imposed on the two sequences the correspondence to P is not good enough for the elements of the sequences to be entries in the ordinary Padé table of P. However, in both cases there are formal power series P^* at ∞ completely determined by P, so that the elements of the sequence are entries in the two-point Padé table for P and P^*. We sketch the situation for T-fractions and work it out in detail for sequences of Schur. The last section contains a discussion of continued fractions whose approximants are the entries along certain diagonals or staircases of a general two-point Padé table.

1 Introduction and T-fractions

The starting point of this investigation was the observation by W.B. Jones that the approximants of T - fractions

$$(1) \quad 1 + d_0 z + \frac{z}{1 + d_1 z} + \frac{z}{1 + d_2 z} + \cdots$$

are entries in a two point Padé table, if all $d_n \neq 0$. This led Jones and the present author to study various aspects of two-point Padé tables. Some of the results of our investigations are presented elsewhere in these Proceedings [4]. Others will be published in forthcoming joint papers [5], [7].

We now review briefly the concept of the two-point Padé table which was introduced by Baker et al [1]. An ordinary (m,n) Padé approximant to a formal power series L

$$(2) \quad L = c_0 + c_1 z + c_2 z^2 + \cdots$$

is a rational function

$$(3) \quad R_{m,n}(z) = \frac{P_{m,n}(z)}{Q_{m,n}(z)} ,$$

where $P_{m,n}$ and $Q_{m,n}$ are polynomials in z of degree not greater than m and n, respectively, which corresponds to L of the highest possible order. This means that

$$(4)\quad LQ_{m,n} - P_{m,n} = d_{m,n} z^{m+n+1} + \text{higher powers of } z .$$

We note that (4) imposes $m+n+1$ linear restrictions on the $m+n+2$ coefficients in the two polynomials $P_{m,n}$ and $Q_{m,n}$. Therefore (4) can always be satisfied.

Before we continue our discussion it is helpful to define for arbitrary formal power series, the statements

$$(5)\quad \chi^{+}(P) = m \quad \text{and} \quad \chi^{-}(P^{*}) = k$$

to mean

$$P = p_m z^m + \text{higher powers of } z ,$$

$$P^{*} = p_k^{*} z^k + \text{lower powers of } z ,$$

respectively.

In the case of the two-point table the conditions to be imposed on the coefficients of $P_{m,n}$ and $Q_{m,n}$ are distributed between L at 0 and

$$(6)\quad L^{*} = c_{\mu}^{*} z^{\mu} + c_{\mu-1}^{*} z^{\mu-1} + c_{\mu-2}^{*} z^{\mu-2} + \cdots$$

at ∞ as evenly as possible (i.e. we approximate alternately at 0 and ∞, starting with 0). Subject to these considerations one can achieve the following correspondence to L at 0 and L^{*} at ∞

$$(7)\quad \chi^{+}(LQ_{m,n} - P_{m,n}) = \left[\frac{m+n}{2}\right] + 1$$

and

$$(8)\quad \chi^{-}(L^{*}Q_{m,n} - P_{m,n}) = \left[\frac{m+n}{2}\right] - m + \mu .$$

Here $[r]$ denotes the largest integer not greater than r.

We shall say that a function $R_{m,n}$, defined in (3), is the (m,n) entry in the <u>two-point Padé table of L and L^{*}</u> if it satisfies (7) and (8). That $R_{m,n}$ is unique can be proved and shall be taken for granted here. We shall also assume that $P_{m,n}$ and $Q_{m,n}$ are relatively prime to each other.

Let L be defined by (2) with $c_0 = 1$ and otherwise arbitrary. It was shown in [10] that there exists a unique T-fraction (1) whose n^{th} approximant A_n/B_n satisfies

$$\chi^+(LB_n - A_n) = n+1 .$$

The degrees of the polynomials A_n and B_n are $n+1$ and n, respectively. The function A_n/B_n is thus, in general, not the $(n+1, n)$ entry in the ordinary Padé table of L. It was felt that this disadvantage was worth accepting in exchange for having all the elements of the continued fraction be linear functions of z.

It was first realized by Perron [8, p. 179] and later elaborated by Waadeland [11] and Jefferson [3] that a T-fraction with all $d_n \neq 0$ satisfies

$$\chi^-(L^* B_n - A_n) = 0$$

for a certain formal power series L^* of the form (6) with $\mu = 1$. Hence A_n/B_n is the $(n+1, n)$ entry in the two-point Padé table of L and L^*. Here L^* is completely determined directly by the T-fraction and thus indirectly by the original L. (In a general two-point Padé table L and L^* are independent of each other.)

We thus obtain an unintended trade off between the linearity conditions on the elements of the continued fraction and a sufficient degree of correspondence to a certain L^* at ∞ to make the nth approximants of the T-fraction $(n+1, n)$ entries of the two-point Padé table of L and L^*.

A very similar property holds for a sequence of rational functions studied by Schur [9] in 1917. Here the trade off is between the approximants being "functions bounded in the unit circle" and a sufficient degree of correspondence to an L^* at ∞ to make the members of the sequence (n,m) entries of the two-point Padé table of L and L^* . In this case $\mu = 0$. As before L^* is completely determined by L .

2 Sequences of Schur

If $f(z)$ is holomorphic and $|f(z)| < I$ for $|z| < 1$ we shall call it a b.u.c. function. To study b.u.c. functions Schur [9] was led to investigate sequences of rational functions defined as follows

$$t_n(w) = \frac{\gamma_n + zw}{1 + \bar{\gamma}_n zw}, \quad |\gamma_n| < 1, \quad n \geq 1 .$$

$$T_n(w) = t_1 \circ \dots \circ t_n(w) = \frac{C_n w + D_n}{E_n w + F_n} .$$

Clearly, for $n \geq 2$

$$T_n(w) = \frac{C_{n-1} \dfrac{\gamma_n + zw}{1 + \bar{\gamma}_n zw} + D_{n-1}}{E_{n-1} \dfrac{\gamma_n + zw}{1 + \bar{\gamma}_n zw} + F_{n-1}}$$

$$= \frac{(C_{n-1} z + D_{n-1} \bar{\gamma}_n z) w + (C_{n-1} \gamma_n + D_{n-1})}{(E_{n-1} z + F_{n-1} \bar{\gamma}_n z) w + (E_{n-1} \gamma_n + F_{n-1})} .$$

For $n = 1$ we have

$$\frac{\gamma_1 + zw}{1 + \bar{\gamma}_1 zw} = \frac{C_1 w + D_1}{E_1 w + F_1} .$$

Hence

$$(9') \quad C_1 = z , \quad D_1 = \gamma_1 , \quad E_1 = \bar{\gamma}_1 z , \quad F_1 = 1 ,$$

and for $n \geq 2$

$$(9'') \quad \begin{aligned} C_n &= C_{n-1} z + D_{n-1} \bar{\gamma}_n z , \\ D_n &= C_{n-1} \gamma_n + D_{n-1} , \\ E_n &= E_{n-1} z + F_{n-1} \bar{\gamma}_n z , \\ F_n &= E_{n-1} \gamma_n + F_{n-1} . \end{aligned}$$

From the recursion relations (9) one easily derives the following identities

$$C_nF_n - D_nE_n = z(1-|\gamma_n|^2)(C_nF_{n-1} - D_{n-1}E_{n-1})$$

$$= z^n \prod_{k=1}^{n} (1-|\gamma_k|^2) ,$$

$$D_nF_{n-1} - F_nD_{n-1} = \gamma_n(C_{n-1}F_{n-1} - D_{n-1}E_{n-1}) ,$$

$$(10)\quad C_nF_{n-1} - E_nD_{n-1} = z(C_{n-1}F_{n-1} - D_{n-1}E_{n-1}) ,$$

$$D_nF_{n-2} - F_nD_{n-2} = (\gamma_n z + \gamma_{n-1})(C_{n-2}F_{n-2} - D_{n-2}E_{n-2}) .$$

It also follows from (9) that C_n, D_n, E_n and F_n are polynomials in z of the form indicated below

$$(11)\quad \begin{aligned} C_n &= \gamma_1\bar{\gamma}_n z + \dots + z^n , \\ D_n &= \gamma_1 + \dots + \gamma_n z^{n-1} , \\ E_n &= \bar{\gamma}_n z + \dots + \bar{\gamma}_1 z^n , \\ F_n &= 1 + \dots + \bar{\gamma}_1\gamma_n z^{n-1} . \end{aligned}$$

We shall be interested in the sequence

$$\{T_n(0)\} = \{D_n/F_n\} .$$

Since $|0| < 1$ each D_n/F_n is a b.u.c. function.

Observe that

$$\frac{D_n}{F_n} - \frac{D_{n-1}}{F_{n-1}} = \gamma_n \prod_{k=1}^{n-1} (1-|\gamma_k|^2) z^{n-1} + \text{higher powers of } z .$$

The series

$$P = \frac{D_1}{F_1} + \sum_{k=2}^{\infty} \left(\frac{D_k}{F_k} - \frac{D_{k-1}}{F_{k-1}} \right)$$

is a well defined formal power series

$$P = c_0 + c_1 z + c_2 z^2 + \dots ,$$

and

$$P - \frac{D_n}{F_n} = \sum_{k=n+1}^{\infty} \left(\frac{D_k}{F_k} - \frac{D_{k-1}}{F_{k-1}} \right) = \gamma_{n+1} \prod_{k=1}^{n} (1 - |\gamma_k|^2) z^n$$

$$+ \text{ higher powers of } z .$$

We then have

$$\lambda^+(PF_n - D_n) \geq n .$$

It is equal to n if $\gamma_{n+1} \neq 0$, since it is known that $F_n(0) = 1$. The rational function D_n/F_n is thus, at least if all $\gamma_n \neq 0$, not the $(n-1, n-1)$ entry in the ordinary Padé table of P. As in the case of T-fractions the disadvantage of not having the maximum possible order of correspondence at 0 is considered less important than the advantage of having all D_n/F_n be b.u.c. functions. For ordinary Padé approximants to P this boundedness condition does not in general hold.

Before proceding to show that $\{D_n/F_n\}$ corresponds to a formal power series at ∞ we note that for each δ with $|\delta| < 1$ the sequence $\{T_n(\delta)\}$ converges to a b.u.c. function f, independent of δ, for all $|z| < 1$. This follows from the theorem of Jones and Thron [6] stated below:

<u>If a sequence $\{R_n\}$ of rational functions is uniformly bounded on every compact subset of a region D containing the origin in its interior and if the sequence corresponds (in the sense that $\lambda^+(P - R_n) \to \infty$) to a formal power series P, then $\{R_n\}$ converges in D to a holomorphic function f whose Taylor series expansion at 0 is P.</u>

Since the $T_n(0)$ are all b.u.c. functions the limit function f is a b.u.c. function and P is thus the Taylor series of the b.u.c. function f.

Conversely, if f is an arbitrary b.u.c. function and P its Taylor series, then a sequence $\{\gamma_n\}$ and hence $\{D_n/F_n\}$ can be determined as follows. Let $f = f_0$ and define recursively

$$f_n(z) = \frac{1}{z}\,\frac{f_{n-1}(z) - \gamma_n}{1 - \bar{\gamma}_n f_{n-1}(z)}\ , \quad \gamma_n = f_{n-1}(0)\ .$$

Then, using Schwarz' Lemma, one can prove that all $f_n(z)$ are b.u.c. functions and that for all $n \geq 1$, $|\gamma_n| < 1$. It is also easy to see that $f(z) = T_n(f_n(z))$ and that $\{T_n(0)\}$ converges to $f(z)$. Thus with each b.u.c. function there is associated a sequence $\{D_n/F_n\}$. By means of these sequences Schur was able to study the behavior of the coefficients of the Taylor series of b.u.c. functions.

Perron [8, p. 179] was the first to observe that a continued fraction similar to the Schur sequence corresponds to a formal power series at ∞. We have

$$\frac{D_n}{F_n} - \frac{D_{n-1}}{F_{n-1}} = \frac{D_n F_{n-1} - F_n D_{n-1}}{F_n F_{n-1}}$$

$$= \frac{\gamma_n z^{n-1} \prod_{k=1}^{n-1} (1 - |\gamma_k|^2)}{(\gamma_n \bar{\gamma}_1 z^{n-1} + \ldots + 1)(\gamma_{n-1}\bar{\gamma}_1 z^{n-2} + \ldots + 1)}$$

$$= \frac{\prod_{k=1}^{n-1} (1 - |\gamma_k|^2)}{\gamma_{n-1}\bar{\gamma}_1^{-2}}\, z^{-(n-2)} + \text{lower powers of } z$$

at least if $\gamma_n \neq 0$ for all $n \geq 1$. Using an argument analogous to the one used to show the existence of a corresponding power series at 0 one obtains a formal power series

$$P^* = c_0^* + c_{-1}^* z^{-1} + c_{-2}^* z^{-2} + \ldots$$

such that

$$\chi^-(P^* F_n - D_n) = 0\ .$$

It then follows that, if all $\gamma_n \neq 0$, then D_n/F_n is the

$(n-1, n-1)$ entry in the two-point Padé table of P and P^*. Here $\mu = 0$.

To express the sequence of Schur as a continued fraction we use the following theorem of Perron [8, p. 6].

For given sequences $\{S_n\}$, $\{T_n\}$ satisfying $T_0 = 1$, $S_n T_{n-1} - S_{n-1} T_n \neq 0$, for all $n \geq 1$, there exists a unique continued fraction

$$t_0 + \underset{n=1}{\overset{\infty}{K}} (s_n/t_n)$$

whose approximants are such that their numerators and denominators are exactly S_n and T_n, respectively. Its elements are as follows:

$$t_0 = S_0, \quad s_1 = S_1 T_0 - S_0 T_1, \quad t_1 = T_1,$$

and for $n \geq 2$

$$s_n = \frac{S_{n-1} T_n - S_n T_{n-1}}{S_{n-1} T_{n-2} - S_{n-2} T_{n-1}}, \quad t_n = \frac{S_n T_{n-2} - S_{n-2} T_n}{S_{n-1} T_{n-2} - S_{n-2} T_{n-1}}.$$

Setting $D_{n+1} = S_n$, $F_{n+1} = T_n$, $n \geq 0$, one obtains $S_0 = \gamma_1$, $T_0 = 1$ and using equations (10)

$$t_0 = \gamma_1, \quad s_1 = D_2 F_1 - D_1 F_2 = \gamma_2 (1 - |\gamma_1|^2) z,$$

$$t_1 = F_2 = 1 + \bar{\gamma}_1 \gamma_2 z$$

and for $n \geq 2$

$$s_n = \frac{\gamma_{n+1}(1 - |\gamma_n|^2)}{\gamma_n}, \quad t_n = \frac{\gamma_{n+1}}{\gamma_n} z + 1.$$

The continued fraction thus becomes

$$\gamma_1 + \frac{\gamma_2 (1 - |\gamma_1|^2) z}{1 + \bar{\gamma}_1 \gamma_2 z} + \frac{-\gamma_3 (1 - |\gamma_2|^2) z/\gamma_2}{1 + (\gamma_3/\gamma_2) z} + \cdots .$$

It occupies the main diagonal of the two-point Padé table of P and P^*. Hamel [2] in 1918 obtained a very similar continued fraction in his work on b.u.c. functions.

3 Continued fractions in the two-point Padé table

In this section we show that for a sufficiently normal table certain sequences of Padé approximants form continued fractions of a surprisingly simple type. We first consider the sequence

$$\{P_{m+\mu,m}/Q_{m+\mu,m}\} \ .$$

It is convenient to set

$$\Gamma_m = P_{m+\mu,m} \ , \quad \Delta_m = Q_{m+\mu,m} \ .$$

Then formulas (7) and (8) lead to

$$\chi^+(L\Delta_m - \Gamma_m) = m+1+v \ ,$$

$$\chi^-(L^*\Delta_m - \Gamma_m) = v \ .$$

Here we have set $v = [\mu/2]$. Let us assume that

(12) $\Delta_k(0) \neq 0$ for all $k \geq 0$.

Then

$$\chi^+(L\Delta_m\Delta_{m-1} - \Gamma_m\Delta_{m-1}) = m+1+v$$

and

$$\chi^+(L\Delta_{m-1}\Delta_m - \Gamma_{m-1}\Delta_m) = m+v \ .$$

It follows that

(13) $\chi^+(\Gamma_m\Delta_{m-1} - \Gamma_{m-1}\Delta_m) = m+v$.

Similarly

$$\chi^-(L^*\Delta_m\Delta_{m-1} - \Gamma_m\Delta_{m-1}) = m-1+v$$

and

$$\chi^-(L^*\Delta_{m-1}\Delta_m - \Gamma_{m-1}\Delta_m) = m+v \ .$$

It follows that

(14) $\chi^-(\Gamma_m\Delta_{m-1} - \Gamma_{m-1}\Delta_m) = m+v$.

From (13) and (14) together one then obtains

(15) $\Gamma_m\Delta_{m-1} - \Gamma_{m-1}\Delta_m = \alpha_m z^{m+v}$.

An analogous argument yields

(16) $\Gamma_m \Delta_{m-2} - \Gamma_{m-2}\Delta_m = \beta_m z^{m-1+v} + \delta_m z^{m+v}$.

Using the theorem of Perron, which we stated at the end of the previous section, together with (15) and (16) we obtain

$$t_0 = \Gamma_0 = P_{\mu,0} = g_0 + \dots + g_\mu z^\mu , \quad s_1 = \alpha_1 z^{1+v} ,$$

$$t_1 = \Delta_1 = h_0 + k_0 z$$

and for $n \geq 2$

$$s_n = -\frac{\alpha_n}{\alpha_{n-1}} z , \quad t_n = \frac{\beta_n}{\alpha_{n-2}} + \frac{\delta_n}{\alpha_{n-2}} z .$$

If in addition to (12) we now assume that all $\alpha_n \neq 0$ then by suitable normalization of the Γ_m and Δ_m we can assume that $\alpha_n = (-1)^n$ for all $n \geq 1$ and thus obtain

$$g_0 + \dots + g_\mu z^\mu + \frac{z^{1+[\mu/2]}}{h_0 + k_0 z} + \frac{z}{h_1 + k_1 z} + \cdots$$

for the continued fraction whose n^{th} approximant is the $(m+\mu, n)$ entry in the two-point Padé table of L and L^* .

For $\mu = 2k+1$ we have $\mu - v = 1 + v$. For these values of μ a simple continued fraction can be obtained for the entries along a certain staircase in the two-point table. Set

$$\Lambda_{2m} = P_{m+\mu,m} , \quad \Lambda_{2m+1} = P_{m+\mu,m+1} ,$$

$$\Lambda_{2m} = Q_{m+\mu,m} , \quad M_{2m+1} = Q_{m+\mu,m+1} .$$

Then from (7) and (8)

$$\chi^+(L M_{2m} - \Lambda_{2m}) = m + 1 + v ,$$

$$\chi^-(L^* M_{2m} - \Lambda_{2m}) = v ,$$

$$\chi^+(L M_{2m+1} - \Lambda_{2m+1}) = m + 2 + v ,$$

$$\chi^-(L^* M_{2m+1} - \Lambda_{2m+1}) = 1 + v .$$

Further, it is easily verified that the degree of Λ_{2m+1} is at most m . Using these facts and the method developed for the previous case one obtains

$$\Lambda_{2m}M_{2m-1} - \Lambda_{2m-1}M_{2m} = \varepsilon_m z^{m+1+v} ,$$

$$\Lambda_{2m+1}M_{2m} - \Lambda_{2m}M_{2m+1} = \eta_m z^{m+1+v} ,$$

$$\Lambda_{2m+1}M_{2m-1} - \Lambda_{2m-1}M_{2m+1} = \kappa_m z^{m+1+v} ,$$

$$\Lambda_{2m+2}M_{2m} - \Lambda_{2m}M_{2m+2} = \alpha_{m+1} z^{m+1+v} .$$

Using Perron's theorem again and assuming

$$\delta_m \neq 0 , \quad \eta_m \neq 0 \quad \text{and} \quad \text{degr } \Lambda_{2m+1} = m , \quad \text{for all} \quad m .$$

one arrives at the continued fraction

$$P_{\mu,0} + \frac{\rho_1 z^{1+[\mu/2]}}{\sigma_1} + \frac{\rho_2 z}{\sigma_2} + \frac{\rho_3}{\sigma_3} + \frac{\rho_4 z}{\sigma_4} + \frac{\rho_5}{\sigma_5} + \cdots$$

for the entries along the staircase of the two-point Padé table, provided μ is odd.

Still incomplete investigations seem to indicate that relatively simple continued fractions may also be obtainable for some, but not all, other diagonals and staircases in two-point tables.

References

1 Baker, G.A., Jr., G.S. Rushbrooke, and H.E. Gilbert, High temperature series expansion for the spin - 1/2 Heisenberg model by the method of irreducible representations of the symmetric group, Physical Review, 135(1964), A1272-A1277.

2 Hamel, G., Eine charakteristische Eigenschaft beschränkter analytischer Funktionen, Math. Annal. 78(1918), 257-269.

3 Jefferson, Thomas H., Some additional properties of T-fractions, Dissertation, University of Colorado, (1969).

4 Jones, William B., Multiple point Padé tables, these Proceedings.

5 Jones, William B., and W.J. Thron, Two-point Padé tables and T-fractions, Bull. Amer. Math. Soc., to appear.

6 Jones, William B., and W.J. Thron, Sequences of meromorphic functions corresponding to a formal Laurent series, submitted.

7 Jones, William B., and W.J. Thron, Results on two-point Padé tables, in preparation.

8 Perron, Oskar, Die Lehre von den Kettenbrüchen, Band II, Stuttgart, Teubner, (1957).

9 Schur, I., Über Potenzreihen die im Innern des Einheitskreises beschränkt sind, J. Reine Angew. Math., 147(1917), 205-232.

10 Thron, W.J., Some properties of continued fractions $1 + d_0 z + K\left(\frac{z}{1 + d_n z}\right)$, Bull. Amer. Math. Soc., 54(1948), 206-218.

11 Waadeland, Haakon, On T-fractions of certain functions with a first order pole at the point of infinity, Det kongelike Norske Videnskabers Selskabs Forhandlinger 40(1967), 1-6.

W.J. Thron
Department of Mathematics
University of Colorado
Boulder, Colorado 80309

Research supported in part by the National Science Foundation under Grant No. MPS 74-22111.

THREE-TERM CONTIGUOUS RELATIONS AND SOME NEW ORTHOGONAL POLYNOMIALS

J. A. Wilson

In this paper, we show how three-term contiguous relations for ${}_2F_1$'s, ${}_3F_2$'s, and ${}_4F_3$'s may be derived, and list the ${}_3F_2$ relations. These relations may be a source for many interesting continued fractions, and the ${}_4F_3$ relations include the recurrence relation for a set of orthogonal polynomials generalizing the classical polynomials.

1 Introduction

Gauss's celebrated continued fraction ([7], [9]) for the ratio ${}_2F_1\left(\begin{matrix}a,b+1\\c+1\end{matrix}; z\right)\Big/{}_2F_1\left(\begin{matrix}a,b\\c\end{matrix}; z\right)$ is derived from the relation

$$(1)\quad {}_2F_1\left(\begin{matrix}a,b+1\\c+1\end{matrix}; z\right) - {}_2F_1\left(\begin{matrix}a,b\\c\end{matrix}; z\right) = \frac{(c-b)az}{c(c+1)}\,{}_2F_1\left(\begin{matrix}a+1,b+1\\c+2\end{matrix}; z\right).$$

While this continued fraction has been well-studied and many special cases written down, very little has been done in the way of new continued fractions for ${}_pF_q$'s. (See however [6].) There are many relations similar to (1), consequences of three-term contiguous relations for ${}_pF_q$'s. A contiguous series to a ${}_pF_q$ is a series obtained by altering one of the parameters by ± 1. Gauss's contiguous relations are three-term linear relations which connect ${}_2F_1\left(\begin{matrix}a,b\\c\end{matrix}; z\right)$ with any two of its contiguous series. Complete lists of Gauss's relations appear in [7] and [1].

Generalizations to ${}_pF_q$'s exist, but in general they are (q+2)-term relations with very complicated coefficients ([3], [8]). In order to get three-term relations - which have a much better chance of tying in with other topics, including continued fractions - it is necessary to put restrictions on the ${}_pF_q$'s.

2 $_3F_2$ relations

For $_3F_2$'s, the condition needed is that $z = 1$ (and for convergence, $\mathrm{Re}(d+e-a-b-c) > 0$) the same condition that allows the $_2F_1$ to be evaluated by Gauss's formula ([2], p. 2) and is needed for the important transformation ([2], p. 98)

$$(2) \quad {}_3F_2\left(\begin{matrix}a,b,c\\ d,e\end{matrix};1\right) = \frac{\Gamma(e)\Gamma(d+e-a-b-c)}{\Gamma(e-a)\Gamma(d+e-b-c)}\,{}_3F_2\left(\begin{matrix}a,d-b,d-c\\ d,d+e-b-c\end{matrix};1\right).$$

Bailey [4] gave a very tedious systematic procedure for deriving relations connecting ${}_3F_2\left(\begin{matrix}a,b,c\\ d,e\end{matrix};1\right)$ with any two contiguous series. We will give the complete list of these relations ((13) through (24)), deriving them by a more efficient procedure which also works for $_2F_1$'s and $_4F_3$'s.

Let $F = {}_3F_2\left(\begin{matrix}a,b,c\\ d,e\end{matrix};1\right)$ and $F_+ = {}_3F_2\left(\begin{matrix}a+1,b+1,c+1\\ d+1,e+1\end{matrix};1\right)$. Denote the ten contiguous series to F by $F(a+), F(a-), \ldots, F(e-)$ and those contiguous to F_+ by $F_+(a+), \ldots, F_+(e-)$.

Note that, by straightforward term-by-term subtraction,

$$(3) \quad F(a+) - F = \frac{bc}{de} F_+$$

$$(4) \quad F(d-) - F = \frac{abc}{(d-1)de} F_+$$

$$(5) \quad F(a-) - F = -\frac{bc}{de} F_+(a-)$$

$$(6) \quad F(d+) - F = -\frac{abc}{d(d+1)e} F_+(d+) .$$

These equations (and the symmetries of the $_3F_2$) make some of the desired relations trivial:

$$(7) \quad a(F(a+) - F) = (d-1)(F(d-) - F) = b(F(b+) - F) = \\ = c(F(c+) - F) = (e-1)(F(e-) - F) ,$$

since all these are equal to $\frac{abc}{de} F_+$. The rest of the contiguous relations may be considered as relations connecting F with pairs of F_+, $F_+(a-)$, $F_+(b-)$, $F_+(c-)$, $F_+(d+)$, and $F_+(e+)$.

Furthermore, all these relations may be generated from a suitably chosen pair of them by combining equations and using the symmetries of the $_3F_2$.

The trick to obtaining such a pair of relations is to use transformation (2). If we apply (2) to each of F, F_+, $F_+(a-)$, and $F_+(d+)$ (getting $_3F_2\left(\begin{matrix}a, d-b, d-c\\ d, d+e-b-c\end{matrix}; 1\right)$, $_3F_2\left(\begin{matrix}a+1, d-b, d-c\\ d+1, d+e-b-c\end{matrix}; 1\right)$, $_3F_2\left(\begin{matrix}a, d-b, d-c\\ d+1, d+e-b-c\end{matrix}; 1\right)$, and $_3F_2\left(\begin{matrix}a+1, d-b+1, d-c+1\\ d+2, d+e-b-c+1\end{matrix}; 1\right)$, respectively) we find that the first three transformed series are connected by the first of equations (7), while the first, third and fourth are connected by (4). In terms of F, F_+, $F_+(a-)$, and $F_+(d+)$, these relations are

(8) $deF - a(d+e-a-b-c-1)F_+ - (d-a)(e-a)F_+(a-) = 0$

(9) $eF - (e-a)F_+(a-) - \dfrac{a(d-b)(d-c)}{d(d+1)} F_+(d+) = 0$.

From (8) and (9) it is easy to generate the following:

(10) $eF - (d+e-a-b-c-1)F_+ + \dfrac{(d-a)(d-b)(d-c)}{d(d+1)} F_+(d+) = 0$

(11) $(e-d) + \dfrac{(d-a)(d-b)(d-c)}{d(d+1)} F_+(d+) - \dfrac{(e-a)(e-b)(e-c)}{e(e+1)} F_+(e+) = 0$

(12) $de(a-b)F + b(d-a)(e-a)F_+(a-) - a(d-b)(e-b)F_+(b-) = 0$.

The contiguous relations now come by substituting (3) - (6) into (8) - (12). We list them, excluding the ones obtained by permuting numerator or denominator parameters:

(13) $a(d+e-a-b-c-1)[F(a+)-F] - (d-a)(e-a)[F(a-)-F] - bcF = 0$

(14) $a[F(a+)-F] - b[F(b+)-F] = 0$

(15) $a(d+e-a-b-c-1)[F(a+)-F] - (d-b)(e-b)[F(b-)-F] - acF = 0$

(16) $ad(d+e-a-b-c-1)[F(a+)-F] + (d-a)(d-b)(d-c)[F(d+)-F] - abcF = 0$

(17) $a[F(a+)-F] - (d-1)[F(d-)-F] = 0$

(18) $(d-a)(e-a)[F(a-)-F] - (d-b)(e-b)[F(b-)-F] + (b-a)cF = 0$

(19) $d(e-a)[F(a-)-F] + (d-b)(d-c)[F(d+)-F] + bcF = 0$

(20) $(d-a)(e-a)[F(a-)-F] - (d-1)(d+e-a-b-c-1)[F(d-)-F] + bcF = 0$

(21) $(d-a)(d-b)(d-c)[F(d+)-F]+d(d-1)(d+e-a-b-c-1)[F(d-)-F]-abcF=0$

(22) $e(d-a)(d-b)(d-c)[F(d+)-F]-d(e-a)(e-b)(e-c)[F(e+)-F] + (d-e)abcF = 0$

(23) $(d-a)(d-b)(d-c)[F(d+)-F] + d(e-1)(d+e-a-b-c-1)[F(e-)-F] - abcF = 0$

(24) $(d-1)[F(d-)-F] - (e-1)[F(e-)-F] = 0$.

3 ${}_4F_3$ relations

To get three-term relations for ${}_4F_3$'s, we need, besides the condition $z = 1$, that the sum of the numerator parameters is one greater than the sum of the denominator parameters, and that one of the numerator parameters is an integer less than or equal to zero, so that the series terminates. These conditions again are natural ones, the same which allow the ${}_3F_2$ to be evaluated by the formula of Pfaff and Saalschutz ([2], p. 9) and are needed for the transformation ([2], p. 56)

$$(25)\quad {}_4F_3\left(\begin{matrix}-n,b,c,d\\ e,f,g\end{matrix};1\right) = \frac{(f-b)_n(g-b)_n}{(f)_n(g)_n}\,{}_4F_3\left(\begin{matrix}-n,b,e-c,e-d\\ e,e+f-c-d,e+g-c-d\end{matrix};1\right).$$

A ${}_{p+1}F_p$ satisfying these conditions is called balanced.

If F is a balanced series, then its contiguous series are not. We define a contiguous balanced series to F to be a balanced series obtained from F by altering two parameters, each by ± 1.

If the techniques applied above to ${}_3F_2$'s are applied to ${}_4F_3$'s, transformation (25) replacing (2), there results a set of relations connecting any balanced ${}_4F_3$ with any two of its contiguous balanced series. We find that they can all be generated from the single relation

$$fg\,{}_4F_3\left(\begin{matrix}a,b,c,d\\e,f,g\end{matrix};1\right) - (f-a)(g-a)\,{}_4F_3\left(\begin{matrix}a,b+1,c+1,d+1\\e+1,f+1,g+1\end{matrix};1\right)$$

$$+\frac{a(e-b)(e-c)(e-d)}{e(e+1)}\,{}_4F_3\left(\begin{matrix}a+1,b+1,c+1,d+1\\e+2,f+1,g+1\end{matrix};1\right) = 0$$

just as the ${}_3F_2$ relations were generated from (8) and (9).

If polynomials p_n (of degrees $n = 0,1,2,\ldots$) are defined by $p_n(t^2) = {}_4F_3\left(\begin{matrix}-n,n+a+b+c+d-1,a-it,a+it\\a+b,a+c,a+d\end{matrix};1\right)$, then one of the ${}_4F_3$ contiguous relations gives a recurrence relation for p_n:

$$A_n(p_{n+1}(x) - p_n(x)) + (a^2+x)p_n(x) + B_n(p_{n-1}(x) - p_n(x)) = 0$$

with $A_n = \dfrac{(n+a+b+c+d-1)(n+a+b)(n+a+c)(n+a+d)}{(2n+a+b+c+d-1)(2n+a+b+c+d)}$ and $B_n = \dfrac{n(n+c+d-1)(n+b+d-1)(n+b+c-1)}{(2n+a+b+c+d-2)(2n+a+b+c+d-1)}$. By a theorem of Favard [5], $\{p_n\}$ forms a set of orthogonal polynomials if $A_nB_{n+1} > 0$ $(n = 0,1,\ldots)$. Results on these polynomials will appear in the author's thesis and future papers. The orthogonality relations have been found and include Racah's orthogonality for the 6-j symbols of angular momentum in quantum mechanics. Limiting cases include the orthogonality relations for the classical orthogonal polynomials.

References

1 Abramowitz, M. and I. Stegun (editors), Handbook of Mathematical Functions, Applied Mathematics Series, vol. 55, National Bureau of Standards, 1964.

2 Bailey, W. N., Generalized Hypergeometric Series, Cambridge University Press, Cambridge, 1935.

3 Bailey, W. N., Associated hypergeometric series, Quart. J. Math. 8 (1937), 115-118.

4 Bailey, W. N., Contiguous hypergeometric functions of the type ${}_3F_2(1)$, Proc. Glasgow Math. Assoc. 2 (1954-1956), 62-65.

5 Favard, J., Sur les polynomes de Tchebicheff, C. R. Acad. Sci. Paris 200 (1935), 2052-2953.

6 Frank, E., A new class of continued fractions for the ratios of hypergeometric functions, Trans. Amer. Math. Soc. 81 (1956), 453-476.

7 Gauss, C. F., Disquisitiones generales circa seriem infinatam $1 + \frac{\alpha\beta}{1 \cdot \gamma} x + \frac{\alpha(\alpha + 1)\beta(\beta + 1)}{1 \cdot 2 \cdot \gamma(\gamma + 1)} xx + \cdots$, Commentationes societatis regiae scientarum Goettingensis recentiores 2 (1813); Werke 3, 123-162.

8 Rainville, E. D., The contiguous function relations for ${}_pF_q$ with applications to Bateman's $J_n^{u,v}$ and Rice's $H_n(\zeta,p,v)$, Bull. Amer. Math. Soc. 51 (1945), 714-723.

9 Wall, H. S., Analytic Theory of Continued Fractions, D. Van Nostrand Company, Inc., New York, 1948, chapter 13.

J. A. Wilson
Department of Mathematics
University of Wisconsin
Madison, Wisconsin 53706

THEORY OF RATIONAL APPROXIMATIONS

ON A PROBLEM OF SAFF AND VARGA CONCERNING BEST RATIONAL APPROXIMATION

Colin Bennett, Karl Rudnick and Jeffrey D. Vaaler

We consider uniform approximation of even, real-valued functions f on [-1,1] by complex-valued rational functions $U = p/q$ with p and q linear, that is, by complex linear fractional transformations $U(x) = (ax+b)/(cx+d)$. We conjecture that there are always best approximants which are symmetric ($U(-x)=\overline{U(x)}$), and for a large class of functions f we determine the degree of symmetric approximation $E_s(f)$. This class includes the functions $f_\alpha(x) = |x|^\alpha$, whenever $\alpha \geq \kappa$, where κ (= 1.4397589...) is an explicitly determined transcendental number. Curiously, the geometry of the extremal transformations is quite different when $0<\alpha<\kappa$, and we obtain only a lower bound for $E_s(f_\alpha)$ in these cases.

1 Introduction

Throughout the paper f will denote an even, real-valued function on the interval [-1,1] satisfying

$$(1.1) \quad 0 = f(0) \leqslant f(x) \leqslant f(1) = 1, \qquad -1 \leqslant x \leqslant 1.$$

Recently, Saff and Varga [2] have considered the uniform approximation of such functions by complex-valued rational functions. Their results show that the complex approximation is often much better than can be attained by using real approximants, and they have posed several interesting problems in this connection. In this paper we shall consider only rational functions of class $R_{1,1}$, that is, linear fractional transformations $U(x) = (ax+b)/(cx+d)$. In the general case when a,b,c and d are complex we shall say that U is complex, when $U(-x) = \overline{U(x)}$ for all x we shall call U symmetric, and when U takes on only (extended) real values we shall say that U is real. For a given function f of the form (1.1), the corresponding degrees of approximation are given by

$$E_R(f) = \inf \left\{ ||U-f||_\infty : U \text{ is real} \right\},$$

$$E_S(f) = \inf \left\{ ||U-f||_\infty : U \text{ is symmetric} \right\},$$

$$E_C(f) = \inf \left\{ ||U-f||_\infty : U \text{ is complex} \right\},$$

where, as usual, $||.||_\infty$ denotes the uniform (Chebyshev) norm over [-1,1]. The following estimates are fairly easy to establish (cf. [1]):

(1.2) $\frac{1}{4} \leq E_C(f) \leq E_S(f) \leq E_R(f) = \frac{1}{2}$.

We conjecture that for any function f the degrees of approximation $E_C(f)$ and $E_S(f)$ coincide, that is, one can approximate as well by symmetric transformations as by arbitrary complex transformations. The remainder of the paper is devoted to determining the <u>degree of symmetric approximation</u> $E_S(f)$.

2 Approximation of $f(x) = x^2$

The results presented here on the approximation of $f(x) = f_2(x) \equiv x^2$ will help motivate the more general results in the succeeding sections. Indeed, we shall see that the geometry of the solution in this special case is characteristic of the solutions for more general f. The degree of real approximation $E_R(f_2)$ is equal to 1/2 by (1.2), but Saff and Varga [2] have obtained the upper bound

$$E_S(f_2) \leq \sqrt{2} - 1 = 0.414...$$

in the symmetric case. We show here that in fact

(2.1) $E_S(f_2) = \sqrt{\frac{4}{27}} \doteq 0.384...$.

An elementary geometric argument (cf. [1]) shows that it suffices to consider symmetric linear fractional transformations U of the form

(2.2) $U(x) = s + r\,\frac{x+it}{x-it}, \qquad -1 \leq x \leq 1,$

where r,s and t are real, and r and t are nonzero. The range of such a transformation is an arc of a circle, with center s and radius r, situated symmetrically with respect to the real axis. The parameter t is determined by the angle the arc subtends at the center of the circle. Note that the transformation obtained from (2.2) by replacing t by -t is simply the complex conjugate $\overline{U}$ of U. Clearly, U and $\overline{U}$ trace out the same arc of the circle but in opposite senses. Hence, since f is real, the transformations U and $\overline{U}$ provide exactly the same degree of approximation to f (i.e., $||U-f||_\infty = ||\overline{U}-f||_\infty$). However, we shall see that the symmetric transformations of best approximation are unique modulo complex conjugation.

From (2.2) we have

$$(2.3)\quad |U(x) - f(x)|^2 = \frac{x^2}{x^2+t^2}\{x^2-r-s\}^2 + \frac{t^2}{x^2+t^2}\{x^2+r-s\}^2 , \quad -1 \leqslant x \leqslant 1 ,$$

and we wish to minimize the supremum on [-1,1] of this quantity over all admissible values of the parameters r,s and t. A computer search led us to consider the values

$$(2.4)\quad r = \frac{2}{9} , \quad s = \frac{5}{9} , \quad t^2 = \frac{1}{3} .$$

If we denote the corresponding transformation in (2.2) by U_2, we find from (2.3) that

$$(2.5)\quad |U_2(x) - x^2|^2 - \frac{4}{27} = \frac{(x^2-1)(x^2 - 1/9)^2}{(x^2 + 1/3)} \leqslant 0 , \quad -1 \leqslant x \leqslant 1,$$

and equality occurs if and only if $x = \pm 1$ or $x = \pm 1/3$. This shows immediately that $E_S(f_2) \leqslant (4/27)^{1/2}$. Hence, to establish (2.1) we have only to show that U_2 provides the <u>best</u> uniform approximation to x^2 on [-1,1]. In fact, we shall show that even on the four-point set {-1, -1/3, 1/3, 1} no other symmetric transformation can attain the degree of approximation $(4/27)^{1/2}$ attained by U_2.

To see this, let U be any transformation of the form (2.2) with associated parameters r,s and t, where r and t are nonzero.

We express these parameters in terms of perturbations ε, η and ξ of the values in (2.4) as follows:

(2.6) $r = \frac{2}{9} + \varepsilon \quad , \quad s = \frac{5}{9} + \eta \quad , \quad t^2 = \frac{1}{3} + \xi \quad .$

Note that since $t \neq 0$ we have

(2.7) $\frac{1}{3} + \xi > 0 \; .$

Now suppose that <u>over the four-point set</u> $\{-1, -1/3, 1/3, 1\}$ the degree of approximation by U does not exceed $(4/27)^{1/2}$. Thus, we suppose that

(2.8) $|U(1/3) - (1/3)^2|^2 \leq 4/27, \quad |U(1) - (1)^2|^2 \leq 4/27.$

Using (2.3) and (2.6) we can rewrite these inequalities in terms of the perturbations ε, η and ξ in the following way:

$$\frac{1}{9}\left\{\frac{2}{3} + \varepsilon + \eta\right\}^2 + (\frac{1}{3} + \xi)\left\{\frac{2}{9} - \varepsilon + \eta\right\}^2 \leq \frac{4}{27}\left\{\frac{1}{9} + (\frac{1}{3} + \xi)\right\} ,$$

$$\left\{\frac{2}{9} - \varepsilon - \eta\right\}^2 + (\frac{1}{3} + \xi)\left\{\frac{2}{3} + \varepsilon - \eta\right\}^2 \leq \frac{4}{27}\left\{1 + (\frac{1}{3} + \xi)\right\} .$$

When the second inequality is multiplied throughout by 1/3 and added to the first, the resulting inequality is

(2.9) $(\frac{2}{3} + \xi)\,\varepsilon^2 - 2\xi\,\varepsilon\,\eta + (\frac{2}{3} + \xi)\,\eta^2 \leq 0.$

The left-hand side of (2.9) is a quadratic form in ε and η whose discriminant $16(1/3 + \xi)/3$ is strictly positive by virtue of (2.7). The form is therefore positive definite and so the only solution to (2.9) is $\varepsilon = \eta = 0$. Hence $r = 2/9$ and $s = 5/9$, and with these values for r and s the original inequalities (2.8) yield $t^2 \geq 1/3$ and $t^2 \leq 1/3$, so $t^2 = 1/3$. Consequently, the only symmetric transformations satisfying (2.8) are U_2 and its conjugate $\bar{U}_2$. This establishes (2.1) as well as the uniqueness, modulo complex conjugation, of the extremal transformation.

3 Geometrical aspects of the solution

In the following diagram the circle is centered at C = 5/9 and has radius 2/9. The points B and $\bar{B}$ are the points of tangency of the tangents through A = 1, and the points B' and $\bar{B}'$ are the points of tangency of the tangents through A' = 1/9. Hence, the diagram is symmetric with respect to the vertical axis through C, and also with respect to the real axis.

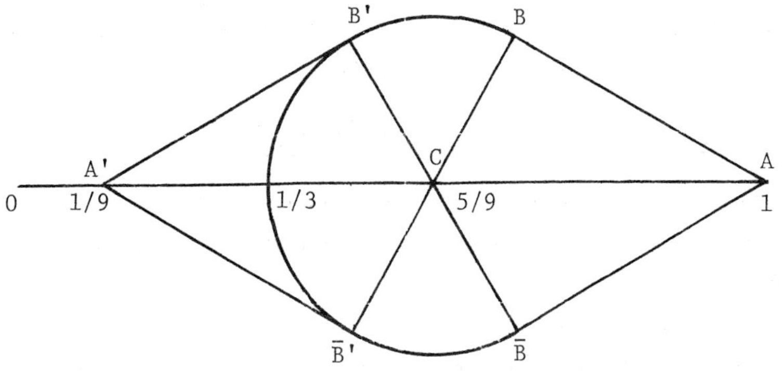

It is not difficult to show that as x varies from 1 to -1 through the interval [-1,1], the extremal transformation $U = U_2$ traces out the arc $BB'\bar{B}$; at the same time, the approximated function $f(x) = f_2(x) = x^2$ traverses the real axis from 1 to 0 and back to 1. In particular, we have B = U(1) and $\bar{B} = U(-1)$. Also, from (2.5) we know that $|U(x) - f(x)|$ attains its norm at x = 1, that is, the length of AB is $(4/27)^{1/2}$. But (2.5) shows that $|U(x) - f(x)|$ also attains its norm at x = 1/3. Since in this case we have f(1/3) = 1/9 = A', it follows that B' is the point U(1/3). Thus, if we write $\omega = 1/3$, then we can express some of the geometrical features of the diagram in terms of U, f and ω as follows:

(I) The line joining the points $U(\omega)$ and $f(\omega)$ is the reflection, in the vertical line through the center of the circle determined by U, of the line joining the points $U(1)$ and $f(1)$.

(II) For each point x in the four-point set $\{-1, -\omega, \omega, 1\}$ the line joining the points $U(x)$ and $f(x)$ is tangent to the circle determined by U.

(III) The function $|U(x) - f(x)|$ attains its norm at each point x of the four-point set $\{-1, -\omega, \omega, 1\}$.

We shall see in the next two sections that properties (I), (II) and (III) characterize the extremal symmetric transformations U for a large class of functions f.

4 A lower bound for $E_S(f)$

Consider an arbitrary function f of the form (1.1). For each ω satisfying $0 < \omega < 1$ we first approximate f on the four-point set

$$Z(\omega) = \{-1, -\omega, \omega, 1\}.$$

For this we select a transformation $U(x;f,\omega)$ which is defined as follows. In the trivial case where $f(\omega) = 1$, we take $U(x;f,\omega) \equiv 1$. If $f(\omega) < 1$, then $U(x;f,\omega)$ is taken to be the unique (modulo complex conjugation) symmetric transformation U determined by the conditions (I), (II) and (III) of Section 3. In this case, $U(x;f,\omega)$ is of the form (2.2) and the associated parameters r, s and t are given by

$$(4.1) \quad \begin{cases} r = r(f,\omega) = \dfrac{(1-\omega)(1-f(\omega))}{2(1+\omega)}, \\ s = s(f,\omega) = \dfrac{1+f(\omega)}{2}, \\ t = t(f,\omega) = \omega^{1/2}; \end{cases}$$

note that r and t are nonzero as required. If we set

$$(4.2) \quad \delta(f,\omega) = \frac{\omega^{1/2}(1 - f(\omega))}{1+\omega},$$

then a simple computation shows that

$$(4.3)\quad |U(x;f,\omega) - f(x)| = \delta(f,\omega), \qquad x \in Z(\omega).$$

The argument involving the quadratic form given in Section 2 can be applied again in this case to show that no other symmetric transformation can approximate f this well on the four-point set $Z(\omega)$ (cf. [1] for details). Hence we have the following result.

THEOREM A. Let f be given, let $\omega \in [0,1]$ be fixed and suppose that U is any symmetric transformation. Then

$$(4.4)\quad \max_{x\in Z(\omega)} |U(x) - f(x)| \geq \delta(f,\omega).$$

If $\omega \in (0,1)$, then equality holds in (4.4) if and only if $U(x) = U(x;f,\omega)$ or $\bar{U}(x) = U(x;f,\omega)$.

If now we define

$$(4.5)\quad \Delta(f) = \sup_{0\leq\omega\leq 1} \delta(f,\omega),$$

then from Theorem A (and (1.2)) we immediately obtain the following lower bound for $E_S(f)$.

THEOREM B. Let f be given and let U be any symmetric transformation. Then

$$(4.6)\quad ||U-f||_\infty \geq \max(1/4, \Delta(f)),$$

and hence

$$E_S(f) \geq \max(1/4, \Delta(f)).$$

In the special case $f(x) = x^2$, the supremum in (4.5) is attained at the (unique) point $\omega = 1/3$, and by (4.2) has value $(4/27)^{1/2}$. This is precisely the degree of approximation $E_S(f_2)$. In the next section we shall see that the lower bound established in Theorem B is exact for a large class of functions f.

5 Exactness of the lower bound

We shall need to impose some restrictions on f.

CONDITION I. *The function* f *is continuous on* $[-1,1]$, *is differentiable on* $(0,1)$, *and* $f'(x) \geq 0$ *for each* $x \in (0,1)$.

It is shown in [1] that the lower bound $\Delta(f)$ can be exact only if the supremum in (4.5) is attained at a *unique* point $\omega = \Omega$ in $(0,1)$. Hence we can restrict our attention to those functions f for which the following condition holds.

CONDITION 2. *There exists a unique point* $\Omega \in (0,1)$, *depending only on* f, *for which* $\Delta(f) = \delta(f,\Omega)$.

This condition is satisfied whenever f is continuous and convex, or when $f(x) = |x|^\alpha$ for any $\alpha > 0$ (cf. [1]). Our main result is as follows.

THEOREM C. *Let* f *satisfy Conditions* 1 *and* 2, *and let* U *be any symmetric transformation. Suppose in addition that the function*

$$(5.1) \quad x \to \frac{(x^2+\Omega)^2 \; f'(x)}{x}$$

is increasing on $(0,1)$. *Then* $1/4 < \Delta(f)$ *and*

$$(5.2) \quad ||U-f||_\infty \geq \Delta(f),$$

with equality in (5.2) *if and only if* $U(x) = U(x;f,\Omega)$ *or* $\bar{U}(x) = U(x;f,\Omega)$. *In particular, we have* $E_S(f) = \Delta(f)$.

In view of the preceding results the only part of Theorem C requiring proof is the assertion that equality holds in (5.2) if $U(x) = U(x;f,\Omega)$. For this it suffices to establish the global inequality

$$(5.3) \quad |U(x;f,\Omega) - f(x)|^2 \leq \Delta(f)^2, \qquad -1 \leq x \leq 1,$$

because, by Theorem A and (4.5), equality is known to occur in (5.3) on the four-point set $Z(\Omega)$. In the special case where $f(x) = x^2$ the inequality is easily proved due to the particularly simple factorization of the left-hand side obtained in (2.5). In the general setting, however, no such factorization is possible and (5.3) is established only after a fairly intricate analysis of the derivative of the left-hand side; complete details are given in [1].

6 Approximation of $f(x) = |x|^\alpha$

For the functions $f(x) = f_\alpha(x) = |x|^\alpha$, $\alpha > 0$, it is easily verified that Conditions 1 and 2 of Section 5 hold for all $\alpha > 0$; we denote the corresponding values of Ω by Ω_α. It is shown in [1] that Ω_α is the unique value of Ω in (0,1) satisfying the equation

$$1 - \Omega - (2\alpha+1)\Omega^\alpha - (2\alpha-1)\Omega^{\alpha+1} = 0.$$

We denote the corresponding transformation in (4.1) (with $f = f_\alpha$ and $\omega = \Omega_\alpha$) by U_α.

When $\alpha \geq 2$, the function f_α satisfies the additional hypothesis (5.1) in Theorem C. Hence, in this case, the degree of approximation (= $\Delta(f_\alpha)$) and the precise extremal transformations (U_α and $\bar{U}_\alpha$) are completely determined.

If $0 < \alpha < 2$, then the hypothesis (5.1) of Theorem C fails to hold for f_α. Nevertheless, it is shown in [1] that the conclusion of Theorem C persists for all $\alpha \geq \kappa$ (=1.4397589...), where κ is the unique solution in $(1,\infty)$ of the equation

$$(2\kappa-1)^{2\kappa-1} = \frac{\kappa}{\kappa-1}$$

(κ is transcendental, cf. [1]). We have the following result.

THEOREM D. <u>Let</u> $f_\alpha(x) = |x|^\alpha$, <u>where</u> $\alpha > 0$, <u>and let</u> U <u>be any symmetric transformation</u>.

(i) <u>If</u> $\kappa \leq \alpha$, <u>then</u> $1/4 < \Delta(f_\alpha)$ <u>and</u>

$$||U-f||_\infty \geq \Delta(f_\alpha),$$

<u>with equality if and only if</u> $U(x) = U(x;f_\alpha,\Omega_\alpha)$ <u>or</u> $\bar{U}(x) = U(x;f_\alpha,\Omega_\alpha)$. <u>In particular</u>, $E_S(f_\alpha) = \Delta(f_\alpha)$.

(ii) <u>If</u> $0 < \alpha < \kappa$, <u>then</u>

$$||U-f||_\infty > \max(1/4, \Delta(f_\alpha)).$$

Notice that the transformations U_α are defined, and have in common the geometrical structure described in Section 3, for <u>all</u> $\alpha > 0$. Furthermore, equality occurs in the inequality

(6.1) $|U_\alpha(x) - |x|^\alpha| \leqslant \Delta(f_\alpha)$

for the four values of x in $Z(\Omega_\alpha) = \{-1, -\Omega_\alpha, \Omega_\alpha, 1\}$, again for all $\alpha > 0$. Now when $\alpha \geqslant \kappa$, the inequality (6.1) holds for all x in $[-1,1]$. Furthermore, if $\alpha > \kappa$, equality occurs if and only if $x \in Z(\Omega_\alpha)$. In the extreme case $\alpha = \kappa$ equality occurs on $Z(\Omega_\alpha)$ but also at the point $x = 0$. However, when $0 < \alpha < \kappa$, the inequality fails at $x = 0$ and so U_α is no longer the best approximant. Thus, it may be the case that for $0 < \alpha < \kappa$ one has to consider "five-point approximations" on sets of the form $\{-1, -\omega, 0, \omega, 1\}$. We conclude with a table of some numerical values for Ω_α and $\Delta(f_\alpha)$.

α	Ω_α	$\Delta(f_\alpha)$
1	0.236	0.300
κ	0.283	0.347
2	0.333	0.385
3	0.404	0.423
4	0.459	0.444
5	0.503	0.457
10	0.637	0.482
100	0.919	0.499

References

1. Bennett, C., Rudnick, K. and Vaaler, J.D., Best uniform approximation by linear fractional transformations (to appear).

2. Saff, E.B. and Varga, R.S., Nonuniqueness of best complex rational approximations to real functions on real intervals, (to appear in J. Approximation Theory).

Colin Bennett
Department of Mathematics
McMaster University
Hamilton, Ontario L8S 4K1
Canada

Karl Rudnick
Department of Mathematics
Texas A & M University
College Station, Texas 77843
U.S.A.

Jeffrey D. Vaaler
Department of Mathematics
University of Texas at Austin
Austin, Texas 78712
U.S.A.

A MINIMIZATION PROBLEM RELATED TO PADÉ SYNTHESIS OF RECURSIVE DIGITAL FILTERS

C. K. Chui, P. W. Smith and L. Y. Su

1 Introduction

Digital signal processing, a field which has its roots in the seventeenth and eighteenth century Mathematics, has become an important modern tool in many fields of Science and Technology. It is concerned with the representation of signals by sequences (of numbers or symbols) and the processing of these sequences. With the rapid advances in digital integrated circuit technology, there is an increase in need of digital filter synthesis for signal processing to suit certain purposes, such as removing interference, saving computer time, modifying the signal to a more easily interpreted form, etc.

Let $H(\omega)$ be a function in $L_2[-\pi,\pi]$. In practice, $H(\omega)$ is usually a filter amplitude characteristic to be synthesized. Consider the Fourier series

$$(1.1) \quad \sum_{n=-\infty}^{\infty} h_n e^{-in\omega}$$

of $H(\omega)$. The Fourier coefficients h_n are usually called the samples of the impulse response of $H(\omega)$. The (formal) Laurent series

$$(1.2) \quad \sum_{n=-\infty}^{\infty} h_n z^{-n}$$

is called the (two-sided) z-transform of the samples h_n. By approximating (1.1) by a partial sum and adding a delay constant to $H(\omega)$, one can also consider a one-sided z-transform

$$(1.3) \quad \tilde{H}(z^{-1}) = \sum_{n=0}^{\infty} \tilde{h}_n z^{-n}$$

of the sample sequence $\{\tilde{h}_n\}$.

Let R_N^M denote the class of all rational functions with degrees no greater than M and N in the numerator and denominator respectively. A realizable digital filter transfer function is a function $G(z^{-1})$ in R_N^M, with variable z^{-1}, of the form

$$G(z^{-1}) = \frac{\alpha_0 + \alpha_1 z^{-1} + \ldots + \alpha_M z^{-M}}{1 - \beta_1 z^{-1} - \ldots - \beta_N z^{-N}}$$

which "approximates" (1.2) or (1.3). If N > 1, it defines a recursive digital filter. If, for example, we wish to match the coefficients $g_0, g_1, \ldots$ of the Maclaurin expansion of $G(z^{-1})$ with the response samples $\tilde{h}_0, \tilde{h}_1, \ldots$ respectively, as far as possible, then $G(z^{-1})$ is the [M/N]-Padé approximant of $\tilde{H}(z^{-1})$. Hence, in this case, $G(z^{-1})$ satisfies the minimization problem:

$$\|\tilde{H}(z^{-1}) - G(z^{-1})\| = \inf\{\|\tilde{H}(z^{-1}) - R(z^{-1})\| : R(z) \in R_N^M\}$$

where $\|\cdot\|$ is the semi-norm, $\|\sum_{n=0}^{\infty} a_n z^{-1}\| = (\sum_{n=0}^{T} w_n |a_n|^2)^{1/2}$, $w_n \geq 0$, for some suitable T, say any $T \leq M + N$ if the power series is normal. For an application of the Padé approximant technique to the synthesis of recursive digital filters, see [3].

The above idea leads to the following minimization problem: Let M be a class of functions in $L_2(T)$, where here and throughout this paper, T denotes the unit circle in the complex plane $\mathbb{C}$. Let $H(z^{-1}) = \sum_{n=-\infty}^{\infty} a_n z^{-n}$ be a function in $L_2(T)$. If we are interested only in the one-sided z-transform, we just set $a_n = 0$ for $n < 0$. The problem is to study the existence, (local) uniqueness, characterization, etc. of an $f^* \in M$ satisfying

(1.4) $\|H-f^*\| = \inf\{\|H - f\| : f \in M\}$,

where $\|\cdot\|$ is some norm (or semi-norm). In this paper, we will only concentrate on the L_2 norm on the unit circle, that is,

$$(1.5)\quad \|f\|_2 = \left(\frac{1}{2\pi}\int_{-\pi}^{\pi}|f(e^{-i\omega})|^2 d\omega\right)^{1/2} .$$

The next section will be devoted to the case where M consists of rational functions with restricted poles. In section 3, we will discuss the case where M is a class of Stieltjes series, and in section 4 some general results on approximation from R_N^M will be given. Some examples will be included in the final section of the paper.

2 Approximation by rational functions with preassigned poles

In this section we will study the problem of best approximation in $L_2(T)$ by rational functions with preassigned poles. The choice of these poles depends on the nature of the filters desired. See for example [6]. Let $\alpha_1, \alpha_2, \ldots$ and $\beta_1, \beta_2, \ldots$ be two sequences of complex numbers that lie in $|z| < 1$ and $|z| > 1$ respectively. These will be the preassigned poles of the approximants. If, for instance, $\alpha_1 = \alpha_{j_1} = \alpha_{j_2} = \ldots$, $1 < j_1 < j_2 < \ldots$, we will treat α_{j_1} as a double pole, α_{j_2} as a pole of order 3, etc. Let R(M,N) be the collection of all rational functions in R_{M+N}^{M+N} with poles $\alpha_1, \ldots, \alpha_M$ and $\beta_1, \ldots, \beta_N$. Hence, R(M,N) is the linear space spanned by

$$1, \frac{1}{\alpha_1 - z}, \ldots, \frac{1}{\alpha_M - z}, \frac{1}{\beta_1 - z}, \ldots, \frac{1}{\beta_N - z},$$

where in case of multiple poles, the corresponding terms have to be replaced by derivatives. More precisely, if, for example, $\alpha_1 = \alpha_{j_1} = \ldots = \alpha_{j_k}$, $1 < j_1 < \ldots < j_k$, then the terms $1/(\alpha_1 - z)$, $1/(\alpha_{j_1} - z), \ldots, 1/(\alpha_{j_k} - z)$ should be replaced by

$$\frac{1}{\alpha_1 - z}, \frac{1}{(\alpha_1 - z)^2}, \ldots, \frac{k!}{(\alpha_1 - z)^{k+1}}$$

respectively. We have the following

THEOREM 2.1. Let f be in $L_2(T)$. For any positive integers m and n, there exists a unique rational function $r^*_{m,n} \in R(m,n)$ such that

$$\|f - r^*_{m,n}\|_2 = \inf\{\|f - r\|_2 : r \in R(m,n)\} .$$

Moreover, if $\{\alpha_m\}$ has at least one limit point in $|z| < 1$ and $\{\beta_n\}$ has at least one limit point in $|z| > 1$, then $\|f - r^*_{m,n}\|_2 \to 0$ as $m,n \to \infty$.

In fact we can say more about the best approximants $r^*_{m,n}$. Let $f \sim \sum_{k=-\infty}^{\infty} c_k e^{ik\theta}$ be the Fourier expansion of f. We set

$$f^+ \sim \sum_{k=0}^{\infty} c_k e^{ik\theta} \text{ and } f^- \sim \sum_{k=-\infty}^{-1} c_k e^{ik\theta} .$$

Then by the M. Riesz theorem (cf. [5 p. 151]), we have $f = f^+ + f^-$ where $f^+(z) = \sum_{k=0}^{\infty} c_k z^k$ and $f^-(z) = \sum_{k=-\infty}^{-1} c_k z^k = \sum_{k=1}^{\infty} c_{-k} z^{-k}$, so that both $f^+(z)$ and $f_1^-(z) \equiv f^-(z^{-1})$ are in the Hardy space H^2. Also, if $r_{m,n}$ is any rational function in $R(m,n)$, we can write $r_{m,n} = r_m + r_n$ where $r_m \in R_m^{m-1}$ has poles at $\alpha_1,\ldots,\alpha_m$ and $r_n \in R_n^n$ has poles at $\beta_1,\ldots,\beta_n$. Here, of course, we have to consider multiple poles as before. By Parseval's equality, we have

$$(2.1)\quad \|f - r_{m,n}\|_2^2 = \|f^+(z) - r_n(z)\|_2^2 + \|f^-(z^{-1}) - r_m(z^{-1})\|_2^2 .$$

Hence, $r_{m,n}$ best approximates f if and only if both r_n best approximates f^+ and r_m best approximates f^-. By a result of Malmquist-Takenaka-Walsh (cf. [7 ,pp. 224-280]), we see that r_n is the unique rational function that interpolates f^+ at 0, $1/\bar{\beta}_1,\ldots, 1/\bar{\beta}_n$ and $\hat{r}_m(z) \equiv r_m(z^{-1})$ the unique rational function that interpolates f_1^- at $\bar{\alpha}_1,\ldots,\bar{\alpha}_m$. Here, of course, if we have multiple poles, we have to consider appropriate derivatives as usual. For instance, if $\alpha_1 = \alpha_{j_1} = \ldots = \alpha_{j_k}$, then $(\hat{r}_m - f_1^-)^{(j)}(\alpha_1) = 0$ for $j = 0,\ldots,k$. To prove convergence, we note that the linear spaces $R(M,N)$ satisfy $R(M,N) \subset R(M,N+1)$ and $R(M,N) \subset R(M+1,N)$, $M,N = 1,2,\ldots$,Hence, by virtue of the Hahn-Banach theorem, it is sufficient to prove that if $f \in L_2(T)$ is orthogonal to each $R(M,N)$, $M,N=1,2,$

..., then $f = 0$ a.e.. To do this, we again write $f = f^+ + f^-$ as above where $f^+(z)$ and $f^-(z^{-1})$ are in the Hardy space H^2. That f is orthogonal to R(M,N) gives by the Cauchy integral formula that $f^+(1/\overline{\beta}_j) = 0$, $j=1,\dots,N$, and $f_1^-(\overline{\alpha}_j) = 0$, $j = 1,\dots,M$, where again appropriate derivatives have to be taken when we have coincident points. For example if $\alpha_1 = \alpha_{j_1} = \dots = \alpha_{j_k}$, $1 < j_1 < \dots < j_k$, then $f_1^-(\overline{\alpha}_1) = f_1^-(\overline{\alpha}_{j_1}) = \dots = f_1^-(\alpha_{j_k}) = 0$ should be replaced by $(f_1^-)^{(j)}(\overline{\alpha}_1)=0$, $j=0,\dots,k$. Since $\{\alpha_j\}$ has at least one limit point in $|z| < 1$ and $\{\beta_j\}$ has at least one limit point in $|z| > 1$, we have $f^- = 0$ and $f^+ = 0$. That is, $f = 0$ a.e.. This completes the proof of the theorem.

In general, instead of the two sequences $\{\alpha_j\}$ and $\{\beta_j\}$, one may also consider triangular sequences $\{\alpha_{n,j}\}$ and $\{\beta_{n,j}\}$, $j=1,..n$, $n = 1,2...$. In this case, however, we do not necessarily have the property that $R(M,N) \subset R(M,N+1)$ and $R(M,N) \subset R(M+1,N)$. A result in this direction will be given in section 5.

3 Approximation by series of Stieltjes

Since the only simple class of formal power series whose diagonal Padé approximants are known to converge nicely is the class of series of Stieltjes (cf. [4]), it is natural to study the possibility of best approximation by such series. Let M be the cone of all functions $g \in L_2(T)$ such that

$$g(z^{-1}) = \int_{-1}^{1} \frac{d\mu(t)}{1-z^{-1}t}$$

where μ is a nonnegative measure on $[-1,1]$ and $|z^{-1}| = 1$. For simplicity, we only consider the approximation of a one-sided z-transform. If a two-sided z-transform is desired, one might want to include those measures μ with larger support and use the idea in section 2 (cf. (2.1)) to uncouple the best approximation problem.

THEOREM 3.1. *Let* $f(z^{-1}) = \sum_{n=0}^{\infty} h_n z^{-n}$ *be in* $L_2(T)$. *There exists a unique* g^* ε M *such that* $\|f - g^*\|_2 = \inf\{\|f - g\|_2 : g \in M\}$.

Since M is a convex set, it is sufficient to prove that M is closed. Let $g_n \in M$ and suppose that $\|g_n - f\|_2 \to 0$ where $f \in L_2(T)$. We set

$$g_n(z^{-1}) = \int_{-1}^{1} \frac{d\mu_n(t)}{1-z^{-1}t}$$

where each μ_n is a nonnegative measure on [-1,1]. Denoting by $|\mu_n|$ the total variation of μ_n, we have, for each $z^{-1} \in T$,

$$|\mu_n| \min\left\{|\alpha| : \alpha \in \mathrm{co}\left\{\frac{1}{1-z^{-1}t}\right\},\ -1 \le t \le 1\right\} \le |g_n(z^{-1})| ,$$

where, as usual, co means convex hull. It is easy to see that $\mathrm{co}\left\{\frac{1}{1-z^{-1}t}\right\}$, $-1 \le t \le 1$, does not contain 0 when z^{-1} is in some neighborhood of i. Since $|g_n| < \infty$ a.e. on T, we can conclude that $\{|\mu_n|\}$ is a bounded sequence. Hence, $\{\mu_n\}$ has a subsequence $\{\mu_{n_k}\}$ which converges to some non-negative measure μ in the weak* topology. In particular, $\{g_{n_k}(z^{-1})\}$ converges to

$$g(z^{-1}) \equiv \int_{-1}^{1} \frac{d\mu(t)}{1-z^{-1}t}$$

a.e. on T. Since $\|g_{n_k} - f\|_2 \to 0$, we have g = f a.e., so that M is closed. This completes the proof of the theorem.

We conclude this section by referring the reader to [1] for another type of approximation by Stieltjes series.

4 Approximation by rational functions with free poles

In this section we consider the problem of best approximation in $L_2(T)$ from R_N^M where the coefficients in both the numerator and denominator are parameters. This is a very nonlinear problem and requires careful analysis. Let $r \in R_N^M$ be in $L_2(T)$ and write

$$r(z) = \frac{a_0 + a_1 z + \ldots + a_M z^M}{1 + b_1(z-1)+\ldots+b_N(z-1)^N} .$$

Consider the mapping $F:\mathbb{C}^{M+N+1} \to R_N^M$ defined by $(a_0,\ldots,a_M,b_1,\ldots,b_n) \mapsto r(z)$. We want to study the set of (local) minima of the functional $\rho(\underline{\alpha}) = \|f - F(\underline{\alpha})\|_2^2$ where $f \in L_2(T)$ and $\underline{\alpha} \in \mathbb{C}^{M+N+1}$. Computing the first and second Fréchet derivatives ρ' and ρ'' of ρ, we obtain

(4.1) $\rho'(\underline{\alpha})(\underline{\beta}) = -2 \operatorname{Re}(F'(\underline{\alpha})(\underline{\beta}), f-F(\alpha))$ and

(4.2) $\rho''(\underline{\alpha})(\underline{\beta},\underline{\gamma}) = 2 \operatorname{Re}(F'(\underline{\alpha})(\underline{\beta}), F'(\underline{\alpha})(\underline{\gamma})) + 2 \operatorname{Re}(F''(\underline{\alpha})(\underline{\beta},\underline{\gamma}), f-F(\underline{\alpha}))$.

If ρ attains a local minimum at $\underline{\alpha}^*$, then $\rho'(\underline{\alpha}^*) = 0$. If $\rho'(\underline{\alpha}^*)=\underline{0}$ and $\rho''(\underline{\alpha}^*)$ is positive definite, then $\underline{\alpha}^*$ is a local minimum for ρ. The following result can be obtained using the techniques in [2.8] and formulae (4.1) and (4.2).

THEOREM 4.1. Let K be a positive integer and M, N be given with $N \geq 1$. Then there exists an $f \in L_2(T)$ so that f has at least K local best approximants from R_N^M.

This theorem implies that it might be extremely difficult to compute the best rational approximant to a given f in $L_2(T)$ since most algorithms only compute local minima. As in [2], we can obtain sharper results if we consider the manifold R_1^ℓ with one real free pole. Set $g_j(z) = z^j$, $j=0,\ldots,\ell-1$, $g_\ell(z) = 1/(z-x)$, and $g_{\ell+1}(z) = 1/(z-x)^2$. Then for any $f \in L_2(T)$, the function

$$\psi(x) = \begin{vmatrix} (g_0,g_0)\ldots\ldots(g_0,g_\ell) & (g_0,f) \\ \ldots\ldots\ldots\ldots & \ldots \\ (g_{\ell+1},g_0)\ldots(g_{\ell+1},g_\ell) & (g_{\ell+1},f) \end{vmatrix}$$

is analytic for $x \in (-1,1)$ and $|x| > 1$. This function must vanish by (4.1) if f is to have a local best approximant with a pole at

x. Since $\psi(x)$ either has isolated zeros or is identically zero in the region of analyticity, we have the following

THEOREM 4.2. Every $f \in L_2(T)$ can have only isolated local best approximants when the approximation is taken in $L_2(T)$ from R_1^{ℓ} with one free real pole.

In proving this theorem, we consider $f = f^+ + f^-$ as in section 2. For $|x| < 1$, it can be shown that if $\psi(x) \equiv 0$, then f^- is a constant multiple of the function $h(z) \equiv \sum_{n=1}^{\infty} (2n)!z^{-2n+1}/(n!2^n)^2$, but this means that $f^- \equiv 0$ since $h \notin L_2(T)$. The argument is similar and somewhat easier if $|x| > 1$.

5 Examples

As an example of best L_2 rational approximation, we consider the case when the restricted poles are equally spaced on a circle. Let $0 \le \rho_n < 1$ and $\alpha_{n,k} = \rho_n\omega_n^k$, $\beta_{n,k} = \rho_n^{-1}\omega_n^k$, $k=1,\ldots,n$, where $\omega_n = e^{i2\pi/n}$. As in section 2, let $R(n,n)$ be the collection of all rational functions in R_{2n}^{2n} with poles at $\alpha_{n,k}$ and $\beta_{n,k}$, $k=1,..,n$. Hence if $\rho_n = 0$, then $R(n,n)$ reduces to the linear span of $\{z^{-n},\ldots,z^{-1},1,z,\ldots,z^n\}$; that is, for $z \in T$, it is the space T_n of all trigonometric polynomials with degree at most n. Let $f \sim \sum_{-\infty}^{\infty} c_k e^{ik\theta}$ be in $L_2(T)$, and as before, write $f = f^+ + f^-$, where $f^+(z) = \sum_{k=0}^{\infty} c_k z^k$ and $f^-(z) = \sum_{k=1}^{\infty} c_{-k}z^{-k}$. Let $r_{n,n}^*$ be the best $L_2(T)$ approximant of f from $R(n,n)$. Then by Theorem 2.1, $r_{n,n}^* = \hat{r}_n + \tilde{r}_n$ where $\hat{r}_n \in R_n^{n-1}$ has poles at $\alpha_{n,k}$, $k=1,\ldots,n$, and $\tilde{r}_n \in R_n^n$ has poles at $\beta_{n,k}$, $k=1,\ldots,n$ such that they respectively best approximate f^- and f^+ among such rational functions. By applying an error formula in Walsh [7,p. 186], one can show that, for $|z| < 1$,

$$(f^+(z) - \tilde{r}_n(z)) + (f^-(z^{-1}) - \hat{r}_n(z^{-1})) =$$

$$\frac{1}{2\pi}\int_0^{2\pi} \frac{f^+(e^{i\theta}) + f^-(e^{-i\theta})}{e^{i\theta} - z} \cdot zM_n(z,\theta)d\theta$$

where

$$M_n(z,\theta) = \frac{z^n - \rho_n^n}{e^{in\theta} - \rho_n^n} \cdot \frac{\rho_n^n e^{in\theta} - 1}{\rho_n^n z^n - 1} \quad .$$

Hence, we have

$$(f^+(z) - \tilde{r}_n(z)) + (f^-(z^{-1}) - \hat{r}_n(z^{-1})) = \sum_{|k|>n} c_k z^k +$$

$$\frac{1}{2\pi}\int_0^{2\pi} \frac{f^+(e^{i\theta}) + f^-(e^{-i\theta})}{e^{i\theta} - z} z\left[M_n(z,\theta) - z^n e^{-in\theta}\right]d\theta \; .$$

For all $|z| \le 1$ and all θ, it is easy to show that

$$\left|\frac{z}{e^{i\theta} - z}\left[M_n(z,\theta) - z^n e^{-in\theta}\right]\right| \le \frac{2n\rho_n^n}{(1-\rho_n^n)^2}$$

That is, we have established the following

THEOREM 5.1. Let $f \in L_2(T)$ and $R(n,n)$, T_n be defined as above. Then

(5.1) $$\inf\{\|f - r\|_2 : r \in R(n,n)\} \le \inf\{\|f - p\|_2 : p \in T_n\} + c_f \frac{n\rho_n^n}{(1-\rho_n^n)^2} \quad ,$$

where $c_f = 2(\|f^+\|_2 + \|f^-\|_2)$.

Hence, if ρ_n does not approach 1 very fast, say $n\rho_n^n \to 0$, then we have a good approximation from $R(n,n)$. Furthermore, the inequality becomes equality if $\rho_n = 0$.

References

1 Barnsley, M. F., Padé approximant bounds for the difference of two series of Stieltjes, J. Math. Phys., 17 (1976), 559-565.

2 Braess, D., On rational L_2-approximation, J. Approx. Th., To appear.

3 Brophy, F. and A. C. Salazar, Considerations of the Padé approximant technique in the synthesis of recursive digital filters, IEEE Trans. Audio Electroaust., AU-21 (1973), 500-505.

4 Chui, C. K., Recent results on Padé approximants and related problems, in Approximation Theory II, G. G. Lorentz, C. K. Chui and L. L. Schumaker, Eds., Academic Press, N. Y. 1976, pp. 79-115.

5 Hoffman, K., Banach Spaces of Analytic Functions, Prentice-Hall, Inc., Englewood Cliffs, N. J., 1962.

6 Shanks, J. L., Recursion filters for digital processing, Geophysics, 32 (1967), 33-51.

7 Walsh, J. L., Interpolation and Approximation by Rational Functions in the Complex Domain, Amer. Math. Soc. Colloq. Publ. Vol. XX, N. Y. 1960.

8 Wolfe, J. M., On the unicity of nonlinear approximation in smooth spaces, J. Approx. Th. 12 (1974), 165-181.

Charles K. Chui*
Philip W. Smith*
L. Y. Su
Department of Mathematics
Texas A&M University
College Station, Texas 77843

* Supported, in part, by the U. S. Army Research Office under Grant Number DAHC04-75-G-0186.

A CONTRIBUTION TO RATIONAL APPROXIMATION ON THE WHOLE REAL LINE

Géza Freud

Estimates are given for the order of approximation of real functions on the whole real line by reciprocals of n-th degree polynomials in terms of the variation of their r-th derivative (order fixed) and the order of decrease for $|x| \to \infty$ of the function f(x) to be approximated.

The set of polynomials the degree of which does not exceed n is denoted by $\mathcal{P}_n$, and the set of trigonometric polynomials with order not more than n by τ_n. Let $k \geqq 1$ be an integer, $F(x) = f(x)\cosh(x^k)$ $(-\infty < x < \infty)$. For an integer $r \geqq 1$ let f(x) be a positive function on the reals which has a generalized derivative $f^{(r)}(x)$ and $f^{(r)}$ has bounded variation over every finite interval. We assume that for a pair $\delta > 0$, $A > 0$, $F(x) > A \exp\{\delta|x|^k\}$ and that the variation of $f^{(r)}$ over $[-\xi,\xi]$, denoted as $V_\xi(f^{(r)})$ does not increase too rapidly.

THEOREM. <u>For every</u> $\rho > 0$ <u>there exists a sequence</u> $\{\pi_n \in \mathcal{P}_n\}$ <u>so that for every</u> $1 \leqq p \leqq \infty$ <u>we have</u>

$$(1)\quad \eta_n^{(p)}(F)=\left\|\frac{1}{F}-\frac{1}{\pi_N}\right\|_p \leqq C(k,\delta,A,\rho)\left\{\frac{(\log n)^{r/k}}{n^{r+(1/p)}} V_{\alpha_n}(f^{(r)})+n^{-\rho}\right\}.$$

<u>Here</u> $C(k,\delta,A,\rho)$ <u>does not depend on</u> n <u>or the choice of</u> F <u>and</u> $\alpha_n = b(\log n)^{1/k}$, b <u>depending on</u> ρ,δ,r <u>only</u>. $\|\cdot\|_p$ <u>is the</u> $L_p(-\infty,\infty)$ - <u>norm</u>. <u>A case of special interest of the Theorem above is if</u> k <u>is odd and</u>

$$F_k(x) = e^{-|x|^k} = 2[1+\exp(-2|x|^k)]^{-1}[\cosh(x^k)]^{-1}$$

in which case we have $r = k$ *and we obtain*

$$\eta_n^{(p)}(F_k) = O(n^{-k-(1/p)}\log n).$$

The special case $p = \infty$ *and* $k = 1$ *of this estimate was proved through a different argument in* [3].

Our device is based on the following two lemmas:

LEMMA 1. *Let* $r \geqq 1$ *be an integer and let* $g(x)$ $(-1 \leqq x \leqq 1)$ *have a generalized derivative* $g^{(r)}$ *of bounded variation. We extend* g *by putting* $g(x) = g(1)$ *for* $x > 1$ *and* $g(x) = g(-1)$ *for* $x < -1$, *resp. Then there exists a sequence* $\{T_n \in P_n\}$ *so that*

(2a) $T_n(x) \geqq g(x)$ $(-\infty < x < \infty)$ *and*

(2b) $$\int_{-1}^{1} \frac{T_n(x) - g(x)}{\sqrt{1 - x^2}}\,dx < c_1(r)V_1(g^{(r)})n^{-r-1} = \varepsilon_n.$$

LEMMA 2. *The* T_n's *above satisfy*

(3) $$0 \leqq T_n(x) - g(x) \leqq c_2(r)V_1(g^{(r)})n^{-r}.$$

Lemma 1 was proved in [2] and [4]. The fact that the polynomials satisfy $T_n \geqq g$ also outside $[-1,1]$ was not stated there but it follows from the construction. We turn to Lemma 2. Let $\mu(x) \in P_{r-1}$ be the first r terms of the MacLaurin expansion of f, $f^* = f - \mu_r$, $T_n^* = T_n - \mu_r \in P_n$ and let $\phi(\theta) = f^*(\cos\theta)$. Then $V_\pi(\phi^{(r)}) < c_3(r)V_1(f^{(r)})$. (2b) is transformed in

(4) $$\int_{-\pi}^{\pi} |T_n^*(\cos\theta) - \phi(\theta)|\,d\theta < c_1(r)V_1(f^{(r)})n^{-r-1}$$

and note that $T_n^*(\cos\theta) \in \tau_n$. There exists by Jackson's theorem a sequence $\{t_n \in \tau_n\}$ approximating ϕ' with an error not exceeding $c_3(r)V_1(f^{(r)})n^{-r}$. By virtue of a theorem of ours (see [1]) this implies

(5) $$\int_{-\pi}^{\pi} |T_n^{*\prime}(\cos\theta) - \phi'(\theta)|\,d\theta < c_4(r)V_1(f^{(r)})n^{-r}.$$

Inequalities (4) and (5) easily imply that $|T_n^*(\cos\theta) - \phi^*(\theta)| < 2[c_4(r) + c_1(r)]V_1(f^{(r)})n^{-r}$ and after transformation $T_n(x) - f(x) < c_2(r)V_1(f^{(r)})n^{-r}$. Q.E.D.

Starting the proof of the main theorem, let $e_n(x) \in P_n$ be the n-th partial sum of the exponential series so that $|e^x - e_n(x)| < c_5 e^{-c_6 n}$ $(|x| < c_7 n)$ and let $\sigma_n(x) = \frac{1}{2}[e_n(x^k) + e_n(-x^k)] \in P_{kn}$. Let $\alpha_n = b_1(\log n)^{1/k}$ and b_1 so large that $\cosh \alpha_n > n^{\rho+100(r+1)+\delta^{-1}}$. We set $g(x) = f(\alpha_n x)$ for $|x| \leqq 1$ and we extend g to every real x as in Lemma 1. Note that $V_1(g^{(r)}) = \alpha_n^r V_{\alpha_n}(f)$. Let $\{T_n(x)\}$ be the sequence as in Lemma 1 and Lemma 2 and $p_n(x) = T_n(x/\alpha_n)$.

By virtue of the lemmas $p_n(x) \geqq f(x)$ and

(6a) $$\int_{-\alpha_n}^{\alpha_n} [p_n(x) - f(x)]dx < c_1(r)V_{\alpha_n}(f^{(r)})(\alpha_n/n)^{r+1}$$

(6b) $$0 < p_n(x) - f(x) < c_2(r)V_{\alpha_r}(f^{(r)})(\alpha_n/n)^r \quad (|x| \leqq \alpha_n).$$

Now let $\pi_N(x) = p_n(x)\sigma_n(x)[1 + n^{-\rho}(x/\alpha_n)^n]$ where n is chosen as the largest integer for which $\pi_N \in P_N$ still holds. By construction π_n approximates $F(x)$ in $[-\alpha_n, \alpha_n]$ and it is small outside this interval. In particular, we have

$$\int_{c_7 n}^{\infty} \frac{dx}{\pi_N(x)} \leqq \frac{c_8}{F(c_7 n)} n^T \int_{c_7 n}^{\infty} (\alpha_n/x)^n dx < c_9 n^{-\rho}.$$

A lengthy but straightforward calculation shows that (1) is satisfied both for p = 1 and p = ∞. The intermediate cases now follow from the elementary inequality

$$||f||_p \leqq (||f||_\infty)^{1-1/p}(||f||_1)^{1/p}.$$

References

1 Czipser, J. and G. Freud, Sur l'approximation d'une fonction periodique et de ses desivees successires par un polynome trigonometrique et par ses derivees successives. Acta Math., (Sweden), 5 (1957), 285-290.

2 Freud, G., Über einseitige Approximation durch Polynome, I. Acta Sci. Math., (Szeged), 6 (1955), 12-28.

3 Freud, G., D.J. Newman and A.R. Reddy, Chebyshev rational approximation to $e^{-|x|}$ on the whole real line. Quart. J. Math. Oxford Ser., to appear.

4 Nevai, G.P., Einseitige Approximation durch Polynome, mit Anwendungen. Acta Math. Acad. Sci. Hungar., 23 (1972), 495-506.

Professor Géza Freud
O.S.U. - Department of Mathematics
Columbus, Ohio 43210

ON THE POSSIBILITY OF RATIONAL APPROXIMATION

P.M. Gauthier

This lecture is intended to introduce the audience to some of the more spectacular results of Alice Roth concerning the existence of rational approximations. We also show the impossibility of characterizing topologically those sets on which Walsh approximation is possible.

1 Runge Approximation

We begin by a well known theorem. Let K be a compact subset of the finite complex plane $\mathbb{C}$.

THEOREM 1.1 (Runge [6]). *Every function holomorphic on K can be uniformly approximated by rational functions.*

The next theorem extends Runge's Theorem to unbounded sets. This result deserves to be much better known than it is. We denote by $\mathbb{C} \cup \{\infty\}$ the extended complex plane.

THEOREM 1.2 (Roth [4]). *Let F be a closed subset of $\mathbb{C}$. Then every function holomorphic on F can be uniformly approximated by functions meromorphic on $\mathbb{C}$. Moreover, if $(\mathbb{C} \cup \{\infty\}) \setminus F$ is connected and locally connected, we may take the approximating functions to be entire.*

Note that in this theorem, it is not assumed that the functions to be approximated are continuous at infinity. Note also that when F is compact, Roth's theorem reduces to Runge's theorems on rational and polynomial approximation respectively.

2 Walsh Approximation

Next we consider a more delicate problem. Namely, we relax the assumption that the functions to be approximated be holomorphic on the set in question. Let $\bar{R}(K)$ denote the uniform limits on K of rational functions whose poles lie outside of K, and denote by A(K) the functions which are continuous on K and holomorphic on the interior K^o of K. J.L. Walsh [7] gave his name to this sort of approximation by showing that if K is a closed Jordan domain, then $\bar{R}(K) = A(K)$. Of course if K^o is empty we may replace A(K) by C(K).

THEOREM 2.1. (Hartogs-Rosenthal [3]). _If K is of measure zero, then_ $\bar{R}(K) = C(K)$.

The interest of this result is considerably enhanced by the following

EXAMPLE 2.2 (Roth [4]). _There exists a nowhere dense compact set K with_ $\bar{R}(K) \neq C(K)$.

Roth's example is her celebrated Swiss cheese. Denote by D(z,r) the open disc of center z and radius r.

$$K = \bar{D}(0,1) \setminus \bigcup_{j=1}^{\infty} D(z_j, r_j)$$

is called a Swiss cheese if the discs $D(z_j, r_j)$ are disjoint and dense in $\bar{D}(0,1)$. Roth showed that some Swiss cheeses have the property that $\bar{R}(K) \neq C(K)$.

A technique of Dolzhenko [1] furnishes a wide variety of sets K for which $\bar{R}(K) \neq A(K)$. Let J be a closed nowhere dense subset of $\bar{D}(0,1)$ having positive area, and let D_j be a sequence of disjoint discs in D(0,1) accumulating at all of J. Then if

$$K = \bar{D}(0,1) \setminus \bigcup_{j-1}^{\infty} D_j,$$

Dolzhenko showed that $\bar{R}(K) \neq A(K)$. When J is a Jordan arc, this example is known as the stitched disc. Cheese connoisseurs may be tempted to call it Morbier cheese.

The following example shows that those sets K for which $\bar{R}(K) = A(K)$ cannot be characterized topologically. This fact is known to experts, but I know of no reference.

THEOREM 2.3. *There is a homeomorphism of the plane* $h : \mathbb{C} \to \mathbb{C}$, *and a compact set* K *such that* $A(K) \neq \bar{R}(K)$ *but* $A(h(K)) = \bar{R}(h(K))$.

Of course we can't replace h by a biholomorphic map since such h preserve rational functions.

To prove the theorem, let K be a stitched disc whose J is a Jordan arc from -1 to +1. Now let ϕ be a homeomorphism of the plane which preserves the unit disc and straightens J into the diameter from -1 to +1. Since the inner boundary of $\phi(K)$ lies on the real axis, we have $\bar{R}(\phi(K)) = A(\phi(K))$ (see [2, p. 238]).

3 Approximating a Given Function

Perhaps more important than Walsh approximation is the problem of approximating a given function on a given set. The following result is known as the fusion lemma. It allows us to approximate two functions simultaneously by a single function. I believe it is one of the most fundamental results in rational approximation.

THEOREM 3.1 (Roth [5]). *Let* K_1, K, *and* K_2 *be compact with* K_1 *and* K_2 *disjoint. There is a positive constant* a *such that if* r_1 *and* r_2 *are rational functions with*

$$|r_1(z) - r_2(z)| < \varepsilon, \quad z \in K,$$

then there is a rational function r *such that for* $j = 1,2$,

$$|r_j(z) - r(z)| < a\varepsilon, \quad z \in K_j \cup K.$$

From the fusion lemma 3.1, one can easily deduce the beautiful localization lemma of Bishop (see [2, p. 51]).

THEOREM 3.2 (Bishop). *Suppose* f *is given on a compact set* K *and suppose for each* $z \in K$, *there is a disc* D(z) *such that*

$$f|_{K \cap \bar{D}(z)} \in \bar{R}(K \cap \bar{D}(z)).$$

Then $f \in \bar{R}(K)$.

To prove the localization lemma, Roth covers K with a rectangular grid and applied the fusion lemma a finite number of times.

For other powerful consequences of the fusion lemma, we refer the reader to [5].

I wish to thank Professor Ted Gamelin for streamlining my proof of Theorem 2.3 and for pointing out that the result is known.

References

1. Dolzhenko, E.P., On approximation on closed regions and on null-sets Dokl. Akad. Nauk SSSR, 143 (1962), 771-774.
2. Gamelin, T.W., Uniform Algebras, Prentice-Hall, Inc., Englewood Cliffs, N.J., 1969.
3. Hartogs, F. and A. Rosenthal, Über Folgen analytischen Funktionen, Math. Ann., 104 (1931), 606-610.
4. Roth, Alice, Approximationseigenschaften und Strahlengrenzwerte unendlich vieler linearer Gleichungen, Comm. Math. Helv., 11 (1938), 77-125.
5. Roth, Alice, Uniform and tangential approximations by meromorphic functions on closed sets, Can. J. Math., 28 (1976), 104-111.
6. Runge, C., Zur Theorie der eindeutigen analytischen Funktionen, Acta Math., 6 (1885), 229-244.
7. Walsh, J.L., Interpolation and Approximation by Rational Functions in the Complex Domain, Amer. Math. Soc., Providence, R.I., 1935.

Département de Mathématiques
Université de Montréal
Montréal, Canada H3C3J7

Research supported in part by NRC of Canada and the Ministère d'Education du Québec.

GEOMETRIC CONVERGENCE OF CHEBYSHEV RATIONAL APPROXIMATIONS ON THE HALF LINE

Myron S. Henry and John A. Roulier

Necessary and sufficient conditions on a real valued function f continuous on $[0, +\infty)$ to insure the existence of a sequence $\{p_n\}_{n=0}^{\infty}$, p_n an algebraic polynomial of degree n or less, and a number $q > 1$ such that

$$\limsup_{n \to \infty} \left(\left\| \frac{1}{f} - \frac{1}{p_n} \right\|_{L_\infty[0, +\infty)} \right)^{\frac{1}{n}} \leq \frac{1}{q}, \text{ are investigated.}$$

In particular, a new necessary condition on f is given which yields a counterexample to the widely held belief that a necessary condition due to G. Meinardus, A. Reddy, G. D. Taylor, and R. S. Varga is also sufficient.

1 Introduction

Let f be a continuous real valued function on $[0, +\infty)$ and define

$$\|f\|_r = \sup\{|f(x)| : 0 \leq x \leq r\} \quad \text{for} \quad r > 0$$

and

$$\|f\|_\infty = \sup\{|f(x)| : 0 \leq x\}.$$

For each nonnegative integer n define π_n to be the algebraic polynomials with real coefficients of degree not exceeding n, and $\pi_{m,n}$ to be the rational functions of the form $r_{m,n}(x) = \frac{p(x)}{q(x)}$ with $p \in \pi_m$, $q \in \pi_n$.

We investigate the following problem:

For which functions $f \in C[0, +\infty)$ does there exist a number $q > 1$ and a sequence of rational functions $\{r_n\}_{n=0}^{\infty}$

such that $r_n \in \pi_{n,n}$, $n = 0, 1, 2, \ldots,$ and

$$(1.1) \quad \limsup_{n \to \infty} \left(\left\| \frac{1}{f} - r_n \right\|_\infty \right)^{1/n} \leq \frac{1}{q}?$$

In particular, for which functions $f \in C[0, +\infty)$ does this happen with $r_n \in \pi_{0,n}$ (i.e. $r_n(x) = \frac{1}{p_n(x)}$, $p_n \in \pi_n$)? The complete answer to this problem is not yet known although many authors in recent years have investigated this. If there exists a $q > 1$ and a sequence $r_n(x) = \frac{1}{p_n(x)}$ $p_n \in \pi_n$ (i.e. $r_n \in \pi_{0,n}$) such that this happens for some f, then we say f has geometric convergence. If (1.1) holds, but with $r_n \in \pi_{n,n}$, then we say that f has weak geometric convergence. Thus we seek to classify all $f \in C[0, +\infty)$ which have geometric convergence, or at least weak geometric convergence.

The first result on this problem established that $f(x) = e^x$ has geometric convergence (see [4]). This result was extended to other functions in [7]. The first in depth study was made by Meinardus, Taylor, Reddy, and Varga in [8]. They obtained a necessary condition as well as a sufficient condition for f to have geometric convergence. Since the appearance of [8], several researchers have suggested that the necessary condition obtained in [8] may also be sufficient. In particular, Roulier and Taylor in [10] obtain a less restrictive sufficient condition and conjecture that the necessary condition in [8] is also sufficient. Blatt in [1] and [2] further weakens the hypotheses of the sufficient condition given in [8]. These results further suggest that the necessary condition in [8] might also be sufficient.

In this paper we obtain a new necessary condition for geometric convergence. This in turn, provides the machinery for constructing a counterexample to the above conjecture. That is, a function f is defined which fails to have geometric convergence, and yet f satisfies the necessary condition obtained in [8]. In addition we obtain a new sufficient condition that is essentially different from those already known. This, in turn, will be used to generate new examples of functions which have geometric convergence but with properties that are somewhat surprising

2 Notation and Previous Results

Let $r > 0$ and $s > 1$ be given, and let $E(r,s)$ denote the unique ellipse in the complex plane with foci at $x = 0$ and $x = r$ and semi-major and semi-minor axes a and b respectively, with $\frac{b}{a} = \frac{s^2 - 1}{s^2 + 1}$. If $f(z)$ is any function analytic inside and on the boundary of $E(r,s)$ define

$$M_f(r,s) = \max\{|f(z)| : z \in E(r,s)\}.$$

The necessary condition obtained in [8] is given in the following theorem.

THEOREM 2.1. *Let f be a real continuous function* $(\not\equiv 0)$ *on* $[0, +\infty)$, *and assume that there exist a sequence of real polynomials* $\{p_n\}_{n=0}^{\infty}$, *with* $p_n \in \pi_n$ *for* $n = 0, 1, \ldots,$ *and* $q > 1$ *such that*

$$(2.1)\quad \limsup_{n \to \infty} \left(\left\|\frac{1}{f} - \frac{1}{p_n}\right\|_\infty \right)^{1/n} \leq \frac{1}{q} < 1.$$

Then, there exists an entire function $F(z)$ *with* $F(x) = f(x)$ *for all* $x \geq 0$, *and* F *is of finite order* ρ. *In addition, for every* $s > 1$, *there exist constants* $K = K(s,q) > 0$, $\theta = \theta(s,q) > 0$ *and* $r_0 = r_0(s,q) > 0$ *such that*

$$(2.2)\quad M_F(r,s) \leq K(\|f\|_r)^\theta \quad \textit{for all} \quad r \geq r_0.$$

3 A New Necessary Condition and a Counterexample

In the following theorem we show that if f has geometric convergence, then f can not oscillate too badly as x approaches $+\infty$.

THEOREM 3.1. *Let* $f \in C[0, +\infty)$ *satisfy*

(3.1) $$\lim_{x \to +\infty} f(x) = +\infty,$$

and let $\{x_j\}_{j=0}^{\infty}$ *be a sequence of real numbers such that*

(3.2) $0 \le x_0 < x_1 < \ldots$

(3.3) $$\lim_{j \to \infty} x_j = +\infty$$

(3.4) $f(x_{2j}) \le f(x_{2j+2})$ *for* $j = 0, 1, 2, \ldots$

(3.5) $f(x_{2j-1}) \le f(x_{2j+1})$ *for* $j = 1, 2, \ldots$

(3.6) $f(x_{2j}) < f(x_{2j-1})$ *for* $j = 1, 2, \ldots$

(3.7) $$\lim_{j \to \infty} \frac{f(x_{2j})}{f(x_{2j-1})} = 0 \quad \textit{and}$$

(3.8) *for any* $r > 1$ $$\lim_{j \to \infty} \frac{f(x_{2j})}{r^j} = 0 .$$

Then f *does not have geometric convergence or even weak geometric convergence.*

Proof. Assume that f does have weak geometric convergence. It follows from (1.1) that there exists $q > 1$, sequence $\{r_n\}_{n=0}^{\infty}$ with $r_n \in \pi_{n,n}$, $n = 0, 1, \ldots,$ and $N_0 \ge 0$ such that $n \ge N_0$ implies

(3.9) $$\left\|\frac{1}{f} - r_n\right\|_{\infty} \le \frac{1}{q^n} .$$

Now (3.1) implies the existence of $r_0 \ge 0$ such that

(3.10) $f(x) \ge 1$ for $x \ge r_0$.

The combination of (3.3) and (3.10) gives the existence of $J_0 \ge 0$ such that

(3.11) $j \ge J_0$ implies $f(x_j) \ge 1$.

Furthermore, the combination of (3.4), (3.7) and (3.11) gives

$$(3.12)\quad \lim_{j\to+\infty} \frac{f(x_{2j-2})}{f(x_{2j-1})} = 0.$$

Choose $J_1 \geq J_0$ ((3.7) and (3.12)) so that $j \geq J_1$ implies

$$(3.13)\quad \frac{f(x_{2j-2})}{f(x_{2j-1})} \leq \frac{1}{2} \text{ and } \frac{f(x_{2j})}{f(x_{2j-1})} \leq \frac{1}{2}.$$

We may now use (3.11) and (3.13) to obtain for $j \geq J_1$

$$\frac{1}{f(x_{2j-2})} - \frac{1}{f(x_{2j-1})} = \frac{1}{f(x_{2j-2})}\left(1 - \frac{f(x_{2j-2})}{f(x_{2j-1})}\right) \geq \frac{1}{2f(x_{2j-2})}$$

(3.14) and

$$\frac{1}{f(x_{2j})} - \frac{1}{f(x_{2j-1)}} = \frac{1}{f(x_{2j})}\left(1 - \frac{f(x_{2j})}{f(x_{2j-1})}\right) \geq \frac{1}{2f(x_{2j})}.$$

It follows from (3.8) that there exists integer $J_2 \geq J_1$ such that $j \geq J_2$ implies both

$$(3.15)\quad \frac{f(x_{2j})}{q^{j/3}} \leq \frac{1}{8} \text{ and } \frac{f(x_{2j-2})}{q^{j/3}} \leq \frac{1}{8}.$$

Note furthermore that if $k \geq j \geq J_2$ then from (3.15) we have

$$(3.16)\quad \frac{f(x_{2j})}{q^{k/3}} \leq \frac{1}{8} \text{ and } \frac{f(x_{2j-2})}{q^{k/3}} \leq \frac{1}{8}.$$

We now use (3.9), (3.14) and (3.16) to observe that for $n \geq N_0$ and

$J_2 \le j \le 3n$

$$r_n(x_{2j-2}) - r_n(x_{2j-1}) = \frac{1}{f(x_{2j-2})} - \frac{1}{f(x_{2j-1})} + r_n(x_{2j-2})$$

$$- \frac{1}{f(x_{2j-2})} + \frac{1}{f(x_{2j-1})} - r_n(x_{2j-1})$$

$$\ge \frac{1}{2f(x_{2j-2})} - \frac{2}{q^n}$$

$$= \frac{1}{2f(x_{2j-2})}\left(1 - \frac{4f(x_{2j-2})}{q^{3n/3}}\right)$$

$$\ge \frac{1}{4f(x_{2j-2})} .$$

That is, if $J_3 \ge \max(J_2, 3N_0)$ then $J_3 \le j \le 3n$ implies

$$(3.17) \quad r_n(x_{2j-2}) - r_n(x_{2j-1}) \ge \frac{1}{4f(x_{2j-2})}$$

In a similar fashion we can show that $J_3 \le j \le 3n$ implies

$$(3.18) \quad r_n(x_{2j}) - r_n(x_{2j-1}) \ge \frac{1}{4f(x_{2j})}$$

It follows from (3.9) and (3.10) that $r_n(x) \ne 0$ and has no poles if $x \ge r_0$ and $n \ge N_0$. It now follows from this, (3.17) and (3.18) that r_n has a relative minimum on each of the intervals

$$(x_{2j-2}, x_{2j}), \quad J_3 \le j \le 3n$$

and a relative maximum on each of the intervals

$$(x_{2j-1}, x_{2j+1}), \quad J_3 \le j \le 3n.$$

Thus r_n has at least $2(3n - J_3 + 1) = 6n - 2J_3 + 2$ relative extrema on $[r_0, +\infty)$. But if we fix J_3 and take n large enough we see that r_n must have at least $5n$ relative extrema on $[r_0, +\infty)$. But this implies that $r_n'(x) = 0$ for at least $5n$ distinct points. Hence, $r_n' \equiv 0$ and r_n is a constant for n sufficiently large. This is a contradiction since f is not a constant.

We now use Theorem 3.1 to construct a function f which satisfies the necessary conditions obtained in Theorem 2.1 but which fails to have geometric convergence or even weak geometric convergence.

EXAMPLE 3.1. Define the entire function

$$F(z) = z + 1 + e^z \sin^2 z;$$

and let f be the restriction of F to the real line;

$$f(x) = x + 1 + e^x \sin^2 x.$$

Then f satisfies both the conclusion of Theorem 2.1 and the hypotheses of Theorem 3.1. Hence, f is the counterexample alluded to in Section 1.

4 A New Sufficient Condition

The following theorem gives a sufficient condition for a function f to have geometric convergence. It is essentially different from the results of Roulier and Taylor [10] and of Blatt [1] and [2]. In order to demonstrate this, an example based on this theorem is given; the example is not obtainable from any of the previously published results.

THEOREM 4.1. *Let* $f \in C[0, +\infty)$ *satisfy*

(4.1) $f(x) \geq \eta > 0$ *on* $[0, +\infty)$,

(4.2) $\lim_{x \to +\infty} f(x) = +\infty$,

(4.3) *there exist real valued functions* h *and* g *such that* h *and* g *are restrictions of entire functions and* $f'(x) = h^2(x) + g^2(x)$,

(4.4) *there exist numbers* $A > 0$, $s > 1$, $\theta > 0$ *and* $r_0 > 0$ *such that* $M_{h^2}(r,s) + M_{g^2}(r,s) \leq A(\|f\|_r)^\theta$ *for* $r \geq r_0$.

<u>Then</u> f <u>has geometric convergence, and the</u> q <u>in</u> (1.1) <u>satisfies</u>

$$q \geq s^{\frac{1}{2(2+\theta)}} > 1.$$

The proof of this theorem appears in [5] and is omitted.

We now employ Theorem 4.1 in conjunction with Theorem 2 in [2] to obtain an example of a function f with geometric convergence which is not obtainable from the previous sufficient conditions.

EXAMPLE 4.1. Let

$$f_1(x) = \frac{1}{8} e^{2x} [2 + \sin(2x) + \cos(2x)]$$

and let

$$f_2(x) = e^{-x}.$$

Let $f(x) = f_1(x) + f_2(x)$. Note that

$$f_1'(x) = (e^x \cos x)^2$$

and

$$f_2'(x) = -e^{-x}.$$

It is easy to see that f_1 satisfies the hypotheses of Theorem 4.1 with

$$h(x) = e^x \cos x \quad \text{and} \quad g(x) = 0.$$

Hence, f_1 has geometric convergence.

It is also easy to see that $f = f_1 + f_2$ satisfies the sufficient conditions of Theorem 2 in [2]. Hence, f has geometric convergence.

Notice, however, that

$$f'(x) = (e^x \cos x)^2 - e^{-x}$$

will assume negative values for arbitrarily large x. Thus there is no r for which f is increasing on $[r, +\infty)$.

The following is an easy corollary to Theorem 4.1.

COROLLARY. *Suppose that f is a positive real valued function on $[0, +\infty)$ and is the restriction of an entire function, and that $\lim_{x \to \infty} f(x) = +\infty$. Assume furthermore that there is an entire function $g(z) = \sum_{j=0}^{\infty} a_j z^j$ such that*

$$(4.5) \quad f'(z) = g(z)\hat{g}(z) \quad \text{where} \quad \hat{g}(z) = \sum_{j=0}^{\infty} \bar{a}_j z^j$$

and $\bar{a}_j$ is the conjugate of a_j, and that there are constants

$$A > 0, \ \theta > 0, \ s > 1 \ \text{and} \ r_0 > 0$$

such that

$$(4.6) \quad M_g(r,s) \le A(\|f\|_r)^\theta \quad \text{for} \quad r \ge r_0.$$

Then f has geometric convergence.

The proof of this also appears in [5] and is omitted.

5 Remarks and Conclusions

Example 4.1 shows that it is possible for a function with geometric convergence to oscillate somewhat. On the other hand Theorem 3.1 shows that such functions can not oscillate too much.

It appears that the complete characterization of functions f with geometric convergence will have to involve the rate of growth of

$$\frac{M_f(r)}{m_f(r)} \quad \text{as} \quad r \to +\infty$$

where $M_f(r) = \|f\|_r$ and $m_f(r) = \inf_{x \ge r} |f(x)|$, as well as the necessary conditions in Theorem 2.1.

Another interesting question is whether f has geometric convergence if it satisfies the necessary conditions in Theorem 2.1 and is increasing on $[r_1, +\infty)$ for some $r_1 \ge r_0$.

References

1. Blatt, H. P., Rationale Tschebyscheff-Approximation über unbeschränkten Intervallen, Habilitationsschrift, Universität Erlangen-Nürnberg, 1974.

2. Blatt, H. P., Geometric convergence of rational functions to the reciprocal of exponential sums on $[0, +\infty)$, in Theory of Approximation with Applications, (A. G. Law and B. N. Sahney, editors), Academic Press, New York, (1976), 111-119.

3. Boas, R. P., Jr., Entire Functions, Academic Press, New York, 1954.

4. Cody, W. J., G. Meinardus, and R. S. Varga, Chebyshev rational approximations to e^{-x} in $[0, +\infty)$ and applications to heat conduction problems, J. Approx. Theory, 2 (1969), 50-65.

5. Henry, M. S. and J. A. Roulier, Geometric convergence of Chebyshev rational approximations on $[0, +\infty)$, J. Approx. Theory (to appear).

6. Meinardus, G., Approximation of Functions: Theory and Numerical Methods, Springer-Verlag, New York, 1967 (translated by L. L. Schumaker).

7. Meinardus, G. and R. S. Varga, Chebyshev rational approximation to certain entire functions in $[0, +\infty)$, J. Approx. Theory, 3 (1970), 300-309.

8. Meinardus, G., A. R. Reddy, G. D. Taylor and R. S. Varga, Converse theorems and extensions in Chebyshev rational approximation to certain entire functions in $[0, +\infty)$, Trans. Am. Math. Soc., 170 (1972), 171-185.

9. Reddy, A. R. and O. Shisha, A class of rational approximations on the positive real axis - a survey, J. Approx. Theory, 12 (1974), 425-434.

10. Roulier, J. A. and G. D. Taylor, Rational Chebyshev approximation on $[0, +\infty)$, J. Approx. Theory, 11 (1974), 208-215.

Myron S. Henry*
Department of Mathematics
Montana State University
Bozeman, Montana 59715

John A. Roulier†
Department of Mathematics
North Carolina State University
Raleigh, North Carolina 27607

*Supported in part by NSF Grant No. MCS 76-06553.

†Supported in part by NSF Grant No. MCS 76-04033.

OPTIMAL APPROXIMATION BY "ALMOST CLASSICAL" INTERPOLATION

F.M. Larkin

Optimal approximation in a Szegö-Hilbert space is shown to be equivalent to the problem of constructing a certain rational interpolating function. This interpolation problem leads to techniques analogous to those developed for classical polynomial interpolation; in particular, forms equivalent to the Lagrange polynomial, the Neville-Aitken method and the Newton polynomial constructed from generalized divided differences, are exhibited. Since polynomial interpolation obtains in the limit as the domain of the functions in the Szegö-Hilbert space becomes infinite, the classical dichotomy of polynomial interpolation and approximation of functions is unified.

1 Introduction

Given a set of distinct abscissae $\{z_j; j=1,2,\ldots,n\}$, and corresponding ordinate values $\{f_j; j=1,2,\ldots,n\}$ sampled from a function $f(\cdot)$, one classical approach to the problem of estimating a functional such as $f(z_0)$ or $\int_a^b f(z)\,dz$ is mathematically equivalent to constructing the unique polynomial $Q_{n-1}(\cdot)$, of degree $<n$, which satisfies

$$Q_{n-1}(z_j)=f_j \;;\quad j=1,2,\ldots,n,$$

and subsequently evaluating $Q_{n-1}(z_0)$ or $\int_a^b Q_{n-1}(z)\,dz$, as required. The interpolation problem has traditionally been treated by the techniques of Lagrange, Newton, Aitken and Neville and the apparatus of divided differences, while the integration problem has given rise to quadrature rules associated with the names of Newton, Cotes, Gauss and many others.

However, the technique of polynomial interpolation seems to have been conceived as a convenient, but somewhat ad hoc, approach to the problem of estimating the values of functionals and, of necessity, the estimation of error bounds on

$|f(z_0)-Q_{n-1}(z_0)|$ and $|\int_a^b (f(z)-Q_{n-1}(z))\cdot dz|$ has been regarded as a separate issue. One purpose of this paper is to show that, far from being ad hoc, the polynomial interpolation approach arises as an unavoidable, natural consequence of optimal, linear approximation in a Szegö-Hilbert space of functions analytic within a large region in the complex plane.

Furthermore, when the region of analyticity is a finite circle centered on the origin, optimal approximation involves a rational interpolation problem bearing a striking resemblance to the polynomial interpolation problem which is its limit as the circle becomes large. In fact, the techniques developed over the centuries for treatment of the polynomial interpolation problem all seem to have extensions applicable to the rational problem, considerably simplifying the task of optimal approximation in the space mentioned. These techniques are developed below; they essentially result from a more thorough consideration of an example used to illustrate a general technique given elsewhere (Larkin, 1970). The style of presentation is discursive, rather than formal, in keeping with the objectives of the paper.

2. The Szegö-Hilbert Space

The space H_r of functions of a complex variable which are analytic within the region $D=\{z:|z|<r\}$ and square-integrable in the Lebesque sense around the perimeter $C=\{z:|z|=r\}$ constitutes a Hilbert space under an inner product defined by

$$(f,g)= \frac{1}{2\pi r}\cdot\int_C f(z)\cdot\overline{g(z)}\cdot|dz| \ ; \ \forall\, f,g\in H_r,$$

and the corresponding natural norm (e.g. Meschkowski, 1962). It may be verified, from the Cauchy residue theorem, that

$$f(w)= \frac{1}{2\pi r}\cdot\int_C \frac{f(z)}{1-\bar{z}w/r^2}\cdot|dz| \ ; \quad \forall\, f\in H_r;\ w\in D,$$

so the Szegö kernel function

$$K(z,\bar{w}) \overset{\text{def}}{=} \frac{1}{1-z\bar{w}/r^2} \; ; \quad \forall z,\ w \varepsilon D$$

which, regarded as a function of either z or w with the other fixed is a member of H_r, is the (unique) reproducing kernel for H_r (Aronszajn, 1950).

i.e. $f(w) = (f(\cdot),\ K(\cdot,\bar{w}));\quad \forall\ f \ \varepsilon\ H_r,\ w\ \varepsilon\ D.$

The ordinate evaluation functional at any point $w \varepsilon D$ is a bounded, linear functional on H_r, since

$$|f(w)| \leq ||f||\cdot||K(\cdot,\bar{w})|| = ||f||[K(w,\bar{w})]^{1/2}$$

so that

$$\frac{|f(w)|}{||f||} \leq \frac{1}{1 - |w|^2/r^2} < \infty \; ;$$

clearly $K(\cdot,\bar{w})$ is its Riesz Representer.

In this space the functions $\{(\frac{z}{r})^k \; ; \; k=0,1,2,\ldots\}$ form a complete orthonormal basis. Also a function $f(\cdot)$, defined by

$$f(z) = \sum_{j=0}^{\infty} c_j z^j \; ; \quad \forall\ z \varepsilon D$$

is a member of H_r if and only if

$$(1) \quad ||f||^2 = \sum_{j=0}^{\infty} r^{2j}|c_j|^2 < \infty$$

Necessity is obvious, and the condition is sufficient for uniform convergence of the series defining $f(\cdot)$; hence $f(\cdot)$ is analytic in D by the Weierstrass theorem on uniformly convergent sequences of analytic functions. H_∞ comprises those entire functions $\{f(\cdot)\}$ for which (1) is true for any finite r; however, under the norm defined H_∞ is not a Hilbert space since

$$||f|| = \infty; \quad \forall\ \text{const.} \neq f\ \varepsilon\ H_\infty$$

3. Optimal Linear Approximation

The ideas in this section are essentially due to Sard (1963) and are included for continuity of presentation.

Suppose we are given a set of ordinate values $\{f_j=f(z_j);j=1, 2,\ldots,n\}$ at distinct abscissae $\{z_j \varepsilon D\}$ and wish to estimate the value of some other bounded, linear functional Lf by means of the rule

$$(2) \qquad Lf \simeq \sum_{j=1}^{n} a_j f(z_j)$$

where the constants $\{a_j;j=1,2,\ldots,n\}$ are to be chosen, independently of the ordinate values, but depending upon the abscissae. Let $g(\cdot)$ be the Riesz representer of L in H_r and consider the error functional

$$Ef \overset{\text{def}}{=} Lf - \sum_{j=1}^{n} a_j f(z_j) = (f, g - \sum_{j=1}^{n} \bar{a}_j K(\cdot,\bar{z}_j)) \ ,$$

the bar denoting complex conjugation.

Thus

$$\frac{|Ef|}{||f||} \leq ||g - \sum_{j=1}^{n} \bar{a}_j K(\cdot,\bar{z}_j)|| < \infty \ ; \quad \forall \ f \ \varepsilon \ H_r .$$

In fact, this bound can be attained by suitable choice of f, so E is a bounded, linear functional and

$$||E|| = ||g - \sum_{j=1}^{n} \bar{a}_j K(\ ,\bar{z}_j)||.$$

If the weights $\{a_j; \ j=1,2,\ldots,n\}$ are chosen so as to minimize $||E||$ rule (2) is said to be optimal with respect to these weights. If the abscissae are chosen so as to minimize the coefficient-optimal value of $||E||$ the rule is said to be optimal with respect to the abscissae.

The function

$$\hat{g}(\cdot) \stackrel{\text{def}}{=} \sum_{j=1}^{n} \bar{a}_j K(\cdot,\bar{z}_j)$$

is the orthogonal projection of $g(\cdot)$ onto the linear manifold K_n spanned by the functions $\{K(\cdot,\bar{z}_j) ; j=1,2,\ldots,n\}$, i.e., $\hat{g} = Pg$ where P denotes this projection operator. Furthermore

$$\sum_{j=1}^{n} a_j f(z_j) = (f, \sum_{j=1}^{n} \bar{a}_j K(\cdot,\bar{z}_j)) = (f,Pg) = (Pf,g)$$

since an orthogonal projection is self-adjoint. Thus, for optimally chosen weights

$$\sum_{j=1}^{n} a_j f(z_j) = (\hat{f},g) = L\hat{f},$$

where

$$\hat{f} = Pf,$$

the orthogonal projection onto K_n of f. In other words, for any L the optimal, linear estimate of Lf can be obtained by finding $\hat{f}$, which is quite independent of L, and then evaluating $L\hat{f}$.

4. Relevance of Interpolation to Optimal Approximation

Since $\hat{f}$ is independent of L, by choosing L to be the ordinate evaluation functional at z_k we see that $\hat{f}(z_k)=f(z_k)$; $k= 1,2,\ldots,n$, i.e. $\hat{f}(\cdot)$ is that linear combination of the representers $\{K(\cdot,\bar{z}_j); j=1,2,\ldots,n\}$ of the ordinate evaluation functionals at the given abscissae which passes through the given points $\{(z_j,f_j); j=1,2,\ldots,n\}$. Thus, all optimal linear approximation problems are made to depend on this single interpolation problem.

5. The Interpolation Problem in Szegö-Hilbert Space

Using the explicit form of the Szegö kernel, the interpolation problem becomes that of constructing a linear combination

$$\hat{f}(w) = \sum_{j=1}^{n} \frac{b_j}{1-w\bar{z}_j/r^2}$$

which satisfies

$$\hat{f}(z_k) = f_k \ ; \quad k = 1,2,\ldots,n.$$

Thus $\hat{f}(\cdot)$ has the rational form

$$\hat{f}(w) = \frac{P_{n-1}(w)}{\prod_{j=1}^{n} (1-w\bar{z}_j/r^2)} \ ; \qquad \forall \ w\varepsilon D,$$

where $P_{n-1}(w)$ is a polynomial in w of degree not greater than n-1. Clearly, as $r \to \infty$, this approaches the problem of constructing the polynomial $Q_{n-1}(\cdot)$ of lowest degree passing through the points $\{(z_j,f_j); \ j=1,2,\ldots,n\}$, confirming for example that the classical, polynomial-based interpolation and quadrature rules are limiting forms of optimal approximation rules formulated in this function space.

It turns out that the analogy between classical polynomial interpolation and the construction of $\hat{f}(\cdot)$ is even closer than the above observation suggests - we now proceed to discuss the details, emphasizing similarities by reference to the classical terminology.

(a) The "Lagrange Form" of $f(\cdot)$

Let $\ell_j(z)$ denote the well-known Lagrange coefficient

$$\ell_j(z) = \frac{\prod_{1=k\neq j}^{n} (z-z_k)}{\prod_{1=k\neq j}^{n} (z_j-z_k)} \ ; \quad \forall \ z \ \varepsilon \ D.$$

Then, by inspection, the function

$$\hat{f}(z) = \frac{\sum_{j=1}^{n} f_j \prod_{k=1}^{n} (1-z_j\bar{z}_k/r^2) \ \ell_j(z)}{\prod_{k=1}^{n} (1-z\bar{z}_k/r^2)} \ ; \quad \forall \ z \ \varepsilon \ D \ ,$$

satisfies the requirements ennumerated in the previous section. In other words, to construct $\hat{f}(z)$ we merely replace $\ell_j(z)$ by

$$\hat{\ell}_j(z) \stackrel{\text{def}}{=} \ell_j(z) \prod_{k=1}^{n} \left(\frac{1-z_j\bar{z}_k/r^2}{1-z\bar{z}_k/r^2}\right) \quad ; \quad j=1,2,\ldots,n; \forall z \in D,$$

in the Lagrange formula

$$Q_{n-1}(z) = \sum_{j=1}^{n} f_j \ell_j(z)$$

Thus $\hat{f}(\cdot)$ exists, and may easily be shown to be unique.

At this point we remark that $\hat{f}(\cdot)$ can be constructed by finding (by any of the classical techniques) its numerator $P_{n-1}(\cdot)$ as that polynomial of degree $\leq$ n-1 which passes through the n points

$$\{(z_j, f_j \prod_{k=1}^{n}(1-z_j\bar{z}_k/r^2)); \ j=1,2,\ldots,n\}.$$

However, this requires a priori knowledge of all the data points $\{(z_j,f_j); \ j=1,2,\ldots,n\}$, whereas the techniques described below can be used to construct a sequence of optimal interpolants, efficiently updating each one as new data becomes available (or as old data becomes outdated).

(b) The "Neville-Aitken Table"

In Aitken's (1932) technique of recursive linear interpolation, as modified by Neville (1934), we construct a table of functions in the following format

$$\begin{array}{lllll}
z_1 & f_1 & & & \\
 & & f_{11} & & \\
z_2 & f_2 & & f_{12} & \\
 & & f_{21} & & f_{13} \\
z_3 & f_3 & & f_{22} & \vdots \\
 & & f_{31} & \vdots & \\
z_4 & f_4 & \vdots & & \\
\cdot & \cdot & & &
\end{array}$$

The general element $f_{jk}(\cdot)$ has the properties that

(i) it is a polynomial of degree $\leq$ k

(ii) $f_{jk}(z_s) = f_s$; s = j,j+1,...,j+k

(iii)

$$f_{jk}(z)=\frac{(z-z_j)f_{j+1,k-1}(z) + (z_{j+k}-z)f_{j,k-1}(z)}{z_{j+k}-z_j}; \quad j,k=1,2,\ldots$$

where

$$f_{jo}(z) = f_j \quad ; \forall\ z \in D \quad ; \quad j = 1,2,\ldots$$

Clearly, by the uniqueness theorem for interpolating polynomials, $f_{1,n-1}(z) \equiv Q_{n-1}(z)$.

An analogous technique for the construction of $\hat{f}(\cdot)$ is based on the observation that property (iii) above shows that $f_{jk}(\cdot)$ is a weighted, linear average of its two nearest neighbors in the preceding column. The appropriate generalization is to construct a similar table

$$\begin{array}{ccccc}
z_1 & \hat{f}_1 & & & \\
 & & \hat{f}_{11} & & \\
z_2 & \hat{f}_2 & & \hat{f}_{12} & \\
 & & \hat{f}_{21} & & \hat{f}_{13} \\
z_3 & \hat{f}_3 & & \hat{f}_{22} & \vdots \\
 & & \hat{f}_{31} & \vdots & \\
z_4 & \hat{f}_4 & \vdots & & \\
\cdot & \cdot & & &
\end{array}$$

where the general element $\hat{f}_{jk}(z)$ has the properties

(iv) it is a rational function whose numerator is a polynomial of degree $\leq k$ and whose denominator is

$$\prod_{s=j}^{j+k} (1-z\bar{z}_s/r^2)$$

(v) $$\hat{f}_{jk}(z_s) = f_s; \quad s=j,\ j+1,\ldots,j+k$$

(vi) $$\hat{f}_{jk}(z) = \hat{f}_{j+1,k-2}(z) + \alpha_{jk}(z)\ \{\hat{f}_{j+1,k-1}(z) - \hat{f}_{j+1,k-2}(z)\} + \beta_{jk}(z)\ \{\hat{f}_{j,k-1}(z) - \hat{f}_{j+1,k-2}(z)\}$$

where

$$\alpha_{jk}(z)=\frac{(z-z_j)(1-z_{j+k}\bar{z}_j/r^2)}{(z_{j+k}-z_j)(1-z\bar{z}_j/r^2)} \; ; \quad \beta_{jk}(z)=\frac{(z_{j+k}-z)(1-z_j\bar{z}_{j+k}/r^2)}{(z_{j+k}-z_j)(1-z\bar{z}_{j+k}/r^2)}$$

and

$$\hat{f}_{j+1,-1}(z)\equiv 0 \; ; \quad \hat{f}_{jo}(z)\equiv \frac{(1-|z_j|^2/r^2)}{(1-z\bar{z}_j/r^2)} \cdot f_j \; ; \quad j,k=1,2,\ldots$$

Here $f_{jk}(\cdot)$ is a weighted, linear average of its three (symmetrically) nearest neighbours in the two preceding columns, reducing to recurrence (iii) as $r\to\infty$.

It is easily verified by induction that properties (iv), (v) and (vi) are consistent, so the table exists; furthermore it is unique. Clearly $\hat{f}(z)\equiv\hat{f}_{1,n-1}(z)$.

(c) "Divided Differences"

A standard classical result (e.g. Milne-Thomson, 1933) is that the divided difference $f[z_j z_{j+1} \ldots z_{j+k}]$ based on the points $\{(z_s,f_s) \; ; \; s=j, j+1,\ldots,j+k\}$ is nothing other than the coefficient a_{jk} of z^k in $f_{jk}(z)$. This can actually be taken as the definition of $f[z_j,z_{j+1}\ldots z_{j+k}]$, and the classical recurrence relation

$$a_{jk}=\frac{a_{j+1,k-1}-a_{j,k-1}}{z_{j+k}-z_j} \; , \quad \text{where } a_{j0} = f_j;\; j,k=1,2,\ldots ,$$

deduced immediately from property (iii) above by equating coefficients of z^k on both sides of the relation.

We now define a generalized divided difference $\hat{f}[z_j,z_{j+1}\ldots z_{j+k}]$, or alternatively $\hat{a}_{jk}$, to be the coefficient of z^k in the polynomial numerator of $\hat{f}_{jk}(z)$, for all relevant j and k. By equating coefficients of z^k on both sides of relation (vi) we find that

$$\hat{a}_{jk} = \frac{(1-z_{j+k}\bar{z}_j/r^2)\hat{a}_{j+1,k-1} - (1-z_j\bar{z}_{j+k}/r^2)\hat{a}_{j,k-1}}{z_{j+k} - z_j} +$$

$$\hat{a}_{j+1,k-2}\,\frac{\bar{z}_{j+k}(1-z_{j+k}\bar{z}_j/r^2) - \bar{z}_j(1-z_j\bar{z}_{j+k}/r^2)}{r^2(z_{j+k}-z_j)}$$

where $\hat{a}_{j,-1} = 0;\quad \hat{a}_{jo} = (1-|z_j|^2/r^2)f_j;\quad j,k = 1,2,\ldots$.

Obviously $\hat{a}_{jk} \to a_{jk}$ as $r\to\infty$, $j,k=1,2,3,\ldots$ Also, by the definition of $\hat{f}[z_j z_{j+1} \ldots z_{j+k}]$, it is a symmetric function of its arguments.

Picking out the coefficients of z^{n-1} in the numerator of the "Lagrange form" of $\hat{f}(z)$ we see that

$$(3)\quad \hat{f}[z_1 z_2 \ldots z_n] = \sum_{j=1}^{n} (1-|z_j|^2/r^2)f_j \cdot \prod_{1=k\neq j}^{n} \left(\frac{1-z_j\bar{z}_k/r^2}{z_j-z_k}\right) =$$

$$\frac{1}{2\pi i}\int_C f(z)\cdot \prod_{k=1}^{n}\left(\frac{1-z\bar{z}_k/r^2}{z-z_k}\right) dz$$

so that

$$|\hat{f}[z_1 z_2 \ldots z_n]| \leq \frac{1}{2\pi}\left\{\int_C |f(z)|^2 |dz|\right\}^{\frac{1}{2}} \cdot \left\{\int_C \prod_{k=1}^{n} \left|\frac{1-z\bar{z}_k/r^2}{z-z_k}\right|^2 \cdot |dz|\right\}^{\frac{1}{2}},$$

by the Schwarz inequality. However,

$$(4)\quad \left|\frac{1-z\bar{z}_k/r^2}{z-z_k}\right| = \frac{1}{r} \quad \text{if} \quad |z| = r$$

so

$$(5)\quad |\hat{f}[z_1 z_2 \ldots z_n]| \leq \frac{\|f\|}{r^{n-1}} .$$

Notice that, although (3) can be written in the form

$$\hat{f}[z_1 z_2 \ldots z_n] = \frac{1}{2\pi r}\int_C f(z)\left\{\frac{1}{\bar{z}r^{2n-2}} \cdot \prod_{k=1}^{n}\left(\frac{\bar{z}-\bar{z}_k}{1-\bar{z}z_k/r^2}\right)\right\} |dz|$$

the function

$$\frac{1}{zr^{2n-2}} \cdot \prod_{k=1}^{n} \left(\frac{z-z_k}{1-z\bar{z}_k/r^2} \right)$$

is not analytic in D, so it cannot be the representer of the bounded, linear functional which evaluates $\hat{f}[z_1,z_2 \ldots z_n]$ from $f(\cdot)$, unless $z_k=0$ for some $k \in \{1,2,\ldots,n\}$.

(d) The "Newton Form" of $\hat{f}(\cdot)$

The quantity $\hat{f}_{1k}(z)-\hat{f}_{1,k-1}(z) = \dfrac{R_k(z)}{\prod_{j=1}^{k+1} (1-z\bar{z}_j/r^2)}$

where $R_k(z)$ must be of the form

$$R_k(z) = \hat{a}_{1k} \prod_{j=1}^{k} (z-z_j)$$

since it is a polynomial of degree $\leq k$, whose coefficient of z^k is $\hat{a}_{1k}$, which vanishes at the abscissae $\{z_s;\ s=1,2,\ldots,k\}$. Hence, summing along the leading diagonal in the generalized Neville-Aitken table, we see that

$$\hat{f}_{1,n-1}(z) = \sum_{s=0}^{n-1} \frac{\hat{a}_{1s}}{1-z\bar{z}_{s+1}/r^2} \prod_{t=1}^{s} \left(\frac{z-z_t}{1-z\bar{z}_t/r^2} \right)$$

or, in a nested product form suitable for computational purposes

$$\hat{f}_{1,n-1}(z) = \left(\frac{1}{1-z\bar{z}_1/r^2} \right) \left\{ \hat{a}_{10} + \left(\frac{z-z_1}{1-z\bar{z}_2/r^2} \right) \left\{ \hat{a}_{11} + \left(\frac{z-z_2}{1-z\bar{z}_3/r^2} \right) \left\{ \hat{a}_{12} + \ldots + \left(\frac{z-z_{n-1}}{1-z\bar{z}_n/r^2} \right) a_{1,n-1} \right\} \ldots \right\} \right\} .$$

Other, similar and equivalent, expressions can be constructed from any sequence of generalized divided differences lying on a connected, non-returning path in the table, starting from any of the elements $\{\hat{a}_{jo};j=1,2,\ldots,n\}$ and terminating at $\hat{a}_{1,n-1}$. Clearly, this "generalized Newton form" degenerates to a classical Newton form for $Q_{n-1}(z)$ as $r \to \infty$.

6. Error of the Optimal Approximation

Recalling the previous discussion of the "Newton form" of $\hat{f}(\cdot)$, and introducing an arbitrary point (z_{n+1}, f_{n+1}) where $f_{n+1} = f(z_{n+1})$, we see that

$$\hat{f}_{1n}(z) - \hat{f}_{1,n-1}(z) = \frac{\hat{a}_{1n}}{1-z\bar{z}_{n+1}/r^2} \cdot \prod_{j=1}^{n} \left(\frac{z-z_j}{1-z\bar{z}_j/r^2} \right) ;$$

$\forall\ z \in D$, so, in particular,

$$\hat{f}_{1n}(z_{n+1}) - \hat{f}_{1,n-1}(z_{n+1}) = \frac{\hat{a}_{1n}}{1-|z_{n+1}|^2/r^2} \cdot \prod_{j=1}^{n} \left(\frac{z-z_j}{1-z_{n+1}\bar{z}_j/r^2} \right) .$$

But, by construction, $\hat{f}_{1n}(z_{n+1}) = f(z_{n+1})$; also z_{n+1} is arbitrary so we can replace it by w to obtain

$$(6) \quad f(w)-\hat{f}(w) = \frac{\hat{f}\,[z_1 z_2 \ldots z_n w]}{1-|w|^2/r^2} \cdot \prod_{j=1}^{n} \left(\frac{w-z_j}{1-w\bar{z}_j/r^2} \right) ; \quad \forall\ w \in D .$$

Hence, using (4) and (5), we find

$$|f(w)-\hat{f}(w)| \leq \frac{||f||}{r^n\,(1-|w|^2/r^2)} \cdot \prod_{j=1}^{n} \left| \frac{w-z_j}{1-w\bar{z}_j/r^2} \right| ; \quad \forall\ w \in D.$$

Thus, for large r and in the absence of any further information about $f(\cdot)$, $\hat{f}(\cdot)$ will be optimal for the purpose of minimax approximation over $[-1,1]$, with respect to the abscissae $\{z_j;\ j=1,2,\ldots,n\}$ if these are chosen to be the zeros of the n-th order Chebyshev polynomial.

More generally, from (3) and (6), using a suffix to indicate the independent variable through which the operation is effected, we find that

$$L_w\{f(w)-\hat{f}(w)\} = \frac{1}{2\pi i} \int_C f(z) \cdot \prod_{k=1}^{n} \left(\frac{1-z\bar{z}_k/r^2}{z-z_k} \right) \cdot G(z) \cdot dz$$

where

$$G(z)=L_w\left\{\left(\frac{1-z\bar{w}/r^2}{z-w}\right)\cdot\frac{1}{(1-|w|^2/r^2)}\cdot\prod_{j=1}^{n}\left(\frac{w-z_j}{1-w\bar{z}_j/r^2}\right)\right\};\ \forall\ z\ \varepsilon\ D.$$

Hence, using (4) and the Schwarz inequality,

$$|L_w\left\{f(w)-\hat{f}(w)\right\}|\ \leq\ \frac{||f||}{\sqrt{2\pi}\cdot r^{n-\frac{1}{2}}}\ \cdot\left\{\int_C |G(z)|^2\cdot|dz|\right\}^{\frac{1}{2}}$$

and, after some algebraic manipulation and use of the Cauchy residue theorem, it may be shown that

$$\int_C |G(z)|^2\cdot|dz|=\frac{2\pi}{r^3}\left|L_w\left\{\frac{\bar{w}}{1-|w|^2/r^2}\cdot\prod_{j=1}^{n}\left(\frac{w-z_j}{1-w\bar{z}_j/r^2}\right)\right\}\right|^2 +$$

$$\frac{2\pi}{r}\sum_{k=0}^{\infty}\left|L_w\left\{\left(\frac{w}{r}\right)^k\prod_{j=1}^{n}\left(\frac{w-z_j}{1-w\bar{z}_j/r^2}\right)\right\}\right|^2 .$$

For example, if

$$Lf = \int_{-1}^{1} f(w)\cdot dw \quad ;\quad \forall\ f\ \varepsilon\ H_r$$

and r becomes large, the quantity $\int_C |G(z)|^2\cdot|dz|$ is minimized by choosing the abscissae $\{z_j;\ j=1,2,\ldots,n\}$ to be the zeros of the n-th order Legendre polynomial. In that case we finally obtain

$$\frac{1}{||f||}\cdot\left|\int_C\left\{f(w)\ -\hat{f}(w)\right\}\cdot dw\right|\ \leq\ 0(r^{-2n})\ .$$

Thus, Gaussian quadrature obtains in the limit as $r\to\infty$ of rules which are optimal with respect to both weights and abscissae, confirming a result given in another paper (Larkin, 1970).

7. References

1 Aitken, A.C., On Interpolation by Iteration of Proportional Parts, Without the Use of Differences, Proc. Roy. Soc. Edinburgh, 53 (1932), 54-78.

2 Aronszajn, N., Theory of Reproducing Kernels, Trans. Amer. Math. Soc., 68 (1950), 337-404. MR 14 #479.

3 Larkin, F.M., Optimal Approximation in Hilbert Spaces With Reproducing Kernel Functions, Math. Comp. 24, #112 (1970), 911-921.

4 Meschkowski, H., Hilbertsche Raume mit Kernfunktion, Springer-Verlag, (1962).

5 Milne-Thomson, L.M., The Calculus of Finite Differences, MacMillan, London, (1933).

6 Neville, E.H., Iterative Interpolation, J. Indian Math. Soc. 20 (1934), 87-120.

7 Sard, A., Linear Approximation, Math. Surveys No. 9, Amer. Math. Soc., Providence, R.I., (1963), MR 28 #1429.

F.M. Larkin
Department of Computing and Information Science
Queen's University
Kingston, Ontario Canada

APPROXIMATION BY INCOMPLETE POLYNOMIALS
(PROBLEMS AND RESULTS)

G. G. Lorentz

This paper surveys some problems and results in a new field, which may be termed approximation by incomplete or lacunary polynomials. Most of the problems raised are still awaiting a complete answer.

In the last years, when the supply of unsolved problems of Approximation Theory has been drying up, very popular have become problems of restricted approximation--problems in which polynomials or other means of approximation are restricted in some way. First to one's mind come here the investigations of D. J. Newman and his collaborators [3, Chapters X, XI] (also by M. v. Golitschek and others) on the Müntz theorem. Also the present author has contributed to problems of this kind, by considering polynomials with positive coefficients in x, $1-x$ in [5], and (in several joint papers with K. Zeller) approximation by monotone polynomials. Recent deep results concerning monotone approximation are due to R. DeVore [2]. In what follows, P_n will always stand for a polynomial of degree $\leq n$. In this paper, we want to consider approximation by "incomplete polynomials"

$$(0.1)\qquad P_n(x) = \sum_{i=1}^{t} a_i x^{k_i}$$

where $0 \leq k_1 < \dots < k_t \leq n$ are integers, and $t \leq n$, in particular by polynomials of the form

$$(0.2)\qquad P_n(x) = \sum_{k=s}^{n} a_k x^k$$

where $s > 0$ may be large. In Section 1 we consider polynomials of best approximation of form (0.2). Section 2 deals with the possibility of approximation by polynomials of types (0.1) and (0.2). The results of Section 1 are in collaboration with K. L. Zeller; in particular, he suggested the use of Bernstein's polynomials (1.1) and the proof of Theorem 2 by means of polynomials (1.2).

1 Best Approximation

LEMMA 1. <u>Let</u> P_n, P_{n+1} <u>be polynomials of best approximation (of degrees not exceeding</u> n <u>and</u> $n+1$, <u>respectively) to</u> $f \in C[a,b]$, <u>and let</u> $P_n \neq P_{n+1}$. <u>Then</u> $P_{n+1} - P_n$ <u>has</u> $n+1$ <u>distinct simple roots, which all lie in the open interval</u> (a,b).

Hence, if f is not a polynomial, it is impossible that for all large n, its polynomials P_n of best approximation are of form (0.2) with $s \geq 2$; and if $0 \notin (a,b)$, this is impossible even for $s \geq 1$. We want to show that s can be arbitrarily large for infinitely many n.

To S. Bernstein one owes the remark that if $b_k \geq 0$, $\sum b_k < +\infty$, then all partial sums of

$$(1.1) \quad f(x) = \sum_{k=1}^{\infty} b_k T_{3^k}(x)$$

(where T_n denotes the Chebyshev polynomial of degree n) are polynomials of best approximation to f. It is important to remark that the condition $b_k \geq 0$ is not essential.

THEOREM 1. <u>Let</u> n_k, p_k, $k = 1,2,\ldots$ <u>be odd integers, with</u> n_k <u>dividing</u> n_{k+1}, <u>let</u> $n_{k+1} \geq n_k p_k + 2$, $k = 1,2,\ldots$, <u>and let</u> $\sum |b_k| < +\infty$. <u>Then the partial sums</u> S_k, $k = 1,2,\ldots$ <u>of the series</u>

$$(1.2) \quad f(x) = b_0 x + \sum_{k=1}^{\infty} b_k T_{n_k}(x)^{p_k}$$

<u>are polynomials of best approximation to</u> f <u>on</u> $[-1,+1]$.

This is because the remainders $f - S_k$, after substitution $x = \cos\theta$, become periodic functions with small periods and odd about certain points. From this we can derive

THEOREM 2. <u>There exists a function</u> $f \in C[-1,+1]$, <u>not a polynomial, for which the polynomials of best approximation have, for infinitely many</u> n, <u>the form (0.2) with</u>
$s = s(n) \geq C(\log n/\log n)$.

One can take

$$f(x) = x + \sum_{k=1}^{\infty} b_k T_{n_k}(x)^{2k-1},$$

where $n_k = 3.\ 5.\ \ldots\ (2k+1)$, and the b_k are selected inductively to satisfy the condition that x is not present in $S_1 = x + b_1 T_{n_1}$, x^3 is not present in S_2, and so on.

The best result obtainable in this way seems to be

THEOREM 3. <u>There exists a function</u> $f \in C[-1,+1]$ <u>with the property that for infinitely many</u> n, <u>its polynomial of best approximation</u> (0.2) <u>has</u> $s(n) \geq C \log n$.

The proof is more computational than that of Theorem 2. It is based on properties of Vandermonde determinants,

$$V = V(x_1,\ldots,x_n) = \begin{vmatrix} 1 & 1 & \cdots & 1 \\ x_1 & x_2 & \cdots & x_n \\ \cdot & \cdot & \cdots & \cdot \\ x_1^{n-1} & x_2^{n-1} & \cdots & x_n^{n-1} \end{vmatrix}.$$

We denote by $V'_{i,k}$, $V''_{i_1,i_2;k_1,k_2}$ the subdeterminants obtained from V by removing row i, or rows i_1 and i_2, and column k, or columns k_1 and k_2, respectively.

LEMMA 2. <u>One has</u>

$$(1.3) \qquad \left|\frac{V'_{ik}}{V}\right| \leq \prod_{\substack{j=1 \\ j\neq k}}^{n} \left|\frac{i+x_j}{x_j - x_k}\right|,$$

$$(1.4)\qquad \left|\frac{V''}{V}\right| \le \prod_{j\ne k_1,\ \ne k_2} \frac{(1+x_j)^2}{|x_{k_1}-x_j|\,|x_{k_2}-x_j|}\,.$$

From this we derive $|V'/V| \le$ Const. $|V''/V| \le$ Const. if $x_k = 9^{k-1}$, $k = 1,\dots,n$.

For the proof of the theorem we put $f(x) = \sum b_k T_{3^k}(x)$. The coefficients of the Chebyshev polynomials are well known (see [10, p. 32]). We have

$$T_{3^k}(x) = \sum_{i=0}^{K} c_{ik} x^{2i+1}, \qquad K = \tfrac{1}{2}(3^k - 1),$$

$$c_{ik} = (-1)^{K+i} \frac{2K+1}{K+i-1} \binom{K+i}{2i+1} 2^{2i+1}$$

$$= (-1)^{K+i} \frac{1}{(2i+1)!} \left(1 + \frac{\alpha_{ik} k^2}{2^k}\right) 3^{(2i+1)k}, \qquad |\alpha_{ik}| \le \text{Const.},$$

the last equation valid for $i \le k$. The determinants of the c_{ik}, for $i = 0,\dots,p$, $k = p,\dots,2p$ are in a certain sense related to Vandermonde determinants. We now write the equations for the b_k, $k = p,\dots,2p$, which assert that the coefficients of x^{2i+1}, $i \le p$ in the sum $\sum_{k=1}^{2p} b_k T_{3^k}$ are zero. By means of Lemma 2 one shows that the determinants of the corresponding linear systems are not zero, and even more importantly, that the solutions b_k will be small: $|b_k| \le$ Const. k^{-2}.

In Theorems 2 and 3, the common zero of the P_n is in the <u>middle point</u> of the interval $[-1,1]$. One wonders where this is necessary. One can formulate several further interesting problems concerning incomplete polynomials of best approximation. For example:

CONJECTURE 1. <u>If all polynomials</u> P_n <u>of best approximation of</u> $f \in C[-1,+1]$ <u>vanish at</u> 0, <u>then</u> f <u>is odd</u>.

In the formula (0.1) let $t = t(n) \le n$ be fixed. We

consider polynomials of best approximation of form (0.1), with a_i, k_i variable, but with fixed t.

CONJECTURE 2. Among all polynomials (0.1), the polynomial of best approximation to x^n of degree $\leq n$ has powers $k_1 = n-t+1$, $k_2 = n-t+2, \ldots, k_t = n$.

This statement is true and easily proved (by means of Cauchy determinants) for the metric L^2. I am happy to report that this conjecture of mine has been meanwhile proved by Borosh, Chui, and Smith [1]. Their interesting proof uses the techniques of the Remez algorithm and the total positivity of determinants of S. Karlin.

2 Possibility of Approximation

Instead of best approximation, we consider here good approximation. We prove several theorems of Weierstrass type for $f \in C[0,1]$ and incomplete polynomials.

A function $s(n)$, $0 < s(n) \leq n$ has the Weierstrass property for polynomials (0.2), if for each $f \in C[0,1]$, $f(0) = 0$, there is a sequence of polynomials of type (0.2) converging uniformly to f.

THEOREM 4. A function $s(n)$ has the Weierstrass property if and only if

$$\frac{s(n)}{n} \to 0. \tag{2.1}$$

Proof. Let the condition be satisfied. We modify the Bernstein polynomials $B_n(x)$ of f to

$$P_n(x) = \sum_{s \leqq k \leqq n} f\left(\frac{k}{n}\right) p_{nk}(x).$$

If $\omega(h)$ is the modulus of continuity of f, then $\|P_n - B_n\| \leq \omega(s(n)/n) \to 0$, hence $P_n \to f$. The necessity of condition (2.1) follows from

THEOREM 5. For each $0 < \theta < 1$, there is a $0 < \delta < 1$ with the following property. If polynomials $P_n(x) = \sum_s^n a_k x^k$, $s \geq \theta n$,

<u>defined for infinitely many</u> n, <u>satisfy</u> $|P_n(x)| \le M$, $0 \le x \le 1$, <u>then</u> $P_n(x) \to 0$ <u>uniformly on</u> $[0,\delta]$.

<u>Proof</u>. We establish this for

$$(2.2) \quad \delta < \Delta_1(\theta) = \max_{0 \leqq \alpha \leqq 1} \alpha \left(\frac{1-\sqrt{\alpha}}{1+\sqrt{\alpha}} \right)^{\frac{1}{\theta}-1} .$$

Let $P_n(x) = x^{\lambda_n} Q_n(x)$, where Q_n are polynomials and $\lambda_n \ge \theta n$. Then for $0 < \alpha < 1$,

$$|Q_n(x)| \le M\alpha^{-\lambda_n}, \quad \alpha \le x \le 1.$$

We use a well-known lemma (see, for example, [10, p. 93]). If for a polynomial P_n of degree $\le n$, one has $|P_n(x)| \le M$ on an interval of length ℓ, then $|P_n(x)| \le M\left(\rho + \sqrt{\rho^2-1}\right)^n$ on the concentric interval of length $\rho\ell$, $\rho > 1$. In the present case,

$$|Q_n(x)| \le M\alpha^{-\lambda_n} \left(\rho + \sqrt{\rho^2-1}\right)^{n-\lambda_n}, \; 0 \le x \le 1+\alpha,$$

where $\rho = (1+\alpha)/(1-\alpha)$. Since $\rho + \sqrt{\rho^2-1} = \dfrac{1+\sqrt{\alpha}}{1-\sqrt{\alpha}}$, this gives

$$|P_n(x)| \le x^{\lambda_n} M\alpha^{-\lambda_n} \left(\frac{1+\sqrt{\alpha}}{1-\sqrt{\alpha}} \right)^{n-\lambda_n}, \quad 0 \le x \le 1.$$

Thus $P_n(x) \to 0$ on $[0,\Delta_1)$, where Δ_1 is found from the condition

$$M\Delta_1^{\lambda_n} \alpha^{-\lambda_n} \left(\frac{1+\sqrt{\alpha}}{1-\sqrt{\alpha}} \right)^{n-\lambda_n} \le 1,$$

and uniformly for $0 \le x \le \delta$, $\delta < \Delta_1$.

Of some interest is the determination of $\Delta(\theta)$, the supremum of numbers δ, for which Theorem 5 is true. We have

$$(2.3) \quad \Delta(\theta) \le \theta.$$

Indeed, let $f(x) = 0$ on $[0,\theta]$, and $f(x) > 0$ on $[\theta,1]$. Then the Bernstein polynomials $P_n(x)$ of f converge to zero exactly on $[0,\theta]$, and are of form (0.2) with $s \ge \theta n$.

CONJECTURE 3. <u>We have</u> $\Delta(\theta) = \theta, \quad 0 < \theta < 1.$

At the Conference, I have been informed by a reliable authority (R. S. Varga) that the majority of participants present at my talk thought this conjecture false. Below, I report upon developments A, B, C after the Conference.

<u>A</u>. First of all, Conjecture 3 is indeed false, for any θ, $0 < \theta < 1$ (C. H. FitzGerald and D. Wulbert for $\theta = 1/2$). The function $x^{-\theta/(1-\theta)}$ is convex, hence its tangent line at a, $0 < a < 1$, lies below the curve:

$$L(x) = \frac{1}{1-\theta} a^{-\frac{\theta}{1-\theta}} - \frac{\theta}{1-\theta} a^{-\frac{1}{1-\theta}} x \leq x^{-\frac{\theta}{1-\theta}} .$$

If $a = \theta$, then $L(1) = 0$. But we select $a = a(\theta)$ so that $L(1) = -1$. Then $0 < a(\theta) < \theta$, and still $|L(x)| \leq x^{-\theta/(1-\theta)}$ on $[0,1]$. Thus, for $t = [n(1-\theta)]$, $s = n - t$, $P_n(x) = x^s L(x)^t$, we have $s \geq n\theta$ and

$$|P_n(x)| \leq x^s x^{-\frac{t\theta}{1-\theta}} = x^{\frac{\epsilon}{1-\theta}} \leq 1, \quad 0 \leq x \leq 1,$$

where $\epsilon = n(1-\theta) - [n(1-\theta)]$, and similarly $P_n(a) = a^{\epsilon/(1-\theta)} \geq$ Const. Since $P_n(a) \not\to 0$, we have $\Delta(\theta) \leq a(\theta) < \theta$. The number $a(\theta)$ can be found from the equation

$$\theta - a = (1-\theta) a^{\frac{1}{1-\theta}}.$$

It follows that

$$\text{(2.4)} \qquad a > \frac{1}{2}\theta \quad \text{and} \quad \frac{a}{\theta} \to \frac{1}{2} \quad \text{for} \quad \theta \to 0.$$

<u>B</u>. We shall prove that $\Delta(\theta) \geq \theta^2$. For this purpose we need a lemma due to Rahman and Schmeisser [8], communicated to the author at the Conference.

LEMMA 3. <u>Let</u> $M(x)$ <u>be a continuous function defined on</u> $[a,b]$, <u>and for a polynomial</u> P_n <u>of degree</u> $\leq n$, <u>let</u> $|P_n(x)| \leq M(x)$ <u>on</u> $[a,b]$. <u>Then for</u> $x < a$,

$$(2.5)\quad \begin{cases} |P_n(x)| \le \dfrac{1}{r^n} \exp\left\{ \dfrac{1}{2\pi} \displaystyle\int_0^{2\pi} \dfrac{(1-r^2)\log M\left(\frac{b-a}{2}\cos t + \frac{b+a}{2}\right)}{1+2r\cos t + r^2} dt \right\}, \\ r = \delta - \sqrt{\delta^2 - 1}, \qquad \delta = \dfrac{b+a-2x}{b-a}. \end{cases}$$

THEOREM 5. <u>One has</u> $\Delta(\theta) \ge \theta^2$ <u>for</u> $0 < \theta < 1$.

<u>Proof</u>. Let $P_n(x) = x^s Q_n(x)$, $s = [\theta n]$ be polynomials of degree $\le n$, which satisfy $|P_n(x)| \le 1$ on $[0,1]$. Then Q_n are polynomials of degree $n-s$, and we apply for the estimate (2.5) with $M(x) = x^{-s}$, $b = 1$, and $0 < a < \theta^2$. The expression in curved brackets in (2.5) is $-s$ times the value at $t = \pi$ of the Poisson transform $\mathscr{P}_r(f,t)$ of the function

$$f(t) = \log\left(\frac{1-a}{2}\cos t + \frac{1+a}{2}\right).$$

We use the formula

$$(2.6)\quad \lim_{r\to 1}(1-r)^{-1}[\mathscr{P}_r(f,t) - f(t)] = -\tilde{f}'(t),$$

where $\tilde{f}$ is the conjugate function of f,

$$(2.7)\quad \tilde{f}(x) = -\frac{1}{2\pi}\int_0^{\pi}[f(x+t) - f(x-t)]\cot\frac{t}{2}\,dt.$$

Since the general term $A_n(t)$ of the Fourier series of f satisfies $|A_n(t)| \le \text{Const.}\ n^{-3}$, formula (2.6) follows by passing to the limit termwise in the representation $\sum A_n(t)(r^n-1)/(1-r)$ of the expression under the limit in (2.6). As $f' \in \text{Lip}\,1$, the differentiation in (2.7) under the integral sign is permissible, by Lebesgue bounded convergence theorem, and gives

$$-\tilde{f}'(x) = \frac{1}{2\pi}\int_0^{\pi}[f'(x+t) - f'(x-t)]\cot\frac{t}{2}\,dt$$

$$-\tilde{f}'(\pi) = \frac{1-a}{\pi}\int_0^{\pi}\frac{\sin t}{(1+a) - (1-a)\cos t}\cot\frac{t}{2}\,dt$$

$$= \frac{2(1-a)}{\pi} \int_0^\infty \frac{du}{(1+u^2)(a+u^2)} = \frac{1-\sqrt{a}}{\sqrt{a}} .$$

Since $f(\pi) = \log a$, we have from (2.6)

$$\mathscr{P}_r(\pi) = \log a + (1-r)\frac{1-\sqrt{a}}{\sqrt{a}} + o(1-r), \quad r \to 1.$$

We apply (2.5) to $Q_n(x)$ and obtain for $0 \leq x \leq a$,

$$|P_n(x)| \leq x^s r^{-(n-s)} \exp(-s\mathscr{P}_r(\pi)),$$

$$\frac{1}{n}\log |P_n(x)| \leq \frac{s}{n}\log x + \frac{n-s}{n}\log\frac{1}{r} - \frac{s}{n}\mathscr{P}_r(\pi),$$

therefore,

$$\text{(2.8)} \quad \lim_{n\to\infty} \frac{1}{n}\log |P_n(x)| \leq (1-\theta)\log\frac{1}{r} + \theta(\log x - \mathscr{P}_r(\pi)),$$

$$\leq \theta\left\{\left(\frac{1}{\theta}-1\right)\log\frac{1}{r} - \left(\frac{1}{\sqrt{a}}-1\right)(1-r) + o(1-r)\right\} .$$

For $0 < x < a$ we have $\delta > 1$, $0 < r < 1$, with $r \to 1$ as $x \to a$. In this case, $\frac{1}{1-r}\log r \to 1$. Since $\sqrt{a} < \theta$, the expression in curved brackets is <0 for all x sufficiently close to a. It follows that for each $a < \theta^2$ and some $\epsilon > 0$, $P_n(x) \to 0$ uniformly on $[a-\epsilon, a]$. By "induction in the continuum" we obtain $P_n(x) \to 0$ on $[0,\theta^2)$.

Remark. Possibly one can obtain an even larger lower bound for $\Delta(\theta)$ by combining the methods of proof of Theorems 4 and 5. For this purpose, one uses the saturation estimate $\|f-\mathscr{P}_n(f)\| \leq (1-r)\|\tilde{f}\|$ instead of (2.6), this gives $\mathscr{P}_n(f,\pi) \leq \log a + (1-r)\|\tilde{f}\|$. Then $\|\tilde{f}\|$ is computed as a function of a. Finally, in (2.8), for a given x, $0 < x < 1$, one selects the most favorable a.

C. Saff and Varga (oral communication) announce that if $\theta = N^{-1}$, where N is a sufficiently large integer, then $\Delta(\theta) \leq 1.83\theta^2$.

Another theorem of Weierstrass type concerns polynomials of type (0.1). We shall say that an integer-valued function $t(n)$, $0 < t(n) \leq n$, $t(n) \to \infty$ <u>has the Weierstrass property for polynomials</u> (0.1) if for each function $f \in C[0,1]$, vanishing at 0, and for each matrix of integers

$$(2.9) \quad 0 \leq k_1^{(n)} < \dots < k_t^{(n)} \leq n, \ n = 1, 2, \dots, \ k_i^{(n)} = k_i$$

there exists a sequence of polynomials (0.1) which converges uniformly to $f(x)$.

THEOREM 6. <u>A function</u> $t(n)$ <u>has the Weierstrass property if and only if</u>

$$(2.10) \quad t(n)/n \to 1.$$

<u>Proof</u>. In view of Theorem 4, we have only to show that the approximation is possible if (2.10) is satisfied. According to a theorem from [3, pp. 122, 126], for given integers k_i $i = 1,\dots,t$, there exists a linear combination P of 1 and the x^{k_i}, which satisfies

$$(2.11) \quad \|f-P\| \leq \rho_n = 2\omega(f, 184\epsilon_n).$$

Here

$$(2.12) \quad \epsilon_n = \sup_{\mathrm{Re}\, z = 1} \left| \frac{1}{z} \prod_{i=1}^{t} \frac{z-k_i}{z+k_i} \right|,$$

or

$$\epsilon_n^2 = \sup_y \left\{ \frac{1}{1+y^2} \prod_{i=1}^{t} \frac{y^2 + (k_i-1)^2}{y^2 + (k_i+1)^2} \right\} = \sup_y \Phi(y),$$

say. Since $f(0) = 0$, the free term of P_n is in absolute value $\leq \rho_n$; hence $\|f - \bar{P}_n\| \leq 2\rho_n$, where $\bar{P}_n$ is obtained from P_n by omitting the free term.

We have to show that $\epsilon_n^2 \to 0$ for $n \to \infty$ and any selection of the integers k_i in (2.9). The function of k, $k \geq 0$

$$\phi(k) = \frac{y^2+(k-1)^2}{y^2+(k+1)^2}$$

decreases for $k \leq Y = (1+y^2)^{1/2}$, and increases for $k \geq Y$. Let $t = t_1 + t_2$, where t_1 is the number of $k_i \leq Y$, and t_2 is the number of $k_i > Y$. Then

$$\Phi(y) \leq \frac{1}{1+y^2} \prod_{k=1}^{t_1} \phi(k) \prod_{k=n-t_2+1}^{n} \phi(k) = \Omega(y)$$

$$= \frac{y^2}{(y^2+t_1^2)(y^2+(t_1+1)^2)} \quad \frac{(y^2+(n-t_2)^2)(y^2+(n-t_2+1)^2)}{(y^2+n^2)(y^2+(n+1)^2)} .$$

We can assume that $t_1 \leq t^{1/2}$, for otherwise

$$\Omega(y) \leq \frac{1}{y^2+t_1^2} \leq \frac{1}{t} .$$

Then $n - t_2 = n - t + t_1 \leq n - t + t^{1/2}$, and therefore, from (2.10), $(n-t_2)/n \to 0$. We have

$$\Omega(y) \leq \frac{1}{y^2} \quad \text{and} \quad \Omega(y) \leq \frac{y^2+(n-t_2)^2}{n^2} .$$

Hence $\Omega(y) < \epsilon$ if $y^{-2} < \epsilon$. If $y^{-2} \geq \epsilon$, then

$$\Omega(y) \leq \frac{\epsilon^{-1}}{n^2} + \left(\frac{n-t_2}{n}\right)^2 < \epsilon$$

for all sufficiently large n. Hence $\Omega(y) \to 0$ uniformly in y for $n \to \infty$.

Interesting are also problems of approximation of functions which depend on n. We quote here

THEOREM 7 (Newman and Rivlin [7]). <u>The monomial</u> x^n <u>is approximable on</u> $C[0,1]$ <u>by polynomials of the form</u> $P_n(x) = \sum_{k=0}^{t} a_k x^k$, $t = t(n)$ <u>if and only if</u>

(2.13) $$\frac{t(n)}{\sqrt{n}} \to +\infty.$$

The "if" part of this theorem can be derived from another result of Newman [3, p. 125]: The degree of approximation $E_\Lambda(x^n)$ of x^n on $C[0,1]$ by linear combination of the powers x^{k_i}, $k_i \in \Lambda$ satisfies

$$(2.14) \quad E_\Lambda(x^n) \leq \prod_{k_i \in \Lambda} \left| \frac{n-k_i}{n+k_i} \right| .$$

Instead of showing this here, we prefer to give a proof of a companion theorem, in which, however, we know only a sufficient condition.

THEOREM 8. The monomial x^n is approximable on $C[0,1]$ by polynomials of the form $\sum_{s(n)}^{t(n)-1} a_k x^k$, if $\left(\sqrt{2}+\epsilon\right) n \leq s(n) < t(n)$ for some $\epsilon > 0$ and if

$$(2.13) \quad n \frac{t(n) - s(n)}{t(n) - n} \to +\infty.$$

Proof. We have to show that $\rho_n \to 0$, where

$$\rho_n = \prod_{k=s}^{t-1} \frac{k-n}{k+n} = \frac{\Gamma(t-n)\Gamma(s+n)}{\Gamma(t+n)\Gamma(s-n)} .$$

Using Stirling's formula we find $\rho_n \leq \text{Const.}\ \bar{\rho}_n$,

$$\bar{\rho}_n = \frac{(t-n)^{t-n}(s+n)^{s+n}}{(t+n)^{t+n}(s-n)^{s-n}} = \left(1-\frac{n}{t}\right)^t\left(1+\frac{n}{t}\right)^{-t}\left(1+\frac{n}{s}\right)^s\left(1-\frac{n}{s}\right)^{-s} \left(1-\frac{t^s-s^2}{t^2-n^2}\right)^n .$$

We show that $\log \bar{\rho}_n \to -\infty$. Taking logarithms and expanding,

$$\log \bar{\rho}_n = 2\left(\frac{n^3}{3s^2}+\frac{n^5}{5s^4}+\frac{n^7}{7s^6}+\ldots\right) - 2\left(\frac{n^3}{3t^2}+\frac{n^5}{5t^4}+\ldots\right) - n\left\{\frac{t^2-s^2}{t^2-n^2}+\frac{1}{2}\left(\frac{t^2-s^2}{t^t-n^2}\right)^2+\ldots\right\} .$$

The sum of the first two expressions is

$$2n \sum_{p=1}^{\infty} \frac{n^{2p}}{2p+1} \left(\frac{1}{s^{2p}} - \frac{1}{t^{2p}} \right) \leq 2n \left(t^2 - s^2 \right) \sum_{p=1}^{\infty} \frac{pn^{2p} t^{2p-2}}{(2p+1) s^{2p} t^{2p}}$$

$$\leq n \frac{t^2 - s^2}{t^2} \sum_{p=1}^{\infty} \left(\frac{n^2}{s^2} \right)^p = n \frac{t^2 - s^2}{t^2} \frac{n^2}{s^2 - n^2} \leq \frac{1}{1+\epsilon} n \frac{t^2 - s^2}{t^2 - n^2} .$$

Therefore,

$$\log \bar{\rho}_n \leq - \frac{\epsilon}{1+\epsilon} n \frac{t-s}{t-n} \frac{t+s}{t+n} \leq -C_1 n \frac{t-s}{t-n} ,$$

where $C_1 > 0$ is a constant.

Concluding, we shall mention the interesting dissertation of M. Hasson [4], which deals with related questions.

References

[1] Borosh, I., C. K. Chui, and P. W. Smith, Best uniform approximation from a collection of subspaces, in print.

[2] DeVore, R. A., (I) Monotone approximation by splines, (II) Monotone approximation by polynomials, in print in SIAM J. Numer. Anal.

[3] Feinerman, R. P., and D. J. Newman, Polynomial Approximation, Williams and Wilkins Co., Baltimore, 1974.

[4] Hasson, M., Comparison between the degrees of approximation by lacunary and ordinary algebraic polynomials, Ph.D. thesis, Queen's University, Kingston, Ont., Canada, Dec. 1976.

[5] Lorentz, G. G., The degree of approximation by polynomials with positive coefficients, Math. Ann. 151 (1963), 239-251.

[6] Lorentz, G. G., and K. L. Zeller, Best approximation by incomplete polynomials, in preparation.

[7] Newman, D. J., and T. J. Rivlin, Approximation of monomials by lower degree polynomials, Aequationes Math. 14 (1976), 451-455.

[8] Rahman, Q. I., and G. Schmeisser, Rational approximation to e^{-x}, in print.

[9] Riess, R. D., and L. W. Johnson, Estimates for $E_n(x^{n+2m})$, Aequationes Math. 8 (1972), 258-262.

[10] Rivlin, T. J., The Chebyshev Polynomials, J. Wiley and Sons, New York, 1974.

G. G. Lorentz
Department of Mathematics
The University of Texas at Austin
Austin, Texas 78712

Research supported in part by Grant MPS75-09833 of the National Science Foundation.

ON THE CARDINALITY OF A SET OF BEST COMPLEX RATIONAL APPROXIMATIONS TO A REAL FUNCTION

Arden Ruttan

Let $\pi_{1,1}^{c}$ be the set of linear fractional transformations. In this note, we show that there exists an even continuous function on $[-1,1]$ with a continuum of best uniform approximations from $\pi_{1,1}^{c}$.

1 Introduction

Let f(x) be a continuous real-valued function defined on the closed interval $[-1,1]$. For any pair (m,n) of non-negative integers, let $\pi_{m,n}^{r} := \{P/Q$: P, Q are polynomials with real coefficients, degree of $P \leq m$, and degree of $Q \leq n\}$. Define $\pi_{m,n}^{c}$ in an analogous manner by admitting complex coefficients. The classical theory of rational approximation asserts that there is a <u>unique</u> best uniform approximation to f on $[-1,1]$ from $\pi_{m,n}^{r}$ (cf. [3, pg. 161]). In a recent paper, [4], E. B. Saff and R. S. Varga, showed that there need not be a <u>unique</u> best uniform approximation to f(x) on $[-1,1]$ from $\pi_{m,n}^{c}$. More specifically, for any function g defined on $[-1,1]$, let

$$\|g\| := \sup_{x \in [-1,1]} |g(x)|, \quad E_{m,n}(g) := \inf_{R \in \pi_{m,n}^{c}} \|g - R\|, \text{ and}$$

$B_{m,n}(g) := \{R \in \pi_{m,n}^{c} : E_{m,n}(g) = \|g - R\|\}$. Then, they showed that whenever f is an even continuous real-valued function on $[-1,1]$ with f monotone and non-constant on $[0,1]$, the cardinality of $B_{1,1}(g)$ is at least 2.

One question they posed in [5] is whether there exists a continuous function f such that the cardinality of $B_{m,n}(f)$ is infinite for some non-negative integers m and n. Surprisingly, the answer to that question is in the affirmative. Indeed, the final result of this note, Theorem 6, shows that there exists an even continuous function f such that $B_{1,1}(f)$ has the cardinality of the continuum.

The heart of this note is a determination of the minimal error of approximation of a function on certain four point sets by elements of $\pi_{1,1}^{c}$, and an explicit representation of the extremal functions. That is, if f is a function on a four point set $A := \{x_1, x_2, x_3, x_4\}$ such that $x_1 > x_2 > x_3 > x_4$, $f(x_1) = f(x_4)$, and $f(x_2) = f(x_3)$, we determine both (cf. Theorem 1)
$\min_{R \in \pi_{1,1}^{c}} \max_{x \in A} |f(x) - R(x)|$ and all $R \in \pi_{1,1}^{c}$ for which the minimal deviation is achieved. Those functions which achieve the minimal error on A have an interesting geometric characterization. Since $\pi_{1,1}^{c}$ is actually the set of linear fractional transformations, the image of the real axis under an element of $\pi_{1,1}^{c}$ is a (possibly degenerate) circle. It turns out that a function R in $\pi_{1,1}^{c}$ achieves the minimal deviation from f on A if and only if the error function f(x) - R(x) is tangent to the R-image of the real axis at $R(x_k)$, $k = 1,2,3,4$ and $|f(x_1) - R(x_1)| = |f(x_2) - R(x_2)| = |f(x_3) - R(x_3)| = |f(x_4) - R(x_4)|$. This is illustrated in Figure 1, where the R-image of the real axis is the circle with center c. As we shall see below, there is a continuum of such functions, and therein lies the key to Theorem 6. By suitably choosing our four point set A and the values of f on A, we can extend f continuously to all of [-1,1] such that a continuum of the extremal functions associated with f on A achieve their maximum deviation from f on [-1,1] only at points of A. Since no function in $\pi_{1,1}^{c}$ can produce a smaller deviation on A, we have a continuum of best approximates to f on [-1,1] from $\pi_{1,1}^{c}$.

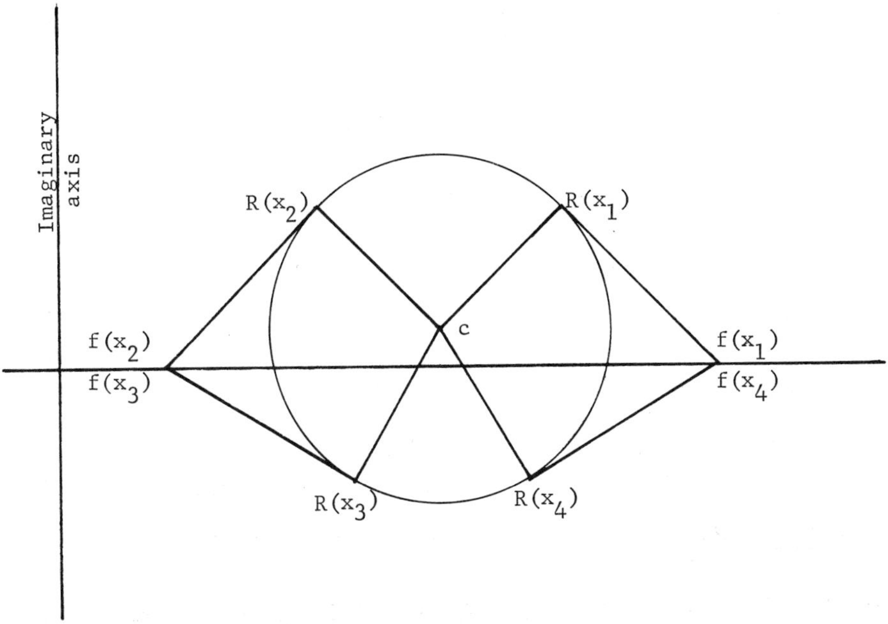

Figure 1

Actually, Theorem 1 is an extension of a result of C. Bennett, K. Rudnick, and J. Vaaler. In [1], they prove the result for symmetric linear fractional transformations (functions in $\pi^c_{1,1}$ which satisfy $R(x) = \overline{R(-x)}$ for all real x). Specifically, they proved

THEOREM A. *Let f be an even real-valued function on* $[-1,1]$ *with* $1 = f(1) = f(-1)$ *and* $0 = f(0) \leq f(x) \leq 1$ *for* $x \in [-1,1]$. *Let* $\omega \in (0,1)$ *be fixed and* $A(\omega) := \{-1,-\omega,\omega,1\}$. *If U is any symmetric linear fractional transformation, then*

$$(1.1)\qquad \max_{x \in A(\omega)} |U(x) - f(x)| \geq \omega^{\frac{1}{2}}(1+\omega)^{-1}(1 - f(\omega)).$$

Equality holds in (1.1) *if and only if* $U(x) = U(x;f,\omega)$ *or* $\overline{U(x)} = U(x;f,\omega)$ *where*

$$\text{(1.2)}\quad U(x;f,\omega) = s + r\,\frac{x+it}{x-it} \quad \text{with}$$
$$r = r(f,\omega) = \frac{(1-\omega)(1-f(\omega))}{2(1+\omega)}$$
$$s = s(f,\omega) = \frac{1+f(\omega)}{2}$$
$$t = t(\omega) = \omega^{\frac{1}{2}}.$$

Their observation that the "tangent condition" held for extremal symmetric linear transformations led us to the more general result.

In the same paper they established

THEOREM B. *Let f be as in Theorem A and let* $\Delta(f) := \sup_{0<\omega<1} \omega^{\frac{1}{2}}(1+\omega)^{-1}(1-f(\omega))$. *If U is any symmetric linear fractional transformation then* $\|U-f\| \geq \max(\frac{1}{4},\Delta(f))$.

THEOREM C. *Let f be as in Theorem A. Further suppose that f is such that:*

(1.3) f is continuous on $[-1,1]$, is differentiable on $(0,1)$ and $f'(x) \geq 0$ for each $x \in (0,1)$;

(1.4) there exists a unique point $\Omega \in (0,1)$, depending only on f, such that $\Delta(f) = \Omega^{\frac{1}{2}}(1+\Omega)^{-1}(1-f(\Omega))$; and

(1.5) the map $x \to (x^2+\Omega)^2 x^{-1} f'(x)$ is increasing on $(0,1)$.

Then, for any symmetric linear fractional transformation U,

$$\text{(1.6)}\quad \|U-f\| \geq \Delta(f) > \tfrac{1}{4},$$

with equality holding in (1.6) if and only if $U(x) = U(x;f,\Omega)$ *or* $\overline{U(x)} = U(x,f,\Omega)$.

THEOREM D. *Let* $f_\alpha(x) = |x|^\alpha$ *where* $\alpha > 0$. *Let* $\varkappa$ *be the unique solution in the interval* $(1,\infty)$ *of the equation* $(2\varkappa-1)^{(2\varkappa-1)} = \varkappa(\varkappa-1)^{-1}$. *Then* f_α *satisfies (1.4) and (1.5). For* $\alpha > 0$ *let* Ω_α *denote the value of* Ω *determined by (1.4). Let U be a symmetric linear fractional transformation. Then*

(1.7) If $\varkappa \le \alpha$, then $\|U - f_\alpha\| \ge \Delta(f_\alpha)$ with equality holding if and only if $U(x) = U(x,f_\alpha,\Omega_\alpha)$ or $\overline{U(x)} = U(x;f_\alpha,\Omega_\alpha)$.

(1.8) If $0 < \alpha < \varkappa$ then

$$\|U - f\| > \max(\tfrac{1}{4},\Delta(f_\alpha)).$$

With our extension of Theorem A, Theorems B, C, and D can be extended to the whole class of linear fractional transformation.

2 Main Results

In this section we state our main results while deferring technical lemmas and proofs to §3.

THEOREM 1. *Let f be a real-valued function defined on a set* $A := \{x_1,x_2,x_3,x_4\}$ *where* $x_1 > x_2 > x_3 > x_4$. *Further, suppose that* $f(x_1) = f(x_4)$ *and* $f(x_2) = f(x_3)$. *Set*

$$K := (x_3-x_1)(x_2-x_4)(x_3-x_4)^{-1}(x_2-x_1)^{-1},$$

and $\delta := \tfrac{1}{2}|f(x_1)-f(x_2)|(K-1)^{\frac{1}{2}}K^{-\frac{1}{2}}$. *Then for any linear fractional transformation* $T(z)$

$$\max_{x\in A} |T(x) - f(x)| \ge \delta. \tag{2.1}$$

Moreover, equality is achieved in (2.1) *if and only if*

(2.2) $T(x_k) = f(x_k) + \delta\beta_k$, $k = 1,2,3,4$, for some complex $\beta_1,\beta_2,\beta_3,\beta_4$ satisfying $|\beta_4| = 1$; $\beta_1 = -\dfrac{a\beta_4 + 1}{a + \beta_4}$ where $a := K^{\frac{1}{2}}(K-1)^{-\frac{1}{2}}$; $\beta_2 = -\overline{\beta}_1$; and $\beta_3 = -\overline{\beta}_4$.

THEOREM 2. *Let f be a real-valued function on* $[-1,1]$. *Suppose that there exist four points* $1 \ge x_1 > x_2 > x_3 > x_4 \ge -1$ *and a linear fractional transformation T such that*

(i) $f(x_1) = f(x_4)$ *and* $f(x_2) = f(x_3)$,

(ii) $\|T(x) - f(x)\| = \tfrac{1}{2}|f(x_1) - f(x_2)|\cdot(K-1)^{\frac{1}{2}}K^{-\frac{1}{2}}$ *where* $K = (x_3-x_1)(x_2-x_4)(x_3-x_4)^{-1}(x_2-x_1)^{-1}$.

Then, $\|T-f\| = E_{1,1}(f)$ *and* $T \in B_{1,1}(f)$.

THEOREM 3. *Let f be an even real-valued function on* $[-1,1]$ *such that* $0 = f(0) \leq f(x) \leq 1 = f(1) = f(-1)$ *for* $x \in [-1,1]$, *and let* $\Delta(f) := \sup_{0 \leq \omega \leq 1} \omega^{\frac{1}{2}}(1+\omega)^{-1}(1 - f(\omega))$. *If U is any linear fractional transformation, then*

$$\text{(2.3)} \quad \|U - f\| \geq \max(\tfrac{1}{4}, \Delta(f)).$$

THEOREM 4. *Let f and* $\Delta(f)$ *be as in Theorem 3. Further suppose that f satisfies (1.3), (1.4), and (1.5). Then for any linear fractional transformation U,*

$$\text{(2.4)} \quad \|U - f\| \geq \Delta(f) > \tfrac{1}{4}.$$

If $U(x) = U(x;f,\Omega)$ *or* $\overline{U(x)} = U(x;f,\Omega)$ *where* $U(x;f,\Omega)$ *is given by (1.2) then equality is achieved in (2.4).*

THEOREM 5. *Let* $f_\alpha(x) = |x|^\alpha$ *where* $\alpha > 0$. *Let* $\varkappa$ *be the unique solution in the interval* $(1,\infty)$ *of the equation* $(2\varkappa-1)^{(2\varkappa-1)} = \varkappa(\varkappa-1)^{-1}$. *Then* f_α *satisfies (1.4) and (1.5). For each* $\alpha > 0$ *let* Ω_α *denote the value of* Ω *determined by (1.4). Let U be any linear fractional transformation. Then*

(2.5) *If* $\varkappa \leq \alpha$, *then* $\|U - f_\alpha\| \geq \Delta(f_\alpha)$ *and equality holds when* $U(x) = U(x;f_\alpha,\Omega_\alpha)$ *or* $\overline{U(x)} = U(x;f_\alpha,\Omega_\alpha)$,

(2.6) *If* $0 < \alpha < \varkappa$ *then*

$$\|U - f\| > \max(\tfrac{1}{4}, \Delta(f_\alpha)).$$

THEOREM 6. *There exists an even continuous real-valued function f on* $[-1,1]$ *such that* $B_{1,1}(f)$ *has the cardinality of the continuum.*

As previously remarked, Theorem 6 answers question 1 raised in [5].

3 Proofs

For any complex number z and real $r > 0$ let $D(z,r) := \{\omega : |\omega - z| < r\}$, and let $\overline{D(z,r)}$ and $D(z,r)^c$ denote respectively the closure of $D(z,r)$ and the complement of $D(z,r)$ in the extended complex plane.

LEMMA 1. Suppose $\hat{a} > 2$ and $\beta \in \overline{D(\hat{a},1)}$. Let T be the linear fractional transformation $T(z) = \frac{z - \beta}{1 - z\bar{\beta}}$. Then for any $z \in D(\hat{a},1)$,

$$(3.1)\quad |T(z)| \le \frac{2}{\hat{a}^2 - 2}.$$

Proof: Let $\hat{K} := \frac{1}{2}(\hat{a}^2 - 2)$, and let S be the linear fractional transformation $S(z) = \frac{z+\beta}{1+z\bar{\beta}}$. Since S is the inverse of T, to establish (3.1), it suffices to show that the S-image of $\overline{D(0,\hat{K}^{-1})}$ contains $\overline{D(\hat{a},1)}$. We first consider the case when $|\beta| > \hat{K}$. Then, the pole of S is contained in $D(0,\hat{K}^{-1})$, and consequently the S-image of the circle $\{z: |z| = \hat{K}^{-1}\}$ is a circle. By the symmetry principle, that circle has center $\hat{c} := S(-\beta\hat{K}^{-2}) = \beta(\hat{K}^2-1)(\hat{K}^2-|\beta|^2)^{-1}$, and the radius of that circle is $\hat{r} := |\hat{c}-S(\hat{K}^{-1})| = \hat{K}(|\beta|^2-1)||\beta|^2-\hat{K}^2|^{-1}$. As the pole of S is contained in $D(0,\hat{K}^{-1})$, the S-image of $D(0,\hat{K}^{-1})$ is $D(\hat{c},\hat{r})^c$. Thus, to establish (3.1) for this case, we must show that $\overline{D(\hat{a},1)} \subseteq D(\hat{c},\hat{r})^c$, or, equivalently, that

$$(3.2)\quad |\hat{c} - \hat{a}| \ge \hat{r} + 1.$$

A calculation shows that (3.2) holds if and only if

$$(3.3)\quad |\beta(\hat{K}^2-1)+\hat{a}(|\beta|^2-\hat{K}^2)|^2 \ge (|\beta|^2-\hat{K})^2(\hat{K}+1)^2.$$

Let $\theta := \arg\beta$. Expanding the left side of (3.3) yields

$$(3.4)\quad |\beta|^2(\hat{K}^2-1)^2+\hat{a}^2(|\beta|^2-\hat{K}^2)^2+2|\beta|\hat{a}(\hat{K}^2-1)(|\beta|^2-\hat{K}^2)\cos\theta.$$

Since $|\beta-\hat{a}| \le 1$, $|\beta|^2+\hat{a}^2-2\hat{a}\,\mathrm{Re}\,\beta \le 1$ and

$$(3.5)\quad \cos\theta \ge (|\beta|^2+a^2-1)(2\hat{a}|\beta|)^{-1}.$$

Consequently, as the coefficient of cos θ in (3.4) is positive, by combining (3.4) and (3.5), we find that the left side of (3.3) satisfies

$$(3.6)\quad |\beta(\hat{K}^2-1)+\hat{a}(|\beta|^2-\hat{K}^2)|^2 \geq |\beta|^2(\hat{K}^2-1)^2+\hat{a}^2(|\beta|^2-\hat{K}^2)^2 + (\hat{K}^2-1)(|\beta|^2-\hat{K}^2)(\hat{a}^2+|\beta|^2-1).$$

Substituting $\hat{a}^2=2\hat{K}+2$, the right side of (3.6) reduces to $(|\beta|^2-\hat{K})^2(\hat{K}+1)^2$, establishing (3.3) and (3.1) for $|\beta| > \hat{K}$.

When $|\beta| < \hat{K}$, the pole of S is not in $D(0,\hat{K}^{-1})$. Therefore, the S-image of $D(0,\hat{K}^{-1})$ is $\overline{D(\hat{c},\hat{r})}$. So, in order to prove (3.1) for this case, we need to show that $|\hat{c}-\hat{a}| \leq \hat{r}-1$, or

$$(3.7)\quad |\beta(\hat{K}^2-1)+\hat{a}(|\beta|^2-\hat{K}^2)|^2 \leq (|\beta|^2-\hat{K})^2(\hat{K}+1)^2.$$

Setting $\theta:=\arg\beta$ and expanding the left side of (3.7) again yields (3.4). But in this case the coefficient of cos θ is negative. Therefore, combining (3.4) and (3.5) gives

$$(3.8)\quad |\beta(\hat{K}^2-1)+\hat{a}(|\beta|^2-\hat{K}^2)| \leq (|\beta|^2-\hat{K})^2(\hat{K}+1)^2$$

which establishes (3.1) for $|\beta| < \hat{K}$.

Finally, when $|\beta| = \hat{K}$, the previous cases and the continuity of $(z-\beta)(1-z\bar{\beta})^{-1}$ as a function of β imply that $\left|\frac{z-\beta}{1-z\bar{\beta}}\right| \leq \frac{2}{\hat{a}^2-2}$.

This together with the above results completes the proof. ∎

Corollary 1. Suppose $\hat{a} > 2$ and $\beta \in \overline{D(\hat{a},1)}$. Let T be the linear fractional transformation $T(z) = \frac{z-\beta}{1-z\bar{\beta}}$. Then $|T(z)| = \frac{2}{\hat{a}^2-2}$ for some $z \in \overline{D(\hat{a},1)}$ if and only if

$$(3.9)\quad |\beta-\hat{a}| = 1 \text{ and } z = \hat{a} - \frac{1-\frac{1}{2}\hat{a}(\hat{a}-\beta)}{\frac{1}{2}\hat{a}-(\hat{a}-\beta)}.$$

Proof: A calculation establishes that if z and β satisfy (3.9), then $|T(z)| = 2(\hat{a}^2-2)^{-1}$. Recalling that a linear fractional transformation of the form $S(\omega) = \frac{1-\bar{\alpha}\omega}{\alpha-\omega}$, $|\alpha| > 1$ maps $D(0,1)^c$

onto $\overline{D(0,1)}$, we see that (3.9) also implies that $z \in \overline{D(\hat{a},1)}$.

To establish necessity, suppose that $|T(z)| = 2(\hat{a}^2 - 2)^{-1}$ for some $z \in \overline{D(\hat{a},1)}$. Then

$$\left|\frac{z - \beta}{1 - \overline{z}\beta}\right| = \left|\frac{z - \beta}{1 - z\overline{\beta}}\right| = \frac{2}{\hat{a}^2 - 2} .$$

Let $R(\omega) = \frac{z - \omega}{1 - \overline{z}\omega}$. Using Lemma 1 with β replaced by z, we have that $|R(\omega)| \leq 2(\hat{a}^2 - 2)^{-1}$ for $\omega \in \overline{D(\hat{a},1)}$. But, by the maximum modulus principle, $|R(\omega)| < 2(\hat{a}^2 - 2)^{-1}$ for $\omega \in D(\hat{a},1)$. Hence it follows that $|\beta - \hat{a}| = 1$. Thus we conclude that the T-image of $\overline{D(\hat{a},1)}$ is a closed disc containing zero as a boundary point. So $|T(\omega)| = 2(\hat{a}^2 - 2)^{-1}$ for at most one point of $\overline{D(\hat{a},1)}$. Consequently, z given by (3.9) is the only point of $\overline{D(\hat{a},1)}$ satisfying $|T(z)| = 2(\hat{a}^2 - 2)^{-1}$. ■

<u>Proof of Theorem 1</u>. If $f(x_1) = f(x_2)$, (2.1) is trivial. So, suppose $f(x_1) \neq f(x_2)$. Without loss of generality we may suppose $f(x_1) > f(x_2)$ since otherwise we may replace T with -T and f with -f.

Suppose (2.1) does not hold for some T. Then there exist complex numbers $\alpha_1,\alpha_2,\alpha_3,\alpha_4$ such that

(i) $|\alpha_k| < 1$ $\quad k = 1,2,3,4$, and

(ii) $T(x_k) = f(x_1) + \delta\alpha_k$, $k=1, 4$ and $T(x_k)=f(x_2) + \delta\alpha_k$, $k=2,3$.

From the invariance of the cross-ratio, it follows that

$$(3.10) \quad \frac{T(x_3)-T(x_1)}{T(x_3)-T(x_4)} \; \frac{T(x_2)-T(x_4)}{T(x_2)-T(x_1)} = K.$$

Using (ii), (3.10) can be written

$$(3.11) \quad \frac{[\delta(\alpha_3-\alpha_1)-(f(x_1)-f(x_2))][\delta(\alpha_2-\alpha_4)-(f(x_1)-f(x_2))]}{[\delta(\alpha_3-\alpha_4)-(f(x_1)-f(x_2))][\delta(\alpha_2-\alpha_1)-(f(x_1)-f(x_2))]} = K.$$

Cancelling δ from the left side of (3.11), we obtain

$$(3.12) \quad \frac{[(\alpha_3-\alpha_1)-2K^{\frac{1}{2}}(K-1)^{-\frac{1}{2}}][(\alpha_2-\alpha_4)-2K^{\frac{1}{2}}(K-1)^{-\frac{1}{2}}]}{[(\alpha_3-\alpha_4)-2K^{\frac{1}{2}}(K-1)^{-\frac{1}{2}}][(\alpha_2-\alpha_1)-2K^{\frac{1}{2}}(K-1)^{-\frac{1}{2}}]} = K.$$

Let $a := K^{\frac{1}{2}}(K-1)^{-\frac{1}{2}}$; $z_k = \alpha_k - a$, $k=2,3$; and $z_k = \alpha_k + a$, $k=1,4$. Then, from (3.12) we find that

$$(3.13) \qquad \frac{(z_3-z_1)(z_2-z_4)}{(z_3-z_4)(z_2-z_1)} = K.$$

Note that $z_1, z_4 \in D(a,1)$, while $z_2, z_3 \in D(-a,1)$. Also observe that $a > 1$ whence $\overline{D(a,1)} \cap \overline{D(-a,1)} = \emptyset$.

Consider the linear fractional transformation $S(z) = \dfrac{z - z_1}{z - z_4}$. Let D_1 be the S-image of $D(-a,1)$, and let D_2 be the K·S-image of $D(-a,1)$. From (3.13) we conclude that $S(z_3) = K \cdot S(z_2)$, and therefore $D_1 \cap D_2 \neq \emptyset$.

Since the pole of S, z_4, is not in $\overline{D(-a,1)}$, D_1 and D_2 are discs. By the symmetry principle the pre-image of the center of D_1 is $P := (\bar{z}_4 + a)^{-1} - a$. Let $c := S(P)$. Then the radius of D_1 is $r := |c - S(1-a)|$. Consequently, the center of D_2 is Kc and the radius of D_2 is Kr. But $D_2 \cap D_1 \neq \emptyset$, so it follows that $|c| + r > K|c| - Kr$. Solving for $r|c|^{-1}$ gives

$$(3.14) \qquad r|c|^{-1} > (K-1)(K+1)^{-1} = (2a^2-1)^{-1}.$$

A calculation shows that

$$(3.15) \qquad r|c|^{-1} = |z_1 - z_4| \cdot |1-(a+z_1)(a+\bar{z}_4)|^{-1}.$$

Combining (3.14) and (3.15) we find

$$(3.16) \qquad \frac{|(z_1+a)-(z_4+a)|}{|1-(a+z_1)(a+\bar{z}_4)|} > \frac{1}{(2a^2-1)}.$$

However (z_1+a), $(z_4+a) \in D(2a,1)$, and therefore (3.16) contradicts Lemma 1. That contradiction establishes (2.1).

To establish the second half of the theorem, first note that if $T(x_k)$, $k=1,2,3,4$ satisfies (2.2) then $T(x_k)$, $k=1,2,3,4$ satisfies (3.10). Hence there exists a linear fractional transformation T which satisfies (2.2). Moreover, the conditions on β_4, $k=1,2,3,4$ imply that $|\beta_k|=1$, $k=1,2,3,4$, and therefore

T achieves equality in (2.1).

Finally, suppose T is a linear fractional transformation for which (2.1) holds with equality. Replacing i) with

i') $|\alpha_k| \le 1 \quad k = 1,2,3,4,$

and following the line of proof used to establish (2.1), we find that (3.14) becomes

$$(3.17) \quad r|c|^{-1} \ge (2a^2 - 1)^{-1}$$

and (3.16) becomes

$$(3.18) \quad \frac{|(z_1+a)-(z_4+a)|}{|1-(z_1+a)(\overline{z}_4+a)|} \ge \frac{1}{2a^2-1} \text{ where } (z_1+a),(z_4+a) \in \overline{D(2a,1)} .$$

By Lemma 1, equality must hold in (3.18). Since $z_4 = a+\alpha_4$ and $z_1 = a+\alpha_1$, Corollary 1 applied to (3.18) yields

$$\alpha_1 = -\frac{a\alpha_4+1}{a+\alpha_4} \quad \text{where } a = K^{\frac{1}{2}}(K-1)^{-\frac{1}{2}}.$$

Next, since equality holds in (3.18), equality holds in (3.17). Therefore $|c| + r = K|c| - Kr$, and consequently $\overline{D}_1$ and $\overline{D}_2$ meet in exactly one point. From the usual properties of linear fractional transformations, we deduce that this condition implies that for fixed α_1, α_4 there is exactly one pair α_2, α_3 for which (3.12) is valid. Since (3.12) holds when $\alpha_2 = -\overline{\alpha}_1$ and $\alpha_3 = -\overline{\alpha}_4$, the proof is complete. ■

Proof of Theorem 2: If there exists a linear fractional transformation U such that $\|f - U\| < \|f - T\|$ then we must have

$$\max_{x \in \{x_1,x_2,x_3,x_4\}} |U(x)-f(x)| < \tfrac{1}{2}|f(x_1)-f(x_2)|(K-1)^{\frac{1}{2}}K^{-\frac{1}{2}}.$$

contradicting Theorem 1. Hence for all linear fractional transformations U, $\|U(x)-f(x)\| \ge \|T(x)-f(x)\|$. Therefore $\|T-f\| = E_{1,1}(f)$ and $T \in B_{1,1}(f)$. ■

Lemma 2: Let f satisfy the hypotheses of Theorem 3, and let U be any linear fractional transformation. Then

$$\|U - f\| > \tfrac{1}{4}.$$

Proof: See Theorem 2.3 of [2]. ■

Corollary 2. For any $\omega \in [0,1]$, let $A(\omega) = \{-1,-\omega,\omega,1\}$. Let f be a real-valued function such that $f(-1) = f(1) = 1$ and $0 \leq f(-x) = f(x) \leq 1$. Let $\delta = (1-f(\omega))\omega^{\frac{1}{2}}(1+\omega)^{-1}$ and $a = \frac{1}{2}(1+\omega)\omega^{-\frac{1}{2}}$. Then for any linear fractional transformation T,

$$\max_{x \in A(\omega)} |T(x)-f(x)| \geq \delta. \tag{3.19}$$

If $\omega \in (0,1)$, equality is achieved in (3.19) if and only if there exists a real number θ such that T satisfies

$$T(z) = T(z,\theta) = c(\theta) + r(\theta)\,\frac{z - \overline{p(\theta)}}{z - p(\theta)}, \tag{3.20}$$

where
$$c(\theta) = \frac{1+f(\omega)}{2} + i\left(\frac{1-f(\omega)}{2a}\right)\left(\frac{a\cos\theta + 1}{\sin\theta}\right)$$

$$r(\theta) = \frac{1-f(\omega)}{2a}\left(\sqrt{a^2-1} - \frac{i(a\cos\theta + 1)}{\sin\theta}\right)$$

$$p(\theta) = \frac{\sqrt{a^2-1} - a - e^{i\theta}}{\sqrt{a^2-1} + a + e^{i\theta}}.$$

Proof: Using Theorem 1 with $\beta_4 = e^{i\theta}$, a tedious calculation yields the result. ■

Proof of Theorem 3. By Lemma 2, $\|U - f\| > \frac{1}{4}$ for any linear fractional transformation U. By Corollary 2,

$$\|U-f\| \geq \max_{x \in A(\omega)} |U(x)-f(x)| \geq \omega^{\frac{1}{2}}(1+\omega)^{-1}(1-f(\omega)).$$

Hence $\|U-f\| \geq \sup_{\omega \in [0,1]} \omega^{\frac{1}{2}}(1+\omega)^{-1}(1-f(\omega)) = \Delta(f)$. Therefore $\|U-f\| \geq \max(\frac{1}{4},\Delta(f))$ for any linear transformation U. ■

Proof of Theorem 4. By Theorem C, $\Delta(f) > \frac{1}{4}$. Hence Theorem 3 implies that

$$\|U - f\| \geq \Delta(f) > \tfrac{1}{4}$$

for any linear fractional transformation U. Theorem C also implies that equality is achieved in (2.4) when $U(x) = (U(x;f,\Omega)$ or $\overline{U(x)} = U(x;f,\Omega)$. ■

Proof of Theorem 5. As in Theorem 4, Theorem 5 is an immediate consequence of Theorem 3 and Theorem D. ■

Lemma 3. Let $y(x)$ be a real-valued function defined on an interval $[x_0, x_1]$. Suppose $\lim_{x \downarrow x_0} y(x) = y(x_0)$ and $y(x_0) \leq y(x) \leq y(x_1)$ for all $x \in [x_0, x_1]$. Then there exists a function $f(x)$ such that $f(x)$ is continuous on $[x_0, x_1]$, $f(x_0) = y(x_0)$, $f(x_1) = y(x_1)$, and $y(x) \leq f(x) \leq y(x_1)$ for all $x \in [x_0, x_1]$.

Proof: For each positive integer n, there exists a δ_n such that $y(x) - y(x_0) \leq 2^{-n}(y(x_1) - y(x_0))$ for $x_0 \leq x \leq \delta_n$. Moreover we can assume that $x_0 < \delta_n < \delta_{n-1} < x_1$, $n = 1, 2, \cdots$. Let f be the piecewise linear function given by

(i) $f(x_0) = y(x_0)$ and $f(x_1) = y(x_1)$,

(ii) $f(\delta_n) = 2^{-n+1}(y(x_1) - y(x_0)) + y(x_0)$ $n = 1, 2, \cdots$,

(iii) $f(x)$ is linear on $[\delta_n, \delta_{n-1}]$,

(iv) $f(x) \equiv y(x_1)$ on $[\delta_1, x_1]$.

Clearly f is continuous and $y(x) \leq f(x) \leq y(x_1)$ for $x \in [x_0, x_1]$ ■

Proof of Theorem 6. Fix an $\hat{x}$ such that $1 > \hat{x} > 0$. Let f be a function defined on $A := \{-1, -\hat{x}, \hat{x}, 1\}$ such that $f(-\hat{x}) = f(\hat{x}) = 0$ and $f(-1) = f(1) = 1$. By Corollary 2, there exists a continuum of linear fractional transformations $T(z, \theta)$ such that

$$\max_{y \in A} |T(y, \theta) - f(y)| = \frac{1}{2a} \text{ where } a = \frac{1 + \hat{x}}{2\sqrt{\hat{x}}}, \tag{3.21}$$

and furthermore no linear fractional transformation can approximate f with a smaller error on A. Thus one way to prove the theorem is to show that f can be extended to all of $[-1,1]$ in such a way that

(3.22) $|T(y,\theta)-f(y)| \le \frac{1}{2a}$ for $y \in [-1,1]$ and $\theta \in [\theta_0, \theta_1]$ for some θ_0, θ_1.

To accomplish this, it is convenient to choose $\hat{x}$ such that $a = 3-\sqrt{3}$, to let $\theta_0 = -\frac{5}{6}\pi$, and to let θ_1 be the unique real number satisfying $-\pi \le \theta_1 \le -\pi/2$ and $\cos\theta_1 = -a^{-1}$. Next, let W be the set of linear fractional transformations given by $T \in W$ if and only if

$$T(z) := T(z,\theta) = c(\theta) + r(\theta)\,\frac{\overline{z - p(\theta)}}{z - p(\theta)}, \quad \theta_0 \le \theta \le \theta_1,$$

where (i) $c(\theta) := \frac{1}{2} + \frac{i}{2a}\left(\frac{a\cos\theta + 1}{\sin\theta}\right)$,

(ii) $r(\theta) := \frac{1}{2a}\left(\sqrt{a^2-1} - i\,\frac{a\cos\theta+1}{\sin\theta}\right)$,

(iii) $p(\theta) := \dfrac{\sqrt{a^2-1} - a - e^{i\theta}}{\sqrt{a^2-1} + a + e^{i\theta}}$.

By Corollary 2, each $T \in W$ satisfies (3.21). In addition, each $T \in W$ has the following properties:

(a) $|T(\hat{x})| = |T(-\hat{x})| = \frac{1}{2a}$,

(b) $|T(1) - 1| = |T(-1) - 1| = \frac{1}{2a}$,

(c) $T(0) = \frac{1}{2}(1 - a^{-1}(a^2-1)^{\frac{1}{2}}) < \frac{1}{2a}$,

(d) $\operatorname{Re} T(\hat{x}) < \frac{1}{2a} < \frac{1}{2}$ and $\operatorname{Re} T(-\hat{x}) < \frac{1}{2a} < \frac{1}{2}$,

(e) $\operatorname{Re} T(1) > 1 - \frac{1}{2a} > \frac{1}{2}$ and $\operatorname{Re} T(-1) > 1 - \frac{1}{2a} > \frac{1}{2}$,

(f) $|T(x)| \le \frac{1}{2a}$ for $|x| \le \hat{x}$,

(g) $|\operatorname{Im} T(x)| \le \frac{1}{2a}$ for $|x| \le 1$,

(h) $|T(x) - \frac{1}{2}| \le \frac{1}{2a}$ for $|x| \le 1$.

Properties (a) and (b) follow from Corollary 2; (c) is just a calculation; (d) and (e) follow from (a) and (b); (f) is a consequence of (a), (c), and the geometry of intersecting circles; (g) is a calculation using $|\operatorname{Im} T(x,\theta)| \le |c(\theta) - \frac{1}{2}| + |r(\theta)|$, x real, $\theta \in [\theta_0, \theta_1]$. Similarly (h) is obtained from $|T(x,\theta) - \frac{1}{2}| \le |c(\theta) - \frac{1}{2}| + |r(\theta)|$, x real, $\theta \in [\theta_0, \theta_1]$. Finally, we note that for $\theta \in [\theta_0, \theta_1]$, $r(\theta)$ and $c(\theta)$ are bounded and $\operatorname{Im} p(\theta)$ is bounded away from zero. Hence it is possible to find a region $R \supseteq [-1,1]$ such that W is uniformly bounded on R.

Let $\hat{x}_1 = \inf\{|x| : -1 \le x \le 1,\ T \in W, \text{ and } \operatorname{Re} T(x) = \frac{1}{2}\}$ and $\hat{x}_2 = \sup\{|x| : -1 \le x \le 1,\ T \in W, \text{ and } \operatorname{Re} T(x) = \frac{1}{2}\}$. Since each $T \in W$ is continuous, (d) and (e) guarantee that $\hat{x}_1$ and $\hat{x}_2$ exist and $\hat{x} < \hat{x}_1 \le \hat{x}_2 < 1$. By (g), we may define for $\theta \in [\theta_0, \theta_1]$

$$y(x,\theta) = \begin{cases} 0, & 0 \le |x| \le \hat{x} \\ \operatorname{Re} T(x,\theta) - \sqrt{(\frac{1}{2a})^2 - \operatorname{Im}^2 T(x,\theta)}, & \hat{x} < |x| < \hat{x}_1 \\ \frac{1}{2}, & \hat{x}_1 \le |x| \le \hat{x}_2 \\ \operatorname{Re} T(x,\theta) + \sqrt{(\frac{1}{2a})^2 - \operatorname{Im}^2 T(x,\theta)}, & \hat{x}_2 < |x| \le 1 . \end{cases}$$

From the definitions of $y(x,\theta)$, $\hat{x}_1$, and $\hat{x}_2$, we conclude that for any $\theta \in [\theta_0, \theta_1]$ we have

$$y(x,\theta) \le \tfrac{1}{2},\ |x| \in [\hat{x}, \hat{x}_1], \text{ and } y(x,\theta) \ge \tfrac{1}{2},\ |x| \in [\hat{x}_2, 1].$$

Observe that for $|x| \in [\hat{x}, \hat{x}_1]$, $y(x,\theta)$ is the smallest solution to the inequality

$$(\operatorname{Re} T(x,\theta) - y)^2 + \operatorname{Im}^2 T(x,\theta) \le (\tfrac{1}{2a})^2 .$$

From (h), it follows that $\frac{1}{2}$ also satisfies that inequality, and therefore

$$(3.23) \qquad |T(x,\theta) - y| \le \frac{1}{2a} \text{ when } y(x,\theta) \le y \le \tfrac{1}{2} \text{ and } |x| \in [\hat{x}, \hat{x}_1].$$

Similarly, we find that

$$(3.24) \qquad |T(x,\theta) - y| \le \frac{1}{2a} \text{ when } \tfrac{1}{2} \le y \le y(x,\theta) \text{ and } |x| \in [\hat{x}_2, 1].$$

Relations (3.23) and (3.24) provide conditions we would like to impose on f. That is, we would like to have that $y(x,\theta) \le f(x) \le \frac{1}{2}$ for $|x| \in [\hat{x},\hat{x}_1]$ and $\theta \in [\theta_0,\theta_1]$, and $\frac{1}{2} \le f(x) \le y(x,\theta)$ when $|x| \in [\hat{x}_2,1]$ and $\theta \in [\theta_0,\theta_1]$. However, it is not obvious that such conditions will insure that f is continuous. To avoid that problem, we will define another function which will allow us to define a continuous f satisfying (3.22). Let

$$y(x) = \begin{cases} \sup\limits_{\theta \in [\theta_0,\theta_1]} \max\{y(x,\theta),\ y(-x,\theta)\} & |x| \in [0,\hat{x}_1) \\[2ex] \inf\limits_{\theta \in [\theta_0,\theta_1]} \min\{y(x,\theta),\ y(-x,\theta)\} & |x| \in [\hat{x}_1,1]. \end{cases}$$

We claim that y(x) has the following properties:

(i) $y(x) = 0 \qquad 0 \le |x| \le \hat{x}$;

(j) $y(x) = \frac{1}{2} \qquad \hat{x}_1 \le |x| \le \hat{x}_2$

(k) $0 \le y(x) \le \frac{1}{2} \qquad |x| \le \hat{x}_1$

(l) $1 \ge y(x) \ge \frac{1}{2} \qquad |x| \ge \hat{x}_2$

(m) $y(x) = y(-x) \quad x \in [-1,1]$

(n) $\lim\limits_{x \downarrow \hat{x}} y(x) = 0$

(o) $\lim\limits_{x \uparrow 1} y(x) = 1.$

Properties (i), (j), (k), (l), and (m) are immediate consequences of the properties of $y(x,\theta)$. To establish property (n), we first note that since W is uniformly bounded and analytic on region R, W is an equicontinuous family at each point of R. In particular, W is an equicontinuous family at $\hat{x}$. Hence $\{\mathrm{Re}\ T(x,\theta)\}_{\theta \in [\theta_0,\theta_1]}$ and $\{\mathrm{Im}\ T(x,\theta)\}_{\theta \in [\theta_0,\theta_1]}$ are equicontinuous at $\hat{x}$, and an elementary argument shows that $\{y(x,\theta)\}_{\theta \in [\theta_0,\theta_1]}$ is equicontinuous at $\hat{x}$. Likewise $\{y(x,\theta)\}_{\theta \in [\theta_0,\theta_1]}$ is equicontinuous at $-\hat{x}$. But for $\theta \in [\theta_0, \theta_1]$, $y(x,\theta) \to 0$ as $x \to \hat{x}$ or $x \to -\hat{x}$, thus for $|x|$ sufficiently close to $\hat{x}$, $y(x,\theta)$ is small independent of θ, whence (n) holds. A similar argument establishes (o).

We are now in a position to define our extension of $f(x)$. Since $y(x)$ satisfies (k) and (n), by Lemma 3 there exists a function $h(x)$ such that $h(x)$ is continuous on $[\hat{x}, \hat{x}_1]$, $h(\hat{x}) = 0$, $h(\hat{x}_1) = \frac{1}{2}$, and $\frac{1}{2} \geq h(x) \geq y(x)$ on $[\hat{x}, \hat{x}_1]$. In a similar manner, we can find a function $g(x)$ such that $g(x)$ is continuous on $[\hat{x}_2, 1]$, $g(\hat{x}_2) = \frac{1}{2}$, $g(1) = 1$, and $\frac{1}{2} \leq g(x) \leq y(x)$ on $[\hat{x}_2, 1]$. For $x \in [-1,1]$ let

$$f(x) = \begin{cases} y(x), & |x| \leq \hat{x} \\ h(|x|), & \hat{x} \leq |x| \leq \hat{x}_1 \\ y(x), & \hat{x}_1 \leq |x| \leq \hat{x}_2 \\ g(|x|), & \hat{x}_2 \leq |x| \leq 1 . \end{cases}$$

The continuity and evenness of f follow immediately from (i), (j), (m), and the definitions of h and g. Moreover (f), (h), (3.23), and (3.24) show that f satisfies (3.22). ∎

References

1 Bennett, C., K. Rudnick, and J.D. Vaaler, On a problem of Saff and Varga concerning best rational approximation (these Proceedings).

2 Bennett, C., K. Rudnick, and J.D. Vaaler, Best uniform approximation by linear fractional transformations, J. Approximation Theory (submitted).

3 Meinardus, G., Approximation of Functions: Theory and Numerical Methods, Springer-Verlag, New York, 1967.

4 Saff, E.B. and R.S. Varga, Nonuniqueness of best approximating complex rational functions, Bull. Amer. Math. Soc. (to appear).

5 Saff, E.B. and R.S. Varga, Nonuniqueness of best complex rational approximation to real functions on real intervals, J. Approximation Theory (to appear).

Arden Ruttan
Department of Mathematics
Kent State University
Kent, Ohio 44242

PHYSICAL APPLICATIONS

THE APPLICATION OF PADÉ APPROXIMANTS TO CRITICAL PHENOMENA

George A. Baker, Jr.

The application of Padé approximants to the field of critical point phenomena has been extremely fruitful and is now quite well developed . It has led to a veritable zoo of different techniques. This paper will begin with a review of the older and better established methods, such as the logarithmic derivative method together with a discussion of error estimation techniques. Such procedures as the rigorous bounding methods will be touched on, and we will also discuss some of the newer methods such as the Gammel-Gaunt-Guttman-Joyce methods, the Padé-Borel methods, and the Fisher two-variable method.

The area of critical phenomena is one of the more fruitful and developed areas of application of Padé approximant methods. In fact, in the solution of such a rich and interesting physical problem as this one, many mathematically interesting techniques, which are themselves of wide and general applicability, have been, and very likely will continue to be developed. The area of critical phenomena is one of the few fields in which the use of Padé approximants has become part of the orthodoxy. I will review a number of the techniques which have been used.

First, to set the stage, let me describe briefly what the problem is. It is convenient to describe it in terms of the Ising model of ferromagnetism. Consider a space lattice (for example the plane square lattice as in Fig.1), and at every point of it place an arrow which can either point up ↑ or down ↓ . We may denote this arrow mathematically by a variable $\sigma_{\vec{i}} = \pm 1$. Define the energy of the system by the Hamiltonian,

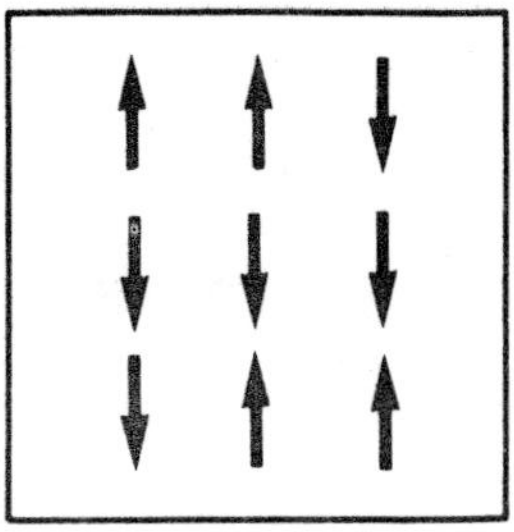

Fig.1. A portion of a plane square lattice

$$(1) \quad \mathcal{H} = -J \sum_{\substack{\text{nearest}\\ \text{neighbors}}} \sigma_{\vec{i}}\,\sigma_{\vec{j}} - H \sum_{\vec{i}} \sigma_{\vec{i}}$$

Then the properties of the Ising may be computed from the partition function,

$$(2) \quad Z = \sum_{\{\sigma_{\vec{i}} = \pm 1\}} \exp\left[-\beta\mathcal{H}\right] \quad ,$$

or alternatively from the free-energy per spin,

$$(3) \quad f = -(\beta N)^{-1} \ln Z \quad ,$$

where N is the number of spins.

Typical thermodynamic quantities of interest would be

$$(4) \quad \begin{aligned} C_H &\propto \left.\frac{\partial^2 f}{\partial \beta^2}\right|_H \propto |T - T_c|^{-\alpha} \;, \\ M &\propto \left.\frac{\partial f}{\partial H}\right|_\beta \propto |T - T_c|^{\beta} \;, \\ \chi &\propto \left.\frac{\partial^2 f}{\partial H^2}\right|_\beta \propto |T - T_c|^{-\gamma} \;, \end{aligned}$$

where C_H is the specific heat at constant magnetic field, M is the spontaneous magnetization, χ is the magnetic susceptibility, and T_c is the critical temperature at which, say the susceptibility first diverses when approached from above. Note is made that

$$(5) \quad \begin{aligned} M &= 0 \qquad T \geq T_c \;, \\ M &> 0 \qquad T < T_c \;. \end{aligned}$$

In the case of two dimensions many of the critical properties are exactly known [11]. For example, $\gamma = 1.75$, $\beta = 1/8$, and $C_H \propto \ln|T-T_c|$. The latter result means, under the usual definition of a critical index,

$$(6) \quad \phi = \lim_{T \to T_c} \frac{\ln \psi(T)}{\ln|T - T_c|} \quad ,$$

that $\alpha = 0$. For higher dimensions $d = 3, 4$, etc. there is no exact solution known, and in particular $d = 3$ is of obvious physical interest.

Even though there is no exact solution for higher dimension, it is possible to obtain exact power series expansions for the free energy and/or its derivatives in both high and lower temperature, and high magnetic field variable. One such expansion may be derived by using the identity,

$$(7)\quad e^{K\sigma_i\sigma_j} = \cosh K\left[1 + \sigma_i\sigma_j \tanh K\right] \quad ,$$

as $|\sigma_i\sigma_j| = 1$. Thus, defining $K = \beta J$, we have for $H = 0$,

$$(8)\quad \sum_{\{\sigma_{\vec{i}} = \pm 1\}} \exp(\Sigma\, K\, \sigma_{\vec{i}}\sigma_{\vec{j}}) = (\cosh K)^N \,(1 + \tanh K\, \Sigma\, \sigma_{\vec{i}}\sigma_{\vec{j}} + (\tanh K)^2\, \Sigma\, \sigma_{\vec{i}}\sigma_{\vec{j}}\sigma_{\vec{k}}\sigma_{\vec{\ell}} + \ldots) .$$

The terms of this expansion are identified with the embeddings of finite graphs on the infinite lattice. Currently many high-temperature, series terms are known. For example, for the susceptibility, χ, for various lattices the series is known through $(\tanh K)^n$ where n is

	triangular	square	honeycombed	face-centered cubic
(9)	16 ,	21 ,	32 ,	12 ,

body-centered cubic	simple cubic	diamond
15 ,	17 ,	22 .

The problem is to design efficient procedures to utilize this information under the hypothesis that the leading behavior, near the critical temperature, is

$$(10)\quad f(x) \approx A(x)\,(1 - x/x_o)^{-\phi} + B(x) \quad ,$$

with A and B regular near $x = x_o$. One seeks estimates for x_o, ϕ, $A(x_o)$, etc. The guiding principle is to try to match the forms of the approximating function to that of the function being approximated. The following procedures have been used [9] :

(i) form the Padé approximants to

(11) $F_1(x) = \frac{d}{dx} \ln f(x) \approx \frac{-\phi}{x-x_o}$.

This procedure gives unbiased estimates of ϕ and x_o , in the sense that the method uses only the exactly known series coefficients and no estimated values of other paramaters ;

(ii) form the Padé approximants to

(12) $F_2(x) = (x_o-x) \frac{d}{dx} \ln f(x) \approx \phi$,

and evaluate them at $x = x_o$. This procedure gives a biased estimate of ϕ given x_o .

(iii) form the Padé approximants to

(13) $F_3(x) = [f(x)]^{1/\phi} \sim A^{1/\phi} (1 - x/x_o)^{-1}$.

This procedure gives biased estimates of x_o and A , given ϕ .

(iv) form the Padé approximants to

(14) $F_4(x) = (1 - x/x_o)^{\phi} f(x) \sim A$,

and evaluate them at $x = x_o$. This procedure gives biased estimates of A, given x_o and ϕ .

(v) form the Padé approximants to

(15) $F_5(x) = \left[\frac{d}{dx} \ln \frac{d}{dx} f(x)\right] \left[\frac{d}{dx} \ln f(x)\right]^{-1} \approx (\phi + 1)/\phi$,

and evaluate at $x = x_o$. This procedure gives biased estimates of ϕ , given x_o ; however as a practical matter, they are relatively insensitive to the choice of x_o .

(vi) critical-point renormalization. If we have two series which behave as

(16) $f(x) = \sum f_j x^j \sim A (1 - x/x_o)^{-\phi} + B$

$g(x) = \sum g_j x^j \sim C (1 - x/x_o)^{-\phi} + D$

near $x = x_o$, and x_o is the closest singularity to the origin, then we can form

(17) $h(x) = \sum \frac{f_j}{g_j} x^j \propto (1-x)^{-(\phi-\psi+1)}$

which has a singular point at $x = 1$. By forming the Padé approximants to

(18) $F_6(x) = (1-x) \frac{d}{dx} \ln h(x)$

and evaluating them at $x = 1$, we may obtain unbiased estimates of $\phi-\psi$. This method has the advantage that no information on the value of x_o is required to get information about the critical indices.

By way of an example of the application of some of these methods, let us consider the high-temperature susceptibility of the spin -1/2 Heisenberg model on the face-centered cubic lattice

(19) $\chi = 1 + 12x + 240x^2 + 6624x^3 + 234720x^4 + 10208832x^5 + 526810176x^6 + 31434585600x^7 + 2127785025024x^8 + 161064469168128x^9 + \ldots$

The results are displayed graphically in Fig.2. We see that these methods are closely correlated over a wide range of values and that the error in γ is an order of magnitude larger than that in x_o.

We next turn to a consideration of the structure of the errors of estimation inherent in the type of problem we have been discussing. Suppose we try to represent a function with a singularity

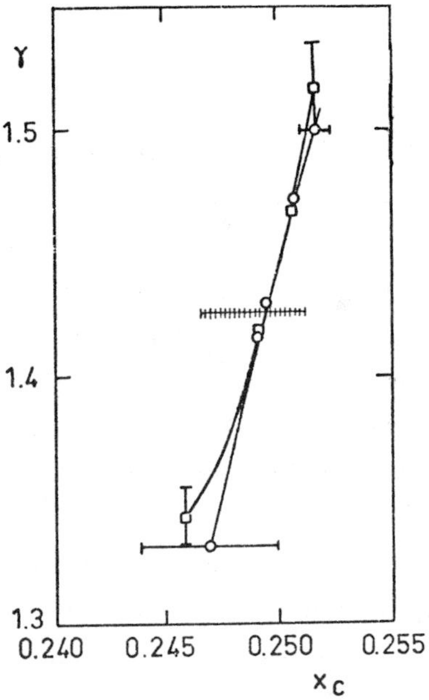

Fig.2. Locus of points in the x_c,γ plane consistent with methods (ii) and (iii). (□) γ from x_c, (o) from γ; barred horizontal indicates γ from method (v).

(20) $f(x) \sim A(1 - Yx)^{-G}$,

by the approximation,

$$(21) \quad a(1-yx)^{-g} \quad .$$

Then we get the equations,

$$(22) \quad a\binom{-g}{j} y^j = A\binom{-G}{j} Y^j (1+\eta_j) \quad , \quad j = J\,,\ J+1\,,\ J+2 \quad ,$$

where the η_j are small percentage errors. Let us expand

$$(23) \quad a = A + \Delta A \quad , \quad g = G + \Delta G \quad , \quad y = Y + \Delta Y \quad .$$

Then on substitution into Eq.(22) we obtain,

$$(24) \quad \begin{aligned} \frac{\Delta Y}{Y} &= (2G+2J+1)\,\eta_{J+1} - (G+J+1)\,\eta_{J+2} - (G+J)\,\eta_J \quad , \\ \Delta G &= (G+J)\left(\eta_{J+1} - \eta_J - \frac{\Delta Y}{Y}\right) \quad , \\ \frac{\Delta A}{A} &= \eta_J - J\,\frac{\Delta Y}{Y} - \left(\sum_{k=0}^{J-1} \frac{1}{G+k}\right) \Delta G \quad . \end{aligned}$$

From Eqs.(24), barring unusual cancellations, we can estimate

$$(25) \quad \begin{aligned} &\frac{\Delta Y}{Y} \simeq J\,O(\eta) \quad , \quad \Delta G \simeq J^2\,O(\eta)\,, \\ &\frac{\Delta A}{A} \simeq J^2 \ln(G+J)\,O(\eta)\,. \end{aligned}$$

This result gives us information about the relative size of the errors ΔY, ΔG and ΔA, but, as accurate estimates of $O(\eta)$ are hard to get, the assessment of the absolute error is more difficult. To date, except in cases where the function being studied belongs to a special class where more information is available, empirical methods which follow those for Taylor series have been used. If two successive Padé approximants are compared, say the $[M+j\,/\,M]$ and the $[M+j-1\,/\,M]$, then for small values of x their difference behaves proportionally to x^{2M+j} , for larger values of x, their difference behaves (except near poles) in a much more mildly varying manner. The apparent error, judged by the fluctuations is often much less in this second region than one would guess by extrapolating the error in the first region to the point under-consideration in the second region. [One relatively conservative approach to error analysis in the second region is to take the larger of the apparent error determined from the first region where we have

$$(26)\quad A\binom{-G}{2M+j} x^{2M+j}\eta \simeq \frac{\det\begin{vmatrix} f_j & \cdots\cdots & f_{M+j} \\ \cdots & \cdots\cdots & \cdots \\ f_{M+j} & \cdots & f_{2M+j}\end{vmatrix}}{\det\begin{vmatrix} f_{j-1} & \cdots & f_{M+j-1} \\ \cdots & \cdots\cdots & \cdots \\ f_{M+j-1} & \cdots & f_{2M+j-1}\end{vmatrix}} ,$$

(successive differences can be used instead to evaluate η), or the apparent error from the second region.] These values can then be used to estimate the order of magnitude of the error.

In another type of approach to the problems of critical phenomena [5,6,13], methods of field theory have been used to generate equations satisfied directly by the critical indices for the Landau-Ginsburg model, which is closely related to the Ising model. These equations have the following form,

$$(27)\quad W(v^*) = 0$$

determines a value v^* of v as a root of (27). Then

$$(28)\quad \gamma = \gamma(v^*) \quad , \quad \eta = \eta(v^*) \quad , \quad \ldots \quad .$$

However, only power series expansions of β, γ, η, etc. are known. Furthermore

$$(29)\quad \beta(v) = \sum \beta_n v^n \quad , \quad \beta_n \propto n!\, a^n n^b ,$$

so that these series expansions are only asymptotic and not convergent. Following the principal of reducing the approximation problem to one that the form of approximation is well suited to handle, we use a generalized Padé-Borel (Padé - Le Roy) summation method. To this end, let us consider

$$(30)\quad \beta(v) = \int_0^\infty t^b e^{-t} f(tv)\, dt \quad ,$$

then

$$(31)\quad f(x) = \sum f_j x^j \quad , \quad f_j \propto a^n \quad ,$$

so that the dominant singularity in f(x) is just a pole at $x = -1/a$. In order to complete the analysis, we must pay for having made the function f(x) regular at the origin, by having to analytically continue f(x) for all positive, real x, $0 \leq x \leq \infty$.

The technique is to Padé approximate f(x). Then, decomposing the Padé approximant in partial fractions, we get

$$(32) \quad f(tv) \simeq \sum_{i=1}^{N} \frac{r_i}{1 + s_i tv} \quad ,$$

so that in this approximation,

$$(33) \quad \beta(v) \simeq \sum_{i=1}^{N} r_i \int_0^\infty \frac{t^b e^{-t}\, dt}{1 + s_i tv} \quad .$$

Since we may immediately identify

$$(34) \quad {}_2F_0(\alpha, 1\, ; -x) = \frac{1}{\Gamma(\alpha)} \int_0^\infty \frac{t^{\alpha-1} e^{-t}\, dt}{1 + tx}$$

and by the elementary remark ${}_2F_0(\alpha, 0\, ; -x) = 1$, the known formula for the confluent version of the continued fraction of Gauss [2] ,

$$(35) \quad \frac{{}_2F_0(\alpha, 1\, ; -x)}{{}_2F_0(\alpha, 0\, ; -x)} = \cfrac{1}{1 + \cfrac{a_1 x}{1 + \cfrac{a_2 x}{1 + \cfrac{a_3 x}{1 + \dots}}}} \quad ,$$

plus the asymptotic expansion at $x \simeq \infty$, gives a rapidly convergent, well controlled method to compute approximation (33). Provided s_i is not real and negative, we can compute these approximations to $\beta(v)$ without difficulty. Baker et al [3], using only the ordinary Padé-Borel method, obtained the high quality results :

$$(36) \quad \begin{array}{llll} v^* = 1.42 \pm 0.01 & , & \gamma = 1.241 \pm 0.002 & , \\ \eta = 0.022 \pm 0.02 & , & \nu = 0.627 \pm 0.01 & , \\ \Delta_1 = \quad = 0.49 \pm 0.01 & & & . \end{array}$$

The theory of series of Stieltjes may also be applied to the investigation of problems of critical phenomena. The famous Lee-Yang [10] theorem for the Ising model, (and now also for many other models) implies that as a function of the magnetic field variable

$$(37) \quad \mu = e^{-2m\,H/kT} \quad ,$$

the free energy has the form

$$(38) \quad F/kT = -mH/kT - \int_0^1 \ln[(1-\mu)^2 + 4\mu y]\, d\phi(y) \quad ,$$

where $d\phi \geq 0$. If one considers the divergence at the critical point of the derivatives

$$(39) \quad \frac{\partial^{2m} F}{\partial H^{2m}} \propto (T-T_c)^{-\gamma_m} \quad ,$$

then the representation (38) allows us to deduce [1] that

$$(40) \quad \gamma_{i+1} - 2\gamma_i + \gamma_{i-1} \geq 0 \quad .$$

We know more about $d\phi$ than is expressed in (38). First, for reference, the only possible singularities of F as a function of μ come when

$$(41) \quad \mu = 1 - 2y \pm 2i\sqrt{y(1-y)} \quad , \quad 0 \leq y \leq 1 \quad ,$$

by elementary algebra. Eq.(41) is however just a description of the unit circle. We know in addition that $d\phi(0) = 0$ (corresponds to $\mu = +1$) for temperatures higher than the critical one, $T > T_c$, and that $d\phi(0) \neq 0$ for $T < T_c$. If we define θ_o (the Yang-Lee angle) as the largest angle for which $F(\mu)$ is regular for any μ with $|\arg \mu| < \theta_o$, then Bessis et al [4] have given bounds on θ_o . The lower bound is derived easily from the work of Ruelle [12]. They obtain the upper bound in the following way. First F is re-expressed in terms of

$$(42) \quad v = \frac{1}{\cosh^2 mH/kT} = \frac{4\mu}{(1+\mu)^2} \quad ,$$

or

$$(43) \quad \mu = \frac{1 - (1-v)^{1/2}}{1 + (1-v)^{1/2}} \quad ,$$

which reduces (38) to

$$(44) \quad \frac{F}{kT} = -\, mH/kT - \ln\left(\frac{1 + (1-v)^{1/2}}{2}\right) - \int_0^{(1+\cos\theta_o)/2} \ln(1-xv)\, d\hat{\phi}(x)$$

where $d\hat{\phi}(x) = d\phi(1-x)$. Now, we see directly that the magnetization is given by

$$(45)\quad M(v) = -\frac{\partial F}{\partial H} = 2m \tanh\frac{mH}{kT}\int_0^{(1+\cos\theta_0)/2}\frac{d\hat{\phi}(x)}{1-xv} \quad ,$$

so that M(−v) is proportional to a series of Stieltjes with a finite radius of convergence. Since the location of the poles of the Padé approximants is on the cut and the poles move towards the end of the cut, the nearest pole provides a converging upper bound for the Lee-Yang angle as

$$(46)\quad \cos\theta_0 \geq \frac{2}{v_1} - 1 \quad .$$

In the course of this analysis they have discovered a remarkable property. To put this property in context, let me first describe what is already known about the coefficients in the expansion of (45). First, the expansion in μ can be thought of as an expansion in the number of over-turned spins from an ordered state. Thought of in this way, it soon becomes apparent that the coefficients are polynomials in $u = \exp(-4K)$. (Notation of Eq.(7)). By (42) the v-series is just a rearrangement of the u series, so its coefficients will be polynomials as well. Bessis et al. [4] have proven, and it also follows from the fact that in the high-temperature series (based on (8)) the coefficient of $(\tanh K)^n$ is a polynomial in $\tanh^2(mH/kT)$ of degree n, that the coefficient of v^ℓ is exactly divisible by $(1-u)^\ell$. The remarkable property is then that the quotient polynomial has all positive terms in every case that they considered. This conclusion has not yet been proven, but if it could be proven it would have important consequences about the analytic structure of the free energy.

In addition to the high-temperature and high-magnetic field series expansions which we have described above, there are also low-temperature series. In these series in $u = \exp(-4K)$, the singularity structure has proven more difficult to deal with. Instead of one nearest singularity which is the most important one, there is a ring of singularities (2 to 6 depending on the lattice considered) of about the same distance and strength from the origin. These singularities appear to be branch points of potentially different exponents. The analysis of such a "ring" of singularities is significantly slowed, simply because enough coefficients are needed to describe each singularity adequately, plus the change in behavior in going between the singularities. In these circumstances, it seems profitable to employ a method which can use more than two parameters directly and independently in the description of each singularity. (The usual Padé method describes each singularity by a location and residue, two parameters.) One alternative is the Gammel-Gaunt-Guttman-Joyce approximants [8]. The idea for these approximants is, in a sense, a natural development of the theory of Padé approximants. The ordinary Padé approximants are formed by first determining two polynomials from a power series by

$$(47) \quad Q_M(x)\ f(x) - P_L(x) = O(x^{M+L+1})$$

and forming the approximant as

$$(48) \quad [L/M]_f = P_L / Q_M \quad .$$

In order to apply the d log Padé method (method (i) above), we impose the equations

$$(49) \quad Q_M(x)\ f'(x) - P_L(x)\ f(x) = O(x^{M+L+1})$$

and again the approximation is given by P_L/Q_M . If we merge Eqs. (47) and (49) we are lead to the scheme where three polynomials are determined by

$$(50) \quad P_L(x)\ f'(x) + Q_M(x)\ f(x) + R_N(x) = O(x^{M+L+N+2})$$

and the approximant is the solution of

$$(51) \quad P_L(x)\ y'(x) + Q_M(x)\ y(x) + R_N(x) = 0 \quad .$$

(A variant of this suggestion which has also been put forward is

$$(52) \quad P_L(x)\ f''(x) + Q_M(x)\ f'(x) + R_N(x)\ f(x) = O(x^{L+M+N+2})$$

because certain functions describing the 2-dimensional Ising model satisfy it exactly.)

What are the characteristics of the solutions of Eq.(51) ? It follows from the standard theory of differential equations that the solutions have singular points where $P_L(x_i) = 0$. At such a point

(53) $$y \approx A_i(x-x_i)^{-\phi_i} + B \quad ,$$

where

(54) $$\phi_i = \frac{Q_M(x_i)}{P_L'(x_i)} \quad ,$$

provided $\phi_i > 0$. By adjusting the coefficients of R_N $(N \geq L)$ we can choose the A_i as we like, independently of each other. Hence we can separately control (if $L \simeq M \simeq N$) x_i , ϕ_i and A_i for each singularity in these approximants. These approximants show a richer structure than do Padé approximants and can represent a wider class of functions. They lack the simplicity of evaluation of the Padé approximants because a differential equation must be solved, but the coefficients of the polynomials P_L , Q_M and R_N are again just the solutions of linear algebraic equations. Of course, if the function being approximated is meromorphic (poles are the only singularities) then more coefficients will be required to obtain an exact representation of each singularity and convergence will be less rapid than with ordinary Padé approximants.

There are other interesting mathematical problems which arise in the theory of critical phenomena. When we consider the effect of an additional field like a staggered-magnetic field (or equivalently an ordinary magnetic field for an antiferromagnet) then one finds not just isolated singular points, but lines of singularities. These lines may meet, producing still other types of singular points. We show in Fig.3 the resulting structure near a tri-critical point. A tri-critical point is at the intersection of 3 critical lines and 3 critical surfaces. Near a tricritical point, one expects physically, that scaling behavior of the form

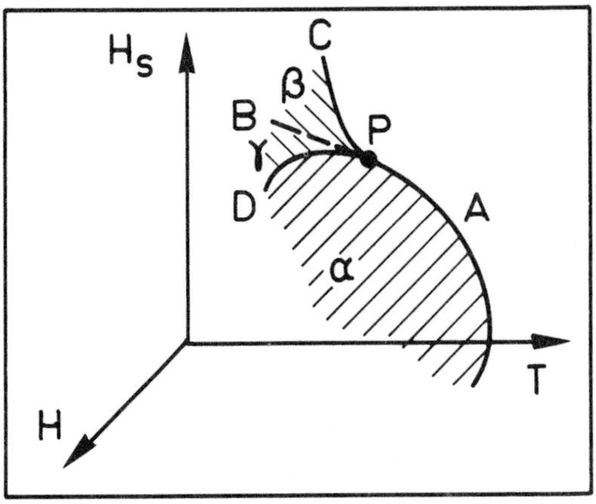

Fig.3. The point P is a tricritical point. A is a line of second-order critical points. B is a line of first-order critical points. C and D are lines of lines of "wing" second-order critical points. α, β, and γ are surfaces of first-order transition points.

$$(55)\quad f(x,y) \simeq (x-x_c)^{-\gamma}\, Z\left(\frac{y-y_c}{(x-x_c)^{\phi}}\right)$$

will occur. The problem is to design approximations so that this behavior may be represented. Fisher [7] has proposed the use of a generalized Gammel-Gaunt-Guttman-Joyce approach to two variables by first noting that f of Eq.(55) exactly satisfies

$$(56)\quad -\gamma f = (x-x_c)\frac{\partial f}{\partial x} + \phi(y-y_c)\frac{\partial f}{\partial y} \qquad ,$$

irrespective of which Z is taken. That function is to be determined by the boundary conditions. He has proposed that one consider the partial differential equation

$$(57)\quad P_{\underset{\sim}{L}}(x,y)\, F(x,y) - Q_{\underset{\sim}{M}}(x,y)\frac{\partial F}{\partial x} - R_{\underset{\sim}{N}}(x,y)\frac{\partial F}{\partial y} = \sum_i O(x^{n_i} y^{m_i}) \ .$$

The known power series expansion of F(x,y) is used to determine the coefficients of the polynomials $P_{\underset{\sim}{L}}$, $Q_{\underset{\sim}{M}}$ and $R_{\underset{\sim}{N}}$. The terms $x^{p_i} y^{\pi_i}$ occurring in the polynomials are the members of the index sets $\underset{\sim}{L}$, $\underset{\sim}{M}$ and $\underset{\sim}{N}$; $(p_i,\pi_i) \in \underset{\sim}{L}$, for example. Some examples of what forms may be expressed exactly by these approximants, are the functions

(58) $$f(x,y) = A(x,y) + B(x,y)$$

where A and B are of the form

(59) $$A_o \exp\left[\frac{a_o(x,y)}{a_{-1}(x,y)}\right] \prod_{j=1}^{J} [a_j(x,y)]^{\alpha_j} ,$$

where the $a_j(x,y)$ are polynomials in x and y for $j = -1, 0, \ldots, J$ and the α_j are arbitrary exponents. Fisher has suggested the trajectory method of solution of the partial differential equations to find the approximants. If a pseudo-time τ is introduced then

(60) $$\dot{F} = \frac{dF}{d\tau} = P_{\underset{\sim}{L}}(x,y)\ F$$
$$\dot{x} = Q_{\underset{\sim}{M}}(x,y) \quad , \quad \dot{y} = R_{\underset{\sim}{N}}(x,y)$$

leads to a solution. It is anticipated that $[x(\tau), y(\tau)]$ will behave like

(61) $$(y-y_c) \propto (x-x_x)^{\phi}$$

near the tricritical point.

Much work remains to be done in this area of application of Padé techniques and I trust that this work will continue, as it has in the past, to enrich both statistical mechanics and mathematics.

References

1 Baker, G.A. Jr., Some rigorous inequalities satisfied by the ferromagnetic Ising model in a magnetic field, Phys. Rev. Lett. 20 (1968), 990-992.

2 Baker, G.A. Jr., Essentials of Padé Approximants, Academic Press, Inc., New York, 1975.

3 Baker, G.A. Jr., B.G. Nickel, M.S. Green and D.I. Meiron, Ising-model critical indices in three dimensions from the Callan-Symanzik equation , Phys. Rev. Lett. 36 (1976), 1351-1354.

4 Bessis, J.D., J.M. Drouffe and P. Moussa, Positivity constraints for the Ising ferromagnetic model, J. Phys. A 9

(1976),2105-2124.

5 Brézin, E., J.C. Le Guillou and J. Zinn-Justin, Field theoretical approach to critical phenomena, in Phase Transitions and Critical Phenomena, C. Domb and M.S. Green, eds., Vol.6, Academic Press, Inc., (to be published).

6 Fisher, M.E., The renormalization group in the theory of critical behavior, Rev. Mod. Phys. 46 (1974), 597-616.

7 Fisher, M.E., Novel two-variable approximants for studying magnetic multicritical behavior, Proc. Int. Conf. on Magnetism, Amsterdam, North-Holland Publishing Co., Amsterdam, (to be published).

8 Gammel, J.L., Review of two recent generalizations of the Padé approximant, pp.3-9, and G.S. Joyce, and A.J. Guttmann, A new method of series analysis, pp.163-168, in Padé Approximants and their Applications, P.R. Graves-Morris, ed., Academic Press, Inc., New York, 1973.

9 Hunter, D.L., and G.A. Baker, Jr., Methods of series analysis. I. Comparison of current methods used in the theory of critical phenomena, Phys. Rev. B 7 (1973), 3346-3376.

10 Lee, T.D., and C.N. Yang, Statistical theory of equations of state and phase transitions. II. Lattice gas and Ising model, Phys. Rev. 87 (1952), 410-419.

11 Mc Coy, B.M., and T.T. Wu, The Two-Dimensional Ising Model, Harvard Univ. Press, Inc., Cambridge, Mass., 1973.

12 Ruelle, D., Extension of the Lee-Yang circle theorem, Phys. Rev. Lett. 26 (1971), 303-304 ; and Some remarks on the location of zeroes of the partition function for lattice systems, Commun. Math. Phys. 31 (1973), 265-277.

13 Wilson, K.G., and J. Kogut, The renormalization group and the expansion, Phys. Repts. 12C (1974), 75-199.

George A. Baker, Jr.*
Service de Physique Théorique
CEN-Saclay
BP n°2, 91190 Gif-sur-Yvette, France

and

Theoretical Division
University of California
Los Alamos Scientific Laboratory
Los Alamos, N.M. 87545, U.S.A.

*Work supported in part by the U.S. E.R.D.A. and in part by the French C.E.A.

VARIATIONAL PRINCIPLES AND MATRIX PADÉ APPROXIMANTS

L. P. Benofy and J. L. Gammel

In this paper we show how accurately a combination of matrix Padé approximants and the Schwinger variational principle sums the Brillouin-Wigner perturbation series for the energies of quantum mechanical bound states.

1 Introduction

Nuttall [6], Alabiso et al [1], Bessis and Villani [3], and Turchetti [7] have discussed the connection between Padé approximants and variational principles. Together with Pierre Mery, we [2] have used this connection to sum the Born series arising in quantum mechanical scattering theory.

The Born series for the tangent of an S (that is, angular momentum L = 0) phase shift satisfies

$$(1.1)\quad \frac{1}{p}\tan\delta = (p|V|p) - \frac{2}{\pi}\int dk(p|V|k)\,\frac{P}{p^2-k^2}\,(k|V|p)$$

$$+ \left[\frac{2}{\pi}\right]^2 \int dkdk'\ (p|V|k)\,\frac{P}{p^2-k^2}\,(k|V|k')\,\frac{P}{p^2-k'^2}\,(k'|V|p)$$

$$- \ldots,$$

where P stands for principal part, and

$$(1.2)\quad (\phi|V|k) = \int dr\ \phi_p(r)V(r)\phi_k(r),$$

where ϕ_p = sin pr is the wave function of an unperturbed incident wave. In [2], we extended Eq. (1.1) in such a way as to define a matrix,

$$(1.3)\quad (a|K|b) = (a|V|b) - \frac{2}{\pi}\int dk\ (a|V|k)\,\frac{P}{p^2-k^2}\,(k|V|b) + \ldots,$$

where it should be noted that the p^2 appearing in the denominators (or propagators) on the RHS of Eq. (1.1) is not changed. The actual phase shift is

$$(1.4)\quad \tan\delta = p(p|K|p).$$

Iterating the integral equation

$$(1.5)\quad (a|K|b) = (a|V|b) - \frac{2}{\pi}\int dk\ (a|V|k)\ \frac{1}{p^2-k^2}\ (k|K|b)$$

yields the Born series. This last Eq. (1.5) is rigorously equivalent to the Schrödinger equation and it is the Schrödinger equation in momentum space.

Matrix Padé approximants have been discussed by Bessis [4] and Turchetti [7]. They are just like ordinary Padé approximants except that the coefficients in the series and the numerator and denominator of the Padé approximant are matrices instead of ordinary numbers. One chooses a finite number of a's and b's, one choice being p, and deals with finite matrices. In the limit of very many a's and b's suitably chosen to approximate a continuum, one gets the exact solution to Eq. (1.5), because Eq. (1.5) is

$$(1.6)\quad K = V + V \cdot G \cdot K,$$

where $\cdot$ means dk, and G is a diagonal matrix. The solution of Eq. (1.6) is

$$(1.7)\quad K = \frac{1}{V - V\cdot G} \cdot V.$$

Since both the numerator and denominator are linear in V, this solution is a [1/1] Padé approximant. But our goal is to get an accurate approximation to K, more particularly its $(p|K|p)$ element, using only a very limited number of a's and b's. In fact, we are able to choose the a's and b's from the set p,q, where q is a single momentum other than p, that is, a single "off-shell" momentum, once the connection with variational principles is made.

The Schwinger variational principle begins with a consideration of

$$(1.8)\quad I = (\phi_b|V|\phi) + (\phi'|V|\phi_a) - (\phi'|V|\phi) - \frac{2}{\pi}\int dk\ (\phi'|V|\phi_k)\frac{1}{p^2-k^2}(\phi_k|V|\phi),$$

and the variational principle itself is

$$(1.9)\quad \delta_\phi I = \delta_{\phi'} I = 0,$$

where δ_ϕ means the variation of I with respect to ϕ.
We have

$$(1.10)\quad \delta_{\phi'} I = V|\phi_a) - V|\phi) - \frac{2}{\pi}\int dk\ V|\phi_k)\frac{1}{p^2-k^2}(\phi_k|V|\phi),$$

and since $(b|K|a) = (\phi_b|V|\phi)$, this Eq. (1.10) gives Eq. (1.5).

The most important thing to notice is that when $\delta_{\phi'} I = 0$, I becomes (for $b = a = p$)

$$(1.11)\quad I = (\phi_p|V|\phi) = (p|K|p) = \frac{\tan\delta}{p}.$$

If there are not sufficiently many variational parameters in ϕ so that ϕ is never sufficiently close to an exact solution to Eq. (1.5), it may be doubted that this equality of I and $\tan\delta$ remains valid. We will return to this question below.

Nuttall [6] showed that the ansatz

$$(1.12)\quad \phi = c_p\phi_p + c_q\phi_q + c_r\phi_r + \ldots,$$
$$\phi' = c_p^*\phi_p + c_q^*\phi_q + c_r^*\phi_r + \ldots,$$

where q,r are momenta other than p and may be chosen arbitrarily, when substituted into I, yields, upon setting $\delta_{c_p} * I = 0$ and $\delta_{c_r} * I = 0$, the [1/1] matrix Padé approximant to K formed by choosing p,q,r,... to be the a's and b's in Eq. (1.3). It is very important to notice that Eq. (1.11) continues to hold even though ϕ is only approximate.

Nuttall also showed that the Cini-Fubini [5] ansatz

$$(1.13)\quad \phi = c_p\phi_p + c_q\phi_q + c_r\phi_r + \dots$$
$$+ c_{p'} \frac{1}{E-H} V\phi_p + c_{q'} \frac{1}{E-H} V\phi_q + c_{r'} \frac{1}{E-H} V\phi_r$$
$$+ c_{p''} \frac{1}{E-H} V \frac{1}{E-H} V\phi_p + c_{q''} \frac{1}{E-H} V \frac{1}{E-H} V\phi_q + \dots$$

out to the (N-1)st order in V with a similar expression for ϕ' using c^*'s, yields upon setting the variation of I with respect to c_p^*, $c_{p'}^*$, $c_{p''}^*$, c_q^*, $c_{q'}^*$, $c_{q''}^*$, ... equal to zero the [N/N] matrix Padé approximant to tan δ. However, these higher order approximants do not concern us here.

Our work is based on the fact that q,r,... may be viewed as variational parameters, so that in view of Eq. (1.11) we may vary q,r,... until the (p,p) element of the matrix Padé approximant, that is, tan δ, becomes stationary with respect to the q,r,..., at which point we should have the best possible approximation to tan δ. In [2] we found that varying just one "off-shell" momentum resulted in an incredibly accurate approximation to tan δ at the stationary point. Figure 1 illustrates this work.

Because of the importance of Eq. (1.11), we give a proof of our assertion that it holds for the approximate ϕ's which we consider.

Let

$$(1.14)\quad \phi = c_p\phi_p + c_q\phi_q, \quad \phi' = c_p^*\phi_p + c_q^*\phi_q.$$

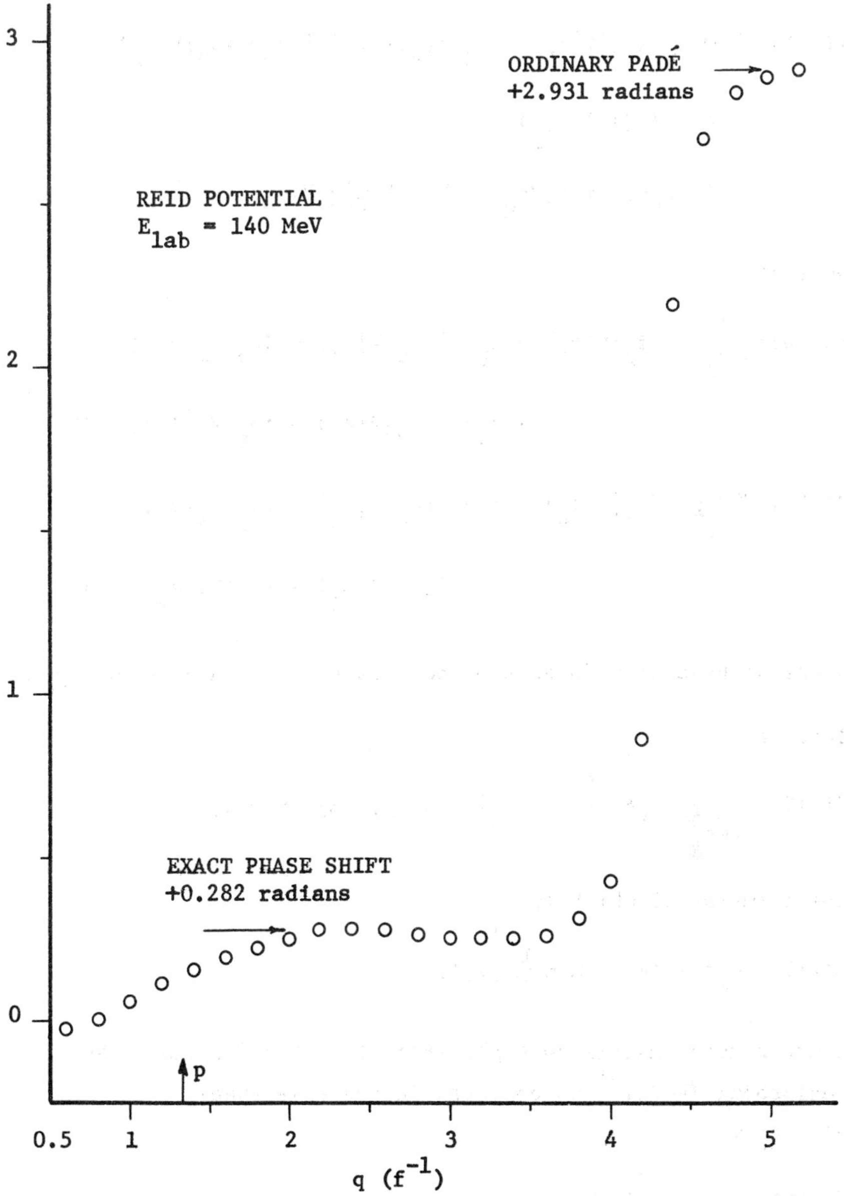

Fig. 1

^1S_o phase shift versus the "off-shell" momentum

Then

$$(1.15)\quad I = c_p\ (\phi_p|V|\phi_p) + c_q\ (\phi_p|V|\phi_q) + c_p{}^*\ (\phi_p|V|\phi_p)$$

$$+ c_q{}^*\ (\phi_q|V|\phi_p)$$

$$+ (c_p{}^*\phi_p + c_q{}^*\phi_q| - V + V\frac{1}{E-H}V|c_p\phi_p + c_q\phi_q),$$

so that

$$(1.16)\quad \frac{\partial I}{\partial c_p} = (\phi_p|V|\phi_p) - c_p{}^*\ [(\phi_p|V|\phi_p) - (\phi_p|V_2|\phi_p)]$$

$$- c_q{}^*\ [(\phi_q|V|\phi_p) - (\phi_q|V_2|\phi_p)] = 0$$

$$(1.17)\quad \frac{\partial I}{\partial c_q} = (\phi_p|V|\phi_q) - c_p{}^*\ [(\phi_p|V|\phi_q) - (\phi_p|V_2|\phi_q)]$$

$$- c_q{}^*\ [(\phi_q|V|\phi_q) - (\phi_q|V_2|\phi_q)] = 0,$$

where we have introduced the short hand notation $V_2 = -V\frac{1}{E-H}V$.

That is,

$$(1.18)\quad \sum_{\ell=p,q} c_\ell{}^*\ (\ell|V - V_2|m) = (p|V|m),\ m = p,q$$

the solution of which is

$$(1.19)\quad c_\ell{}^* = (\phi_p|V \times \frac{1}{V-V_2}|\phi_\ell),$$

where x means matrix multiplication in a 2 x 2 space. The reciprocal $(V-V_2)^{-1}$ is computed in the same space.
Similarly,

$$(1.20)\quad c_\ell = (\phi_\ell|\frac{1}{V-V_2} \times V|\phi_p).$$

Thus

$$(1.21)\quad \phi = \sum_{\ell=p,q} \phi_\ell \; (\phi_\ell|\frac{1}{V-V_2} \times V|\phi_p),$$

and

$$(1.22)\quad (\phi_p|V|\phi) = (\phi_p|V \times \frac{1}{V-V_2} \times V|\phi_p)$$

is the (p|p) element of the [1/1] matrix Padé approximant to K, as asserted.

Then,

$$(1.23)\quad I = \sum_{\ell=p,q} \{(\phi_p|V|\phi_\ell)\; c_\ell + c_\ell^* \;(\phi_\ell|V|\phi_p)\}$$

$$- \sum_{\ell=p,q} \sum_{m=p,q} c_\ell^* \;(\phi_\ell|V-V_2|\phi_m)\; c_m,$$

$$(1.24)\quad I = \sum_{\ell=p,q} (\phi_p|V|\phi_\ell)\;(\phi_\ell|\frac{1}{V-V_2} \times V|\phi_p)$$

$$+ \sum_{\ell=p,q} (\phi_p|V \times \frac{1}{V-V_2}|\phi_\ell)\;(\phi_\ell|V|\phi_p)$$

$$- \sum_{\ell=p,q} \sum_{m=p,q} (\phi_p|V \times \frac{1}{V-V_2}|\phi_\ell)\;(\phi_\ell|V-V_2|\phi_m)$$

$$\cdot\; (\phi_m|\frac{1}{V-V_2} \times V|\phi_p),$$

$$= 2(\phi_p|V \times \frac{1}{V-V_2} \times V|\phi_p)$$

$$- (\phi_p|V \times \frac{1}{V-V_2} \times (V-V_2) \times \frac{1}{V-V_2} \times V|\phi_p),$$

$$= (\phi_p|V \times \frac{1}{V-V_2} \times V|\phi_p),$$

also as asserted.

This important result cannot be restated too often: once the matrix Padé approximant is formed and tan δ calculated from its (p|p) element, tan δ itself should be made stationary with respect to any variational parameters which remain. In particular, the "off-shell" momenta may be viewed as variational parameters, and tan δ should be made stationary with respect to them.

2 The Brillouin-Wigner Perturbation Theory

The Brillouin-Wigner perturbation series to the energy of a bound quantum mechanical system is very similar to the Born series in scattering theory,

$$E = E_p^{(o)} + (p|V|p) + \sum_k (p|V|k) \frac{1}{E-E_k} (k|V|p) + \ldots \tag{2.1}$$

where |p) is an energy eigenstate associated with some "unperturbed" Hamiltonian H_o which together with a "perturbation" V yields the Hamiltonian H of the system under consideration: $H_o|p> = E_p^{(o)}|p>$ and $H = H_o + V$. We imagine the system contained in a big box, so that the energy levels are discrete, and $\sum$ replaces $\int$ in the Born series. There are other, more important, differences. First, the equation for E is implicit; that is, E appears on both sides of the equation. This difference is resolved by calling the E on the RHS of Eq. (2.1) E_{input} or E_{in}. For a given E_{in} we must sum the RHS of Eq. (2.1), and this is exactly the same problem as summing the Born series. Ultimately, E_{in} must be adjusted until the sum, which we call E_{out}, is equal to E_{in}. Second, there is no specific connection between E_{in} and p^2 as in the case of the Born series for which $E_{in} = p^2$. If one wants the lowest energy level, he takes p to be the lowest energy level of the unperturbed system from which he starts. Energy levels will not cross as the strength of the perturbing interaction increases. The ϕ_a in the Brillouin-Wigner case are just the wave functions of the unperturbed

Hamiltonian.

We follow the same method as in the case of the Born series. We extend Eq. (2.1) into a matrix equation,

$$(2.2)\quad (a|E|b) = (a|V|b) + \sum_k (a|V|k) \frac{1}{E_{in}-E_k} (k|V|b) + \ldots,$$

$$E_{out} = (p|E|p),$$

and notice that everything is perfectly analogous to the Born series case.

Our technique is to choose one or several "off-shell" momenta q,r,... and form the [1/1] matrix Padé approximant for a given E_{in}. We vary q,r,... until E_{out} is stationary against variations in these, and use this as the best E_{out}. We then vary E_{in} until the best E_{out} is equal to E_{in}. It may be shown that the order of varying q,r,... and making $E_{out} = E_{in}$ is immaterial. For fixed q,r,... we may vary E_{in} until $E_{out} = E_{in} = E$ and then vary q,r,... until E is stationary.

3 Example

We begin with the problem of one particle confined to a one-dimensional box, so that

$$(3.1)\quad \phi_p = \sqrt{\frac{2}{L}}\ \sin \frac{p\pi x}{L},$$

$$(3.2)\quad E_p = \frac{\hbar^2}{2M} \frac{p^2\pi^2}{L^2},$$

where L is the size of the box, M the mass of the particle, and p = 1 evidently corresponds to the lowest energy level. We imagine the particle to be a nucleon so that $2M/\hbar^2 = 0.04824$ $\text{Mev}^{-1}\ f^{-2}$. We then put a perturbing potential at an edge of the box (see Fig. 2), of depth $V_o = 34$ MeV.

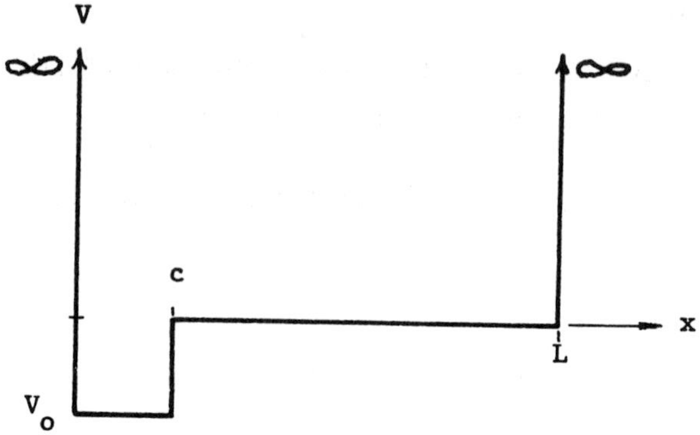

Fig. 2

and range c = 1.5 f. The exact binding energies for various box sizes may be calculated from elementary quantum mechanics. These values are shown in Table I.

Table I

Particle in a Box Acted Upon by Square-Well Potential

V_o = -34 MeV c = 1.5f

L =	E_{exact} =	$E_{calculated}$ =	q =
6f	-2.2312 MeV		2
15f	-2.5423 MeV	-2.5424 MeV	6
30f	-2.5428 MeV	-2.5428 MeV	12
60f	-2.5428 MeV	-2.5431 MeV	24

6f box is too small for variational work, the spacing between energy levels having become so large that the plateau in Fig. 1 disappears.

In Fig. 3, we show E_{out} as a function of q for various E_{in}, for a box size of 30f. (This box size is so large that the box itself perturbs the energy very little.) The existence of a stationary point is not in doubt, there being in fact a very

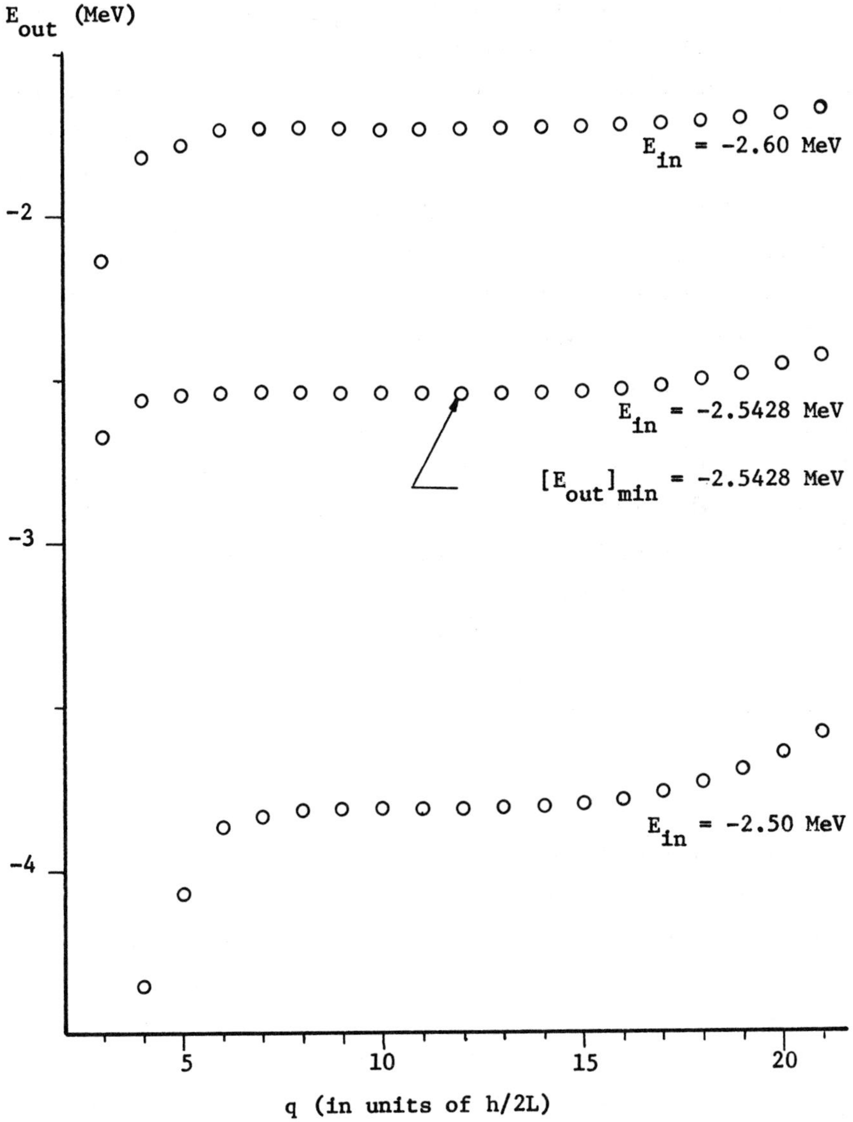

Fig. 3

E_{out} versus "off-shell" momentum q for nucleon in a box of 30f and acted upon by a square-well potential, as shown in Fig. 2. V_o = -34 MeV, c = 1.5f.

broad, flat plateau. Consistency of $E_{out} = E_{in} = E$ is obtained for E = -2.5428 MeV. The exact energy is E = -2.5428 MeV, so that the method is impressively accurate. The minimum along the plateau is the point which we actually chose as "the" stationary point. This is not an important point because the plateau is so flat; a similar point arose in the work on the Born series in the scattering case and the problem of why this prescription is so incredibly accurate is not resolved as yet. Results for other box sizes are summarized in Table I.

We considered a more complicated shape for the potential; namely, a potential with a repulsive core of depth $V_1 = +666.7$ MeV and radius $c_1 = 0.45$f, outside of which was an attractive potential of depth $V_2 = -31.75$ MeV extending to $c_2 = 1.95$f. The exact binding energy is E = -3.8064 MeV for a box size of L = 6f. The approximate binding energy calculated according to our method is E = -3.776 MeV. With L = 30f the approximate binding energy we calculate, E = -4.004 MeV, should be compared to the exact value, E = -4.057 MeV.

We compared our work to Turchetti [7], who searches for a pole in the $(p|p)$ element of the matrix Padé approximant; that is, a zero of the denominator. For a fixed E_{in}, he varies V_o until a pole in this element of the matrix Padé approximant is found. The exact value of this V_o depends on q (the "off-shell" momentum), and he searched until the V_o resulting in a pole was stationary against variations in q. Looking at the nucleon in a box perturbed by the earlier square-well potential, for $E_{in} = -2.5428$ MeV in a 30f box, his method gives $V_o = -34.336$ MeV versus the exact -34 MeV. When it is required that $E_{out} = E_{in}$ rather than that $E_{out} = \infty$, and the same technique employed, one gets very much closer to the exact $V_o = -34$ MeV.

As an example of a harder problem, no easier than the three body problem of nuclear physics (but limited to one dimension), we imagine two particles in a box interacting via a local potential of such depth and range as to imitate the deuteron problem of nuclear physics. The depth of the local potential is $V_o = -7.9529$ MeV, the range is c = 1.5f (that is, $|x_1 - x_2| \leq 1.5f$), resulting in a deuteron binding energy of 2.226 MeV). The local potential is viewed as a perturbation.

The first problem is to number the states of a several particle system. The unperturbed states are direct products of single particle wave functions

$$(3.3)\quad \phi_{p_1}(x_1)\,\phi_{p_2}(x_2) = \sqrt{\frac{2}{L}}\,\sin\frac{\pi p_1 x_1}{L}\,\sqrt{\frac{2}{L}}\,\sin\frac{\pi p_2 x_2}{L},$$

and the unperturbed energies are

$$(3.4)\quad E_{p_1,p_2} = \frac{\hbar^2}{2M}\frac{p_1^2\pi^2}{L^2} + \frac{\hbar^2}{2M}\frac{p_2^2\pi^2}{L^2}.$$

The matrix elements,

$$(3.5)\quad (k_1k_2|V|p_1p_2)$$

$$= \int\int \phi_{k_1}(x_1)\,\phi_{k_2}(x_2)\,V\,(x_1-x_2)\,\phi_{p_1}(x_1)\,\phi_{p_2}(x_2)dx_1dx_2,$$

are small between states which do not satisfy one of the several conditions

$$(3.6)\quad |p_1 \pm p_2| = |k_1 \pm k_2|.$$

In calculating the second Born approximation from Eq. (2.1) one sums over two k's; namely, k_1 and k_2 and does not neglect any matrix elements of V in the sum; however, in choosing "off-shell" states in forming the matrix Padé approximant, one is guided by Eq. (3.6). The unperturbed state of lowest energy is one for which both particles have p = 1. Thus

(3.7) $|p_1 \pm p_2| = 0$ or 2.

So, in considering what sort of "off-shell" states to use in forming the matrix Padé approximant, it is plain that one concludes the most likely candidates are those for which

(3.8) $|q_1 \pm q_2| = 0$ or 2.

The diagram in Fig. 4 is intended to clarify the situation.

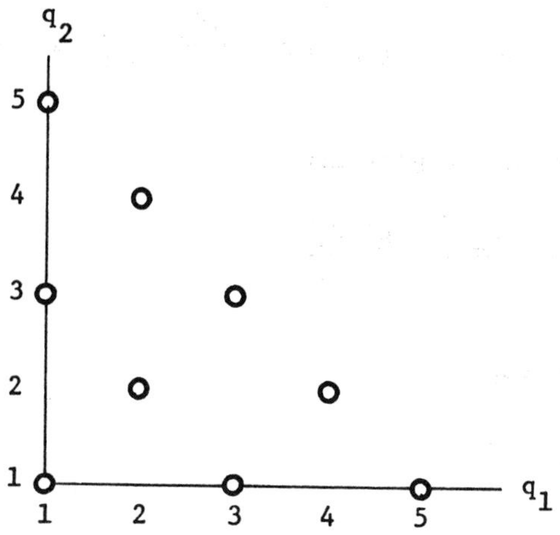

Fig. 4

We have done calculations using the four lowest energy states in Fig. 4, and the lowest nine, and the lowest 16 (adding, at each stage, an anti-diagonal in Fig. 4, as it turns out), forming 4 x 4 [1/1] matrix Padé approximants, 9 x 9 [1/1] matrix Padé approximants, and 16 x 16 [1/1] matrix Padé approximants, respectively, for various box sizes. In these calculations the "off-shell" momenta were not varied, but we did require $E_{in} = E_{out}$, of course. The results are shown in Table II.

Table II

V_o = -7.9529 MeV   c = 1.5f   L = 30f

No. of states	1*	4	9	16
E	-2.0554 MeV	-2.0689 MeV	-2.0701 MeV	-2.0704 MeV

*ordinary (non-matrix) Padé approximant.

We then used only one "off-shell" state, in fact taking $q = q_1 = q_2$ and formed 2 x 2 [1/1] matrix Padé approximants. We varied q until $E_{out} = E_{in} = E$ was stationary with respect to q. The results are shown in Table III.

Table III

V_o = -7.9529 MeV   c = 1.5f

L =	$E_{calculated}$
30f	-2.0579 MeV
60f	-2.1826 MeV
120f	-2.2107 MeV

It is seen that the asymptotic result, E = -2.226 MeV, will be reached as the box size approaches ∞, and by adding the lowest possible value of the kinetic energy of a deuteron confined to the box size tabulated, the result can be understood, the more so the larger the box.

Figure 5 shows how E_{out} varies as a function of q for a certain E_{in}. Again, the wide plateau manifests itself; in fact, it is almost embarrassingly wide and flat.

4 Discussion and Conclusion

It must be noted that the Brillouin-Wigner perturbation theory is useless. As the box size increases, the matrix elements all tend toward zero, so that it is impossible to get the correct binding energy.

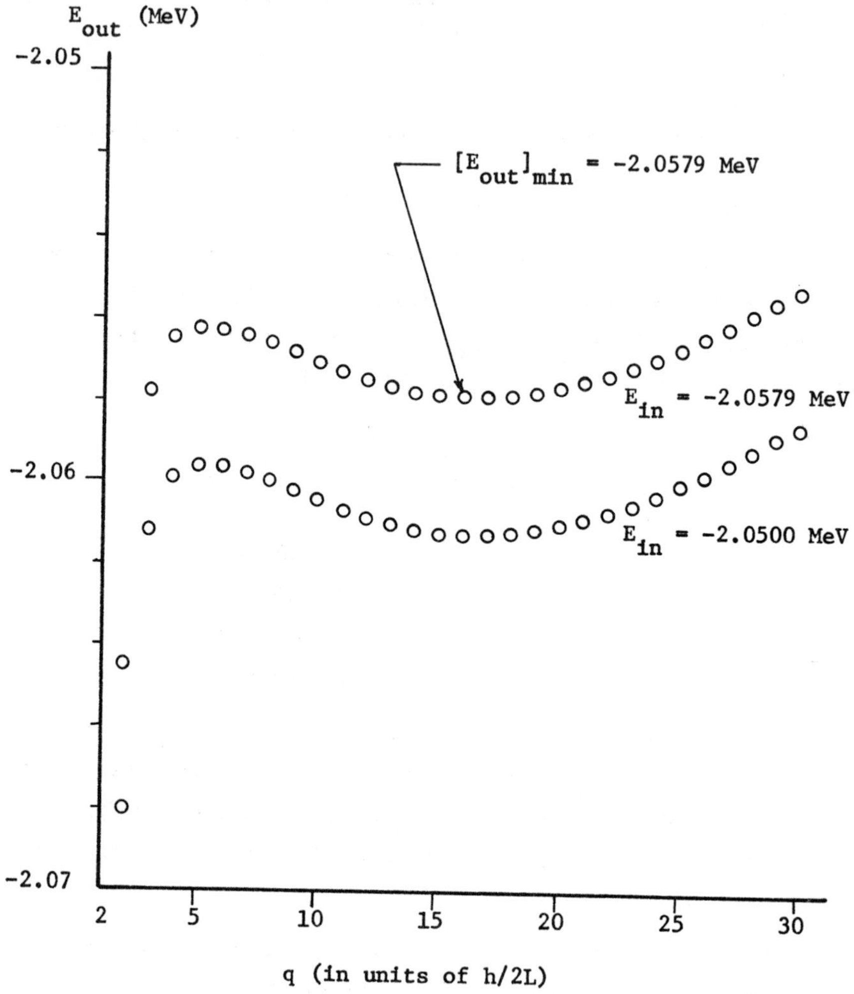

Fig. 5

E_{out} versus "off-shell" momentum q for two nucleons in a box of 30f and interacting through a square-well potential having depth V_o = -7.9529 MeV and range c = 1.5f.

It remains to apply the technique to realistic problems. We intend to calculate the several body problems in one-dimension. The results will be useful as a test of many body methods in quantum mechanics. It would be of interest to extend the method so that it would apply to the many body problem directly, or to employ techniques similar to the Rayleigh-Schrodinger perturbation series.

Acknowledgments

We are grateful to Dr. G. McCartor, especially for his help in locating a crucial error in our calculations, and to Mr. E. Bernardi for calculating some of the exact ground state energies and for checking our single particle results through independent calculations. We also gratefully acknowledge the indispensable services of the Saint Louis University Computer Center.

References

1 Alabiso, C., P. Butera, and G. M. Prosperi, Nucl. Phys. B31, 141 (1971); B42, 493 (1972); B46, 593 (1972).

2 Benofy, L. P., J. L. Gammel, and P. Mery, Phys. Rev. D13, 3111 (1976).

3 Bessis, D. and M. Villani, J. Math. Phys. 16, 462 (1975).

4 Bessis, D., in Padé Approximants, edited by P. R. Graves-Morris, Institute of Physics, London, 1973, pp. 19-44.

5 Cini, M. and S. Fubini, Nuovo Cimento 10, 1695 (1953); 11, 142 (1954).

6 Nuttall, J., in The Padé Approximant in Theoretical Physics, edited by G. A. Baker, Jr., and J. L. Gammel, Academic Press, New York, 1970, pp. 219-230.

7 Turchetti, G., Variational Principles and Matrix Approximants in Potential Theory, University of Bologna preprint, 1976, and private communication.

L.P. Benofy
Department of Physics
St. Louis University
St. Louis, Missouri 63103

J.L. Gammel
Department of Physics
St. Louis University
St. Louis, Missouri 63103

OPERATOR PADÉ APPROXIMANTS FOR THE
BETHE-SALPETER EQUATION OF NUCLEON-NUCLEON SCATTERING

J. Fleischer and J.A. Tjon

The method of the Operator Padé Approximants (OPA's) has been studied as a mean of solving the Bethe-Salpeter equation (BSE) of Nucleon-Nucleon scattering. Accurate solutions have been obtained before by the application of the ordinary Padé method. Comparing our results with these solutions, we obtain perfect agreement for the $^3S_1$ with ten external momenta in the OPA and for the $^1S_0$ with only two by considering the off-shell momentum as variational parameter.

1 Introduction

In momentum space, the BSE for states of definite total angular momentum, isospin, and parity is [1]

$$(1.1)\quad \Phi(p,p_0,\alpha) = G(p,p_0,\alpha;\hat{p},o,\varkappa) - \frac{i}{2\pi^2}\int dq\,dq_0 \sum_{\beta,\gamma} G(p,p_0,\alpha;q,q_0,\beta)S(q,q_0,\beta\gamma)\Phi(q,q_0,\gamma),$$

with $\hat{p} = \sqrt{E^2 - m^2}$ (m the nucleon mass) and $\alpha, \beta, \gamma = 1,2...8$ so that eight channels couple. For the meaning of labels 1,...,8 see [2] . In equation (1.1) G is the kernel and S the two-nucleon propagator.

Equation (1.1) has been solved by iteration and summation of the perturbation series for the physical elements by ordinary Padé approximants. All approximants after the [5/5] give the same result in general. The coupling constants contained in the kernel G, which is a superposition of one-boson exchanges, were fitted such that reasonable agreement of our phase shifts with the experimental ones were obtained [3,4].

In this paper we want to explore further the applicability of the OPA's. Equation (1.1) is formally solved by

$$(1.2)\quad \Phi = G\,[G - B]^{-1}G,$$

where B stands for the box graph (1st iteration, see also [3]). All of these quantities are now supposed to be matrices in the direct product of spin space and discretized off-shell momenta.

2 OPA for the $^3S_1$

What exactly these discretized off-shell momenta are supposed to be cannot be answered in a unique manner. If one selects as off-shell momenta the same points as have been used in the integration procedure of iterating the BSE, equations (1.1) and (1.2) are identical. Since 200 points are ordinarily used to solve the BSE [1], the matrices in (1.2) would become so huge (taking into account the dimension 8 of the "spin space") that matrix inversion would become impossible. In ref. [3] it has been shown, however, that for the OPA's to give reasonable results, only a few points (of the order of 5) out of the complete set of 200 meshpoints are already sufficient for the Isospin I=1 partial waves. The most difficult case to be dealt with in elastic NN-scattering, however, is the $^3S_1 + {}^3D_1$ state, which has isospin I=0. Since recently we have been able to find an adequate "potential" to describe the I=0 NN partial waves in terms of the BSE [4], the study of the OPA's for this case also is an important problem.

Without any special presumptions of how to distribute the off-shell points in the $p - p_4$ plane ($p_0 = ip_4$ after the Wick rotation), one should realize that in solving the BSE, about half of the points have been used in the region $0 \leq p \leq 2\hat{p}$ and about half of them in $2\hat{p} < p < \infty$. We take this as a guide and also observe that there is less structure in the p_4-direction than in the p-direction, which can be taken into account by distributing the higher points on a straight line with $p_4 \cong \frac{2}{3}\, p$.

We thus arrive at the distribution of points shown in Fig. 1.

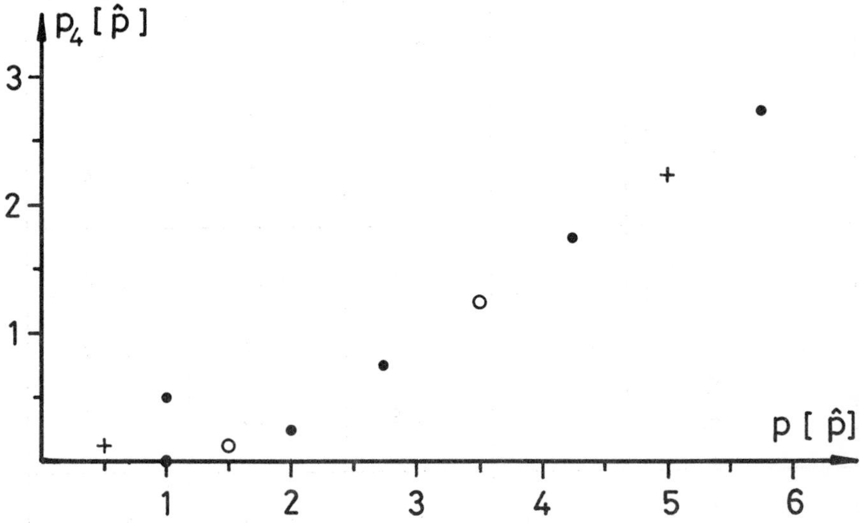

Fig. 1
Distribution of off-shell points
in the p-p_4 plane

Since for I=0 slower convergence than for I=1 has to be expected, we begin already with 6 points (thick dots in Fig. 1). Three of these have $o \leq p \leq 2\hat{p}$ and three $2\hat{p}<p$. Then we add two more points (circles), again one with $p<2\hat{p}$ and one with $2\hat{p}<p$ and finally use all ten points shown in Fig. 1. As one sees from table 1, the results are very bad with six points, but they are qualitatively alright with 8 points, and they are almost exact with ten points.

E_{Lab} (MeV)	PW	$[1/1]_6$	$[1/1]_8$	$[1/1]_{10}$	BSE
50	$^3S_1$	112.4	49.5	58.4	59.2
	$^3D_1$	- 6.8	- 7.1	- 6.9	- 6.9
	ε_1	5.9	0.5	1.9	1.8
100	$^3S_1$	28.6	31.0	38.0	38.8
	$^3D_1$	- 13.9	-13.6	-12.9	-12.9
	ε_1	- 1.8	- 0.9	1.3	1.3
150	$^3S_1$	67.2	32.7	24.9	25.8
	$^3D_1$	66.9	-13.9	-17.2	-17.1
	ε_1	18.2	5.9	0.96	0.9
200	$^3S_1$	38.0	21.0	15.1	15.9
	$^3D_1$	- 6.9	-14.0	-20.4	-20.1
	ε_1	17.7	6.6	0.65	0.6

Table 1
Results for OPA's ($[1/1]_i$ with i = 6,8 and 10 momenta as in Fig. 1) for various energies (E_{Lab}) and coupled partial waves (PW) in comparison with the exact BSE result.

The computer time used to solve the BSE for the $^3S_1 + {}^3D_1$ state in the described manner (with ten points) is about a factor of two less than used by the ordinary Padé method. This is one advantage. The main point, however, is that one hopes for applicability of the OPA's in summing the perturbation theory of a renormalizable strong interaction field theory. Whether, and to what extent a different analytic behavior in terms of the coupling constant spoils our result for the one-loop approximation in the case of actual field theory remains to be seen.

3 Variational OPA for the ^1S_o

A variational procedure to solve the Schrödinger equation has been proposed on the basis of the OPA [5] and various contributions to this question have been made at this conference. In this paper we performed a calculation as closely related to [5] as possible. With one off-shell momentum to be varied parallel to the p-axis ($p_4 = \frac{m}{40}$) we calculate the OPA for the ^1S_o. The behavior of the ^1S_o phase shift for $p \gtrsim \frac{m}{2}$ is similar to what is observed in the case of the Schrödinger equation [5] (see Fig. 2), except that the minimum gives the correct result (within 0.2^o). Stability against small variations of p_4 was observed.

For the $^3S_1 + {}^3D_1$ state, we have not thoroughly investigated the variational OPA's. Calculations performed in the manner described for the ^1S_o state yielded only qualitatively accurate results for the $^3S_1 + {}^3D_1$ state.

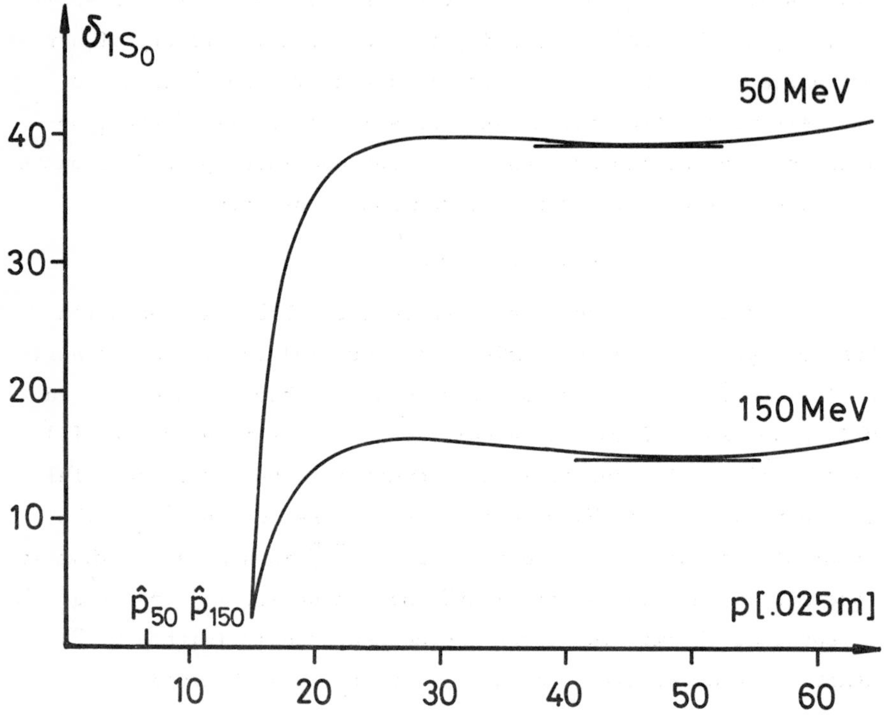

Fig. 2
OPA for the $^1S_0$ - the off-shell momentum considered as variational parameter. E_{Lab} = 50 MeV and 150 MeV with $\hat{p}_{50}$ and $\hat{p}_{150}$ the corresponding on-shell momenta. The horizontal lines at the minima are the exact phase shifts.

References

1 Fleischer J. and J.A. Tjon, Nuclear Physics B84 (1975), 375 - 396.

2 Kubis, J.J., Phys. Rev. D6 (1972), 547 - 564.

3 Fleischer J. and J.A. Tjon, to be published in Phys. Rev. D.

4 Fleischer J. and J.A. Tjon, unpublished.

5 Benofy L.P., J.L. Gammel and P. Mery, Phys. Rev. D13, (1976) 3111 - 3114.

J. Fleischer
Fakultät f. Physik der
Universität Bielefeld
Universitätsstr. 1
D-48 Bielefeld
Germany

J.A. Tjon
Instituut voor Theoretische
Fysica der Rijksuniversiteit
Sorbonnelaan 4
De Uithof - Utrecht
The Netherlands

SERIES SUMMATION METHODS

J.L. Gammel

Eugene Feenberg and his collaborators have shown that certain Padé approximants to the Brillouin-Wigner perturbation series have interesting invariance properties. This paper contains an account of this work and other work of Feenberg which I presented on the occasion of his retirement as Wayman Crowe professor of physics at Washington University in St. Louis.

1 Introduction

In a series of papers [4] Eugene Feenberg and his collaborators described a refinement of the Brillouin-Wigner perturbation scheme which improves the accuracy and convergence of the resulting series for the energy. They established the connection of the method with certain approximants to a continued fraction expansion of the energy (that is, they established the connection with certain Padé approximants), the connection with variational principles, and the invariance of the method under a certain transformation of H_o and the interaction V which they call the μ transformation (see below). Invariance under transformations is a very important property of Padé approximants. It is well known that the [N/N] Padé approximants are invariant under transformations of the form

(1) $x = A\omega/1 + B\omega$,

which helps explain why the regions in which the Padé approximants are accurate are not just circles containing the origin.

A problem considered by Feenberg and his collaborators is that of the lowest eigenvalue of Mathieu's equation

(2) $(-\frac{d^2}{dx^2} + s\cos^2 x)\psi = E\psi$,

with periodic boundary conditions, $\psi(0) = \psi(2\pi)$, a problem about which one can find almost everything in Morse and Feshbach [6]. One finds in particular that the Brillouin-Wigner perturbation scheme yields

$$(3) \quad E = \frac{s}{2} + \frac{s^2}{8(E-4)} + \frac{s^3}{16(E-4)^2} + \frac{s^4}{32(E-4)^2}\left\{\frac{1}{E-4} + \frac{1}{4}\frac{1}{E-16}\right\} + \ldots .$$

A Padé-approximatist will have no trouble in seeing what to do. The right hand side is a power series in s, and he will form Padé approximants to it. For example, the [1/1] Padé approximant yields

$$(4) \quad E = \frac{\frac{s}{2}}{1 - \frac{s}{4(E-4)}},$$

and solve these for E rather than solving the truncated power series for E,

$$(5) \quad E = \frac{s}{2} + \frac{s^2}{8(E-4)}.$$

He will find this game rewarding for reasons which I come to below.

2 The Method of Feenberg

Eugene Feenberg did not come on this scheme in this way. He and his coworkers modified the Brillouin-Wigner perturbation series for the ground state wave function ψ,

$$(6) \quad \psi = \psi_0 + \sum_a{}' \psi_a \frac{V_{ao}}{E-E_a} + \sum_{a,b}{}' \psi_b \frac{V_{ba}V_{ao}}{(E-E_b)(E-E_a)} + \ldots ,$$

by putting arbitrary coefficients G in front of the $\sum'$ (by the way, $\sum'$ means that a,b,c,... can never be 0)

$$(7) \quad \psi = \psi_0 + G_1 \sum_a{}' \psi_a \frac{V_{ao}}{E-E_a} + G_2 \sum_{a,b}{}' \psi_b \frac{V_{ba}V_{ao}}{(E-E_b)(E-E_a)} + \ldots ,$$

and determining the G's by taking advantage of the Rayleigh-Ritz variational principle to make stationary

$$(8)\quad E = \left(\psi^{(n)}|H + V|\psi^{(n)}\right) \Big/ \left(\psi^{(n)}, \psi^{(n)}\right) ,$$

where $\psi^{(n)}$ is just the power series Eq. (7) truncated at the nth term. In Young, Biedenharn, and Feenberg, it is proved that this refinement proposed by Goldhammer and Feenberg is to be interpreted as yielding alternate approximants to a continued fraction expansion for the energy. As is well known [8], these alternate approximants are just certain Padé approximants. John Nuttall notes [7] this work as the earliest connection of Padé approximants with variational principles.

3 Numerical Data

Let me show you the numerical data, in part calculated by Young, Biedenharn, and Feenberg, for s = 4. For s = 4, Eq. (4) yields E = 1.43845. The [2/1] approximant based on Eq. (3) yields E = 1.55051, and the [2/2] E = 1.54446. In fact, if we use the data of Young, Biedenharn, and Feenberg we may prepare a Padé Table of [M/N] Padé results,

(9)

N ↓ \ M →	1	2	3
0	2	1.26795	2.00000
1	1.43845	1.55051	1.54429
2		1.54450	1.54487

where the entries given by the method of Goldhammer and Feenberg are put inside of the solid lines. Young, Biedenharn, and Feenberg noted that the entries just above the diagonal are all greater than the exact eigenvalue 1.54486 and the ones two above the diagonal are all less. The important connection with the whole method with series of Stieltjes, noted, I think, especially among the physicists by Baker, and on which the luck of the Padé-approximatist depends in this case and in many other cases, is responsible for the fact that these entries are respectively decreasing upper bounds and increasing lower bounds. I will return to all of this below.

4 The μ Transformation

The μ transformation noted by Feenberg and his collaborators is

$$(10)\quad H_o' = H_o + (\mu - 1)(H_o - E),$$
$$V' = V - (\mu - 1)(H_o - E).$$

These leave E unchanged because

$$(11)\quad H_o' + V' = H_o + V.$$

Also, the eigenfunctions of H_o and H_o' are the same, so that the matrix elements of V and V' are easily gotten one from the other. Also, they imply

$$(12)\quad E - H_o' = \mu(E - H_o),$$

and this causes in the Brillouin-Wigner perturbation theory a uniform change of scale in the energy denominators. Clearly, E ought to be invariant against all of this in view of Eq. (11).

It is trivial to check explicitly that this is so in particular cases. For example, the [2/1] approximant to

$$(13)\quad E = -(\mu - 1)E + \left[\frac{s}{2} + (\mu - 1)E\right] + \left[\frac{s^2}{8\mu(E-4)}\right] + \left[\frac{s^3}{16\mu^2(E-4)^2} + (\mu - 1)\frac{s^2}{8\mu^2(E-4)}\right],$$

is

$$(14)\quad E = -(\mu - 1)E + \left[\frac{s}{2} + (\mu - 1)E\right] + \left[\frac{s^2/8\mu(E-4)}{\frac{1}{\mu} - \frac{s^2}{2\mu(E-4)}}\right],$$

from which μ disappears, as desired.

Young, Biedenharn, and Feenberg gave a general proof of this property. Bolsterli and Feenberg studied a near invariance under a uniform shift of the zeroth order eigenvalues.

5 Connection of the Method With Series of Stieltjes

Finally, I want to discuss the connection with series of Stieltjes. The Brillouin-Wigner expression,

$$(15)\quad E = (\psi|sv + s^2 v \frac{p}{b} v + s^3 v \frac{p}{b} v \frac{p}{b} v + \ldots|\psi),$$

where p projects away the ground state compressed of H_o, and $b = E - H_o$, can be rearranged to read

$$(16)\quad E = s(\sqrt{v}\psi|1 + s\sqrt{v} \frac{p}{b} \sqrt{v} + b^2(\sqrt{v} \frac{p}{b} \sqrt{v})^2 + \ldots|\sqrt{v}\,\psi),$$

provided $\sqrt{v}$ exists, as it does if $v \geq 0$, which is the case for the Mathieu equation. Then $\sqrt{v} \frac{p}{b} \sqrt{v}$ is manifestly Hermitian and negative definite provide $E < E_1$, where, of course $H_o\psi_1 = E_1\psi_1$. Let ϕ_n denote the eigenfunctions of $(- \sqrt{v} \frac{p}{b} \sqrt{v})$,

$$(17)\quad (- \sqrt{v} \frac{p}{b} \sqrt{b})\phi_n = \lambda_n\phi_n .$$

Since the ϕ_n are complete and orthonormal,

$$(18)\quad \sqrt{v}\,\psi = \sum_n c_n\phi_n ,$$

so that finally

$$(19)\quad E = \sum_m \sum_n |c_n|^2 \quad (-\lambda_n)^m s^m .$$

Let $d\phi(t)$ be zero except at $t = \lambda_n$, and at these points let $d\phi(t) = |c_n|^2 > 0$. Then

$$(20)\quad E = s \int_0^\infty \frac{d\phi(t)}{1 + st}$$

so that Eq. (3) is s times a series of Stieltjes in s provided $E < 4$.

The [N/N] Padé approximants to the Brillouin-Wigner expansion with one factor of s removed then form a decreasing sequence of upper bounds to the actual value of the right hand side, so that

(21) $\frac{E}{s} \leq$ [N/N] to right hand side,

also the [N-1/N] Padé approximants form an increasing sequence of lower bounds,

(22) $\frac{E}{s} \geq$ [N-1/N] to right hand side.

Actually, at this point, there is a slight flaw. The entries below and outside the solid lines in Eq. (9) are actually the best lower bounds obtainable from a certain number of terms in the expansion, but the results obtained are not invariant under Feenberg's μ transformation. The [N+1/N] Padé approximants are lower bounds, however.

At this point it is only a matter of graph drawing to understand why Young, Biedenharn, and Feenberg obtained alternating upper and lower bounds to E. The necessary graph is shown in Fig. 1.

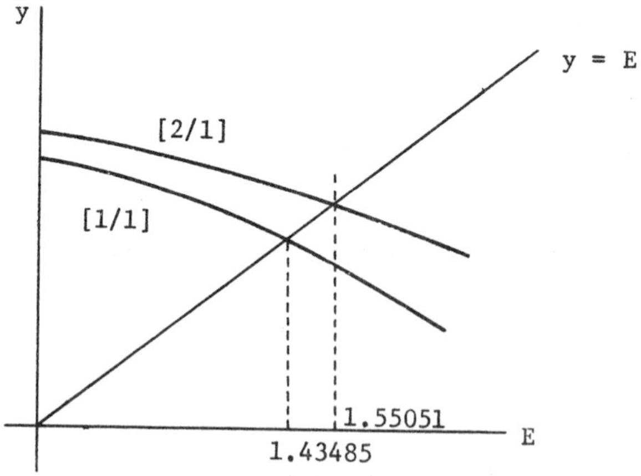

Fig. 1. The ordinate y is the Padé approximant to the Brillouin-Wigner expansion of E evaluated at s = 4. The Padé approximants plotted do not have the factor s removed, so that the [1/1] is really the [0/1] and the [2/1] the [1/1]. Since the [2/1] must be greater than the actual value of y for all E, the exact value of E must lie below 1.55051, and since the [1/1] must be less than the actual value of y for all E, the exact value of E must lie above 1.43485.

6 Other Contributions of Feenberg

Feenberg [3] has done other work on perturbation theory. This work is reviewed in Morse and Feshbach (page 1010 et. seq.). It begins with the observation that in, for example, the fifth term of the Brillouin-Wigner series,

$$(23)\quad \sum_{pqrs}{}' \frac{V_{op}V_{pq}V_{qr}V_{rs}V_{so}}{(E-E_p)(E-E_q)(E-E_r)(E-E_s)},$$

V^2_{ab} appears when p = r = a and q = s = b. Feenberg pointed out that it should be possible to obtain formulas in which such repetitions do not occur in view of the results of the next paragraph.

Following Morse and Feshbach, let the problem to be solved be

$$(24)\quad H\psi + \lambda V\psi = E\psi.$$

Then

$$(25)\quad H\psi_a = E_a\,\psi_a,$$

and

$$(26)\quad \int \psi_a\,\psi_b = \delta_{ab}.$$

Try expanding

(27) $\psi = \sum_p c_p \psi_p,$

and put this into Eq. (24)

(28) $\sum_p c_p (E_p - E)\psi_p + \lambda \sum_p c_p V\psi_p = 0,$

so that

(29) $(E - E_p) c_p = \lambda \sum_p c_q \int \psi_p V\psi_q \equiv \lambda \sum_q c_q V_{pq},$

or

(30) $\sum_q \left[(E-E_q) \delta_{pq} - \lambda V_{pq}\right] c_q = 0.$

For this to have a solution, it is necessary that

(31) $\det \left[(E - E_q) \delta_{pq} - \lambda V_{pq}\right] = 0.$

In the expansion of this secular determinant, no repetition of V_{pq} occurs.

Feenberg suggests that a solution of Eq. (30) which has $c_n = 1$ for some n and is such that all other c's → 0 as λ → 0 be obtained by rewriting Eq. (30) in the form

(32) $E = E_n + \lambda V_n + \lambda \sum_{p\neq n} c_p V_{np},$

$$(E - E_p) - \lambda V_{pp}) c_p = \lambda V_{pn} + \lambda \sum_{q\neq n \text{ or } p} c_q V_{pq}, \quad p \neq n.$$

To avoid repetitions of a matrix element in solving this by iteration, one uses for c_q,

(33) $(E - E_q - \lambda V_{qq})c_q = \lambda V_{qn} + \lambda c_p V_{qp} + \lambda \sum_{r\neq npq} c_r V_{qr},$

and so on

(34) $(E - E_r - \lambda V_{rr}) c_r = \lambda V_{rn} + \lambda c_p V_{rp} + \lambda c_q V_{rq}$

$$+ \lambda \sum_{s\neq npqr} c_s V_{rs}.$$

Were we dealing with a finite secular determinant, the last such equation would not contain a sum. One solves this last equation, substitutes the result into the next to last equation, and so on back to the beginning. For an infinite secular determinant, one resorts to mathematical induction. The details are given in Morse and Feshbach. In the case of the Mathieu equation, the method leads to the exact continued fraction. However, on page 1017, Morse and Feshbach appear to me to calculate $b_o = 1.5$ for $s = 4$ instead of the much more accurate value obtainable from the result given there.

7 Recent Developments

Feenberg's method has been of much utility recently in Samuel Bowen's work [2] on the Kondo problem. Bowen has also generalized the method for application to quantum field theory and all sorts of model Hamiltonians used in many body physics. The work has some rough similarity to approximations which Bessis [1] has called "operator Padé approximants", and to Masson's [5] Padé approximants in Hilbert space.

One expects still further progress in these lines of approximation initiated by Feenberg.

References

1. Bessis, D., Padé Approximants, edited by P.R. Graves-Morris, Institute of Physics (London and Bristol, 1973).
2. Bowen, Samuel, JMP 16 620 (March 1975).
3. Feenberg, E., Phys. Rev. 74, 206 and 664 (1948).
4. Goldhammer, P. and E. Feenberg, Phys. Rev. 101, 1233 (1956) and Phys. Rev. 105, 750 (1957); Bolsterli, M. and E. Feenberg, Phys. Rev. 101, 1349 (1955); Young, R.C., L.C. Biedenharn, and E. Feenberg, Phys. Rev. 106, 1151 (1957); Feenberg, E., Phys. Rev. 103, 1116 (1956).
5. Masson, D., cf. his articles in The Padé Approximant in Theoretical Physics, edited by Baker and Gammel, Academic Press (New York, 1970); Padé Approximants and Their Applications, edited by P.R. Graves-Morris, Academic Press (New York, 1973).

6. Morse and Feshbach, Methods of Theoretical Physics, Vol. II, p. 1008 et seq., McGraw-Hill (New York, 1953).

7. Nuttall, John, Padé Approximants and Their Applications, edited by P.R. Graves-Morris, Academic Press (New York, 1973).

8. Wall, H.S., Continued Fractions, D. Van Nostrand (New York, 1948).

J.L. Gammel
Department of Physics
Saint Louis University
St. Louis, Missouri 63103

A VARIATIONAL APPROACH TO OPERATOR AND MATRIX PADÉ APPROXIMANTS. APPLICATIONS TO POTENTIAL SCATTERING AND FIELD THEORY

P. Mery

We define operator and matrix Padé approximants. We emphasize the fact that these approximants can be derived from the Schwinger variational principle. In potential theory, we shall show, using this variational property, that the matrix Padé approximants can reproduce the exact solution of the Lippman-Schwinger equation with arbitrary accuracy taking into account only the first two coefficients in the Born expansion. The analytic structure of this variational matrix Padé approximant (hyper Padé approximant) is discussed.

1 Introduction

The convergence properties of Padé approximants in potential scattering have been studied for a long time [5,18,20,26]. It has been shown that if one takes into account a very high number of coefficients in the Born expansion the result is arbitrarily accurate.

Very important progress was made after it was shown that the Padé approximants can be derived from variational principles [9,16,23].

Using the Rayleigh-Ritz variational principle, Padé approximants have been derived which make it possible to increase the accuracy of the calculation of bound states and resonances for the two body problem, the three body problem, and the Bethe-Salpeter equation [1,2,3,4].

Another type of approximation has been defined by using the variational property of the operator and matrix Padé approximants [10,11,27,28]. It has been shown that such an approximation makes it possible to get incredibly accurate results in potential scattering [7,8,12,21].

We present here the variational Matrix Padé approximant

method and prove its convergence in the case of potentials which do not change sign. Numerical tests show that one also gets convergence for any type of potential [7,8,12,21].

This Padé method (hyper Padé) is not a rational fraction approximation. It allows for the possibility of deeper analytic structure such as an algebraic cut.

As a consequence the variational matrix Padé approximant competes with other standard methods for solving classical Schrodinger-like problems. Furthermore it can be extended to field theory and may make it possible to improve the description of strong interaction particle physics using a minimal number of parameters [13,22].

2 The Operator Padé Approximant

2.1 Definition

Let $T(\alpha)$ be an operator analytic in α which can be expanded around $\alpha = 0$ in a formal power series according to

$$(2.1)\quad T(\alpha) = T_0 + \alpha T_1 + \alpha^2 T_2 + \ldots + \alpha^n T_n + \ldots ,$$

the T_i being non commuting operators. Because of this non commuting character one can define several types of Padé approximants [14] such as left, right, or mixed Padé approximants,

$$(2.2)\quad T(\alpha) - {}^{L}Q_M^{-1}(\alpha)\, {}^{L}P_N(\alpha) = O(\alpha^{N+M+1}),$$

$$(2.3)\quad T(\alpha) - {}^{R}P_N(\alpha)\, {}^{R}Q_M^{-1}(\alpha) = O(\alpha^{N+M+1}),$$

$$(2.4)\quad T(\alpha) - {}^{M}P_{N'}(\alpha)\, {}^{M}Q_M^{-1}(\alpha)\, {}^{M}P_{(N-N')}(\alpha) = O(\alpha^{N+M+1}),$$

respectively. One can prove that these types of approximants are equal [12,14,21], and so we shall define the operator Padé approximant to $T(\alpha)$ by

$$(2.5)\quad [N/M]_{T(\alpha)} = {}^{R}P_N(\alpha)\, {}^{R}Q_M^{-1}(\alpha) = {}^{L}Q_M^{-1}(\alpha)\, {}^{L}P_N(\alpha)$$
$$= {}^{M}P_{N'}(\alpha)\, {}^{M}Q_M^{-1}(\alpha)\, {}^{M}P_{N-N'}(\alpha).$$

The operator Padé approximants (O.P.A.) have the same covariance properties as the usual approximants. Furthermore one can prove that the O.P.A. to a direct sum of operators is the direct sum of the O.P.A. to each operator, i.e.

$$(2.6)\quad T(\alpha) = \bigoplus_i T_i(\alpha) \Rightarrow [N/M]_{T(\alpha)} = \bigoplus_i [N/M]_{T_i(\alpha)}.$$

2.2 The Variational Property Of The Operator Padé Approximant

Consider the Lippman-Schwinger functional

$$\begin{aligned} F(V,V^+) &= I - i\int_{t_0}^{t} dt'[V^+(t',t)H_I(t') + H_I(t')V(t',t_0)] \\ (2.7)\qquad &+ i\int_{t_0}^{t} dt'\; V^+(t',t)H_I(t')V(t',t_0) \\ &+ i^2\int_{t_0}^{t} dt' \int_{t_0}^{t} dt''\; \Theta(t''-t')V^+(t'',t)H_I(t'')V(t',t_0), \end{aligned}$$

where H_I is the interaction Hamiltonian and V and V^+ are arbitrary time dependent operators.

The Lippman-Schwinger variational principle [19] states that

$$(2.8)\quad \delta F(V,V^+) = 0 \Rightarrow V=U_I,\; V^+=U_I^+,\; F_{stat}=U_I(t,t_0),$$

where $U_I(t,t_0)$ is the evolution operator between times t and t_0. Letting $t_0 \to -\infty$ and $t \to +\infty$ one gets the S-scattering operator.

Using the Cini-Fubini ansatz [17], i.e. choosing for trial operator V a linear combination (with operator coefficients) of the first N-terms of the Taylor expansion of $U_I(t,t_0)$, one can show that the formal solution of the Lippman-Schwinger variational principle is the [N/N] O.P.A. to $U_I(t,t_0)$ [14]. Again if $t_0 \to -\infty$ and $t \to +\infty$ the [N/N] O.P.A. to the S-scattering operator is a solution of the Lippman-Schwinger variational principle.

2.3 The Operator Padé Approximant and the Lippman-Schwinger Equation

Let $K(E)$ be a scattering operator which is the solution of the Lippman-Schwinger equation for a Hamiltonian $H = H_0 + \lambda V$,

(3.1) $K(E) = \alpha V + \alpha V G_0(E) K(E),$

with

(3.2) $G_0(E) = \frac{1}{2}[(E + i\varepsilon - H_0)^{-1} + (E - i\varepsilon - H_0)^{-1}].$

The exact solution of this equation can be written in a formal way as

(3.3) $K(E) = \alpha V[V - \alpha V\, G_0(E) V]^{-1} V,$

which is nothing but the [1/1] O.P.A. to K(E),

(3.4) $K(E) = [1/1]_{K(E)}.$

Such a result can be extended to semi-relativistic equations such as the Blankenbecler-Sugar equation [25] or to the Bethe-Salpeter equation in the ladder approximation.

3 The Variational Matrix Padé Approximant

3.1 Definition

All the results mentioned in the previous section are formal results because the operators we are interested in act in an infinite dimensional Hilbert space. To compute an O.P.A. we need to invert an operator and this can be done only by discretizing the Hilbert space so that in practice we can compute only a matrix Padé approximant.

Using the Lippman-Schwinger principle one can prove for potential scattering that the discretization points must be chosen in a variational way. We shall call such an approximation a variational matrix Padé approximant (V.M.P.A.).

The problem is to look at the accuracy of this method and for that we shall study potential scattering. As has already been mentioned, were the number of discretization points infinite, we would get the exact solution to the Lippman-Schwinger equation. Let us now look at what happens in a realistic case.

We shall call q_i the discretization points of our Hilbert space; q_0 is the center of mass momentum (on-shell point). We shall consider states $|q_i\rangle$ with defined angular momentum and energy $\frac{q_i^2}{2m}$; we shall call E_L the set of $|q_i\rangle$; i.e., our discretized Hilbert space with (L-1) off-shell points is

$$(3.5)\quad E_L = \{|q_i\rangle;\ i = 0,\ldots,L-1\}.$$

$K_L(E)$ will be the restriction of the scattering operator K(E) to the discretized Hilbert space E_L.

3.2 The Quasi-Potential Theorem

Let V be a positive and regular potential ($V(r) = O(r^{-2+\varepsilon})$ for $r \to 0$ and $V(r) = O(r^{-3-\varepsilon})$ for $r \to \infty$ with $\varepsilon > 0$). Let $E_{N,L}$ be the following space

$$(3.6)\quad E_{N,L} = \{(\sqrt{V}G_0(E)\sqrt{V})^m\sqrt{V}|q_i\rangle;\ m=0,\ldots,N-1;\ i=0,\ldots,L-1\},$$

and $P_{N,L}$ be the projector over the space $E_{N,L}$.

The [N/N] matrix Padé approximant to K_L is the solution of a Lippman-Schwinger equation with a non-local potential $V_{N,L} = \sqrt{V}\,P_{N,L}\sqrt{V}$. The proof of this theorem can be found in references [12,21]. This theorem has some physical significance because it asserts that it is possible to construct a quasi potential, i.e., a non-local potential reproducing a given number of coefficients of the perturbative expansion.

3.3 The Variational Matrix Padé Approximant in the Case of Potentials Which Do Not Change Sign

If we increase L we increase the number of vectors in the space $E_{N,L}$; i.e., for a given order of the approximation [N/N] we increase the number of off-shell points so that

$$(3.7)\quad E_{N,1} \subset E_{N,2} \subset \cdots \subset E_{N,L} \subset \cdots \subset H,$$

where H is the infinite dimensional Hilbert space.

The above inequalities imply

$$(3.8)\quad P_{N,1} < P_{N,2} < \cdots < P_{N,L} < \cdots < I,$$

where $P_{N,L}$ projects over the space $E_{N,L}$. I is the identity over the Hilbert space.

In the case of positive potentials (V > 0) we have

$$(3.9)\quad V_{N,1} < V_{N,2} < \dots < V_{N,L} < \dots < V,$$

and as a consequence of the Feynman-Hellmann theorem we get

$$(3.10)\quad \delta_{N,1} > \delta_{N,2} > \dots > \delta_{N,L} > \dots > \delta,$$

where $\delta_{N,L}$ is the exact phase shift computed with the $V_{N,L}$ potential

$$(3.11)\quad tg(\delta_{N,L}) = -q_0 < q_0 | [N/N]_{K_L} | q_0 >.$$

The phase shift δ computed with the potential V is given by

$$(3.12)\quad tg\,\delta = - q_0 < q_0 | K_L | q_0 > \quad \forall\ L.$$

The above relations show how to choose the off-shell momenta in a variational way; namely

$$(3.13)\quad \sup_{q_1,\dots,q_{L-1}} \delta_{N,L} \geq \delta.$$

Equation (3.10) shows the convergence to the exact solution of the phase shifts computed from the V.M.P.A.

It was known for a long time that convergence can be obtained by increasing the order of the approximation; this requires the computation of a large number of coefficients of the perturbative expansion. Such convergence can also be obtained for a given order of approximation by increasing the size of the matrix, i.e., by increasing the dimensionality of the discretized Hilbert space. The latter is a far easier task from a practical point of view. The numerical examples given below will illustrate the fact that convergence can be obtained using only a very small number of off-shell points.

Let us notice that i) the same kind of results can be obtained for $V < 0$ with opposite inequalities, and ii) an identical proof can also be given in the case of singular potentials using the regulator as a variational parameter [15].

In the case of potentials which change sign one must notice that accurate results cannot be obtained from the ordinary [1/1] Padé approximation; indeed, in this case, a change of sign can result only when the denominator changes sign and this means that the phase shift goes to 90° rather than 0°. We have no rigorous proof of convergence of the V.M.P.A. in the case of such potentials. However, numerical examples show convergence to the exact solution and a conjecture has been put forth [7] that the correct choice of off-shell momentum is that q_i corresponding to the extremum closest to the on-shell value q_0.

3.5 Numerical Examples

The practical way to compute variational matrix Padé approximants can be found in references [7,8,12,21]. The V.M.P.A. method has first been applied to study square wells potential. In the case of a single well it has been proved [24] that with one off-shell momentum, the variational choice of this one momentum is the momentum inside the well, and that the V.M.P.A. reproduces numerically the exact result. With two square wells [7] studies of the variational choice of the off-shell point in the case of a potential which changes sign lead to the conjecture presented above.

The table summarizes our result [12,21] in the case of an exponential potential $V = - V_0 \, e^{-\mu r}$. The table illustrates the convergence of the method.

We have used the same notation as in reference [6] in which convergence resulting from increasing the order of the approximation was studied. In the present work we always compute a [1/1] Padé approximant including one or two off-shell momenta. In the case of a weakly attractive or repulsive potential only one off-shell momentum results in very accurate results. Using two off-shell points, we can describe very strongly attractive potentials, for example, a potential with two bound states. The method has also been applied to study double combinations of two exponential potentials which change sign, and the results confirm our conjecture for the choice of the discretization points [12, 21].

4 The Analytic Structure of the Variational Matrix Padé Approximant

The matrix Padé approximant is a rational fraction approximation, but this is no longer true of the variational matrix Padé approximant (hyper Padé approximant).

Indeed, let us consider for example the computation of a phase shift through a [1/1] matrix Padé approximant to the K-matrix with one off-shell point, i.e.,

$$(4.1)\quad \delta(\alpha,q_0,q) = \text{arc tg}[-q_0 < q_0|[1/1]_{K_2}|q_0>].$$

$\delta(\alpha,q_0,q)$ is a rational fraction in the coupling constant α. When we consider the V.M.P.A. we eliminate the off-shell momentum q through

$$(4.2)\quad \frac{\partial\delta(\alpha,q_0,q)}{\partial q} = 0 \Rightarrow q = q(\alpha,q_0).$$

Then $\delta[\alpha,q_0,q(\alpha,q_0)]$ is a function with a more complicated analytic structure than a rational fraction.

As an example consider a square well

$$\begin{aligned} V &= \alpha V_0, && \text{if } r < r_0,\\ &= 0\;, && \text{if } r > r_0. \end{aligned}$$

It has been proved [24] that the [1/1] V.M.P.A. reproduces exactly the $^1S_0$ phase shift, i.e., the function

$$\text{tg}\delta = \frac{q_0\sin(\sqrt{q_0^2+\alpha V_0}\,r_0)\cos(q_0r_0)-\sqrt{q_0^2+\alpha V_0}\cos(\sqrt{q_0^2+\alpha V_0}\,r_0)\sin(q_0r_0)}{q_0\sin(\sqrt{q_0^2+V_0}\,r_0)\sin(q_0r_0)+\sqrt{q_0^2+V_0}\cos(\sqrt{q_0^2+V_0}\,r_0)\cos(q_0r_0)}.$$

5 Conclusion

As we have seen the variational matrix Padé approximant is a very powerful method to study any Schrodinger like problem. Using the variational property of the approximation, we have shown that for potential scattering the method makes it possible to reproduce the exact results with any required numerical accuracy from only the first two coefficients of the perturbative expansion.

In field theory this method can be applied in the same way and may make it possible to describe strong interaction physics such as, for example, nucleon-nucleon scattering.

For this kind of problem the Green's function in the center of mass system can be parametrized as shown in the figure.

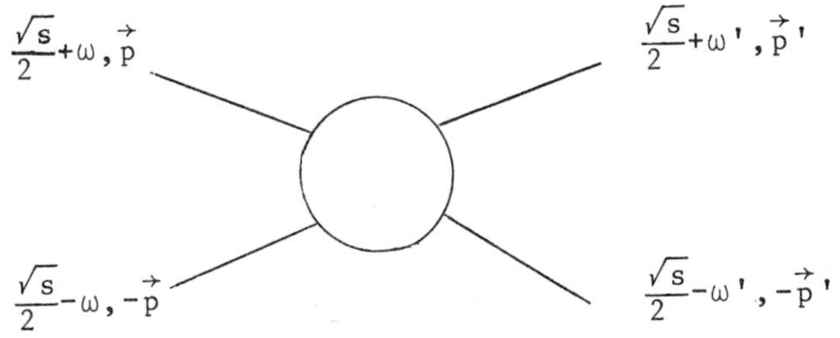

If we neglect spin,the partial wave expansion of this Green's function reads

$$(5.1)\quad G(s,\omega,\omega',p,p',\cos\theta) = \sum_J G^J(s,\omega,\omega',p,p')\, P_J(\cos\theta),$$

where θ is the scattering angle in the center of mass system. The partial wave coefficients can be computed by

$$(5.2)\quad G^J(s,\omega,\omega',p,p') = \langle\omega,p|G^J(J)|\omega',p'\rangle.$$

The perturbative expansion of $G^J(s)$ is given by Feynman's diagrams

$$(5.3)\quad G^J(s) = \alpha G^{(1)J}(s) + \alpha^2 G^{(2)J}(s) + \ldots .$$

The [1/1] Padé approximant to $G^J(s)$ is then

$$(5.4)\quad [1/1]_{G^J(s)} = \alpha G^{(1)J}(s)\, [G^{(1)J}(s) - \alpha G^{(2)J}(s)]^{-1}\, G^{(1)J}(s).$$

Although we cannot prove any rigorous result we can hope that a variational choice of the discretization points can also be made in field theory. In this case we could have for the first time a good description of the strong interaction physics with a minimal number of parameters.

Table

$^1S_0$ phase shift for an exponential potential $V(r) = -V_0 e^{-\mu r}$

(phase shifts are given in degrees)

Notation $g = 2\frac{m}{\mu}\frac{V_0}{\mu}$; $\nu = 2i\frac{\sqrt{2mE}}{\mu}$.

g and ν are dimensionless.

$m = 938$ MeV = nucleon mass

$\mu = 140$ MeV = pion mass

Case g = -8

$-i\nu$	Ordinary Padé	Variational 2x2 Matrix Padé	Exact solution
.1	- 7.610	- 9.237	- 9.242
.5	-34.962	- 43.734	- 43.771
1.	-56.975	- 76.726	- 76.861
2.	-71.029	-110.868	-111.801

Case g = -1

$-i\nu$	Ordinary Padé	Variational 2x2 Matrix Padé	Exact solution
.1	- 3.506	- 3.572	- 3.574
.5	-15.452	-15.794	-15.805
1.	-22.620	-23.281	-23.308
2.	-21.557	-22.254	-22.308

Case g = +1

$-i\nu$	Ordinary Padé	Variational 2x2 Matrix Padé	Variational 3x3 Matrix Padé	Exact solution
.1	14.459	16.054	16.145	16.157
.5	35.882	38.088	38.210	38.228
1.	32.005	33.525	33.618	33.633
2.	22.051	22.809	22.873	22.885

Case g = +8

$-i\nu$	Ordinary Padé	Variational 2x2 Matrix Padé	Variational 3x3 Matrix Padé	Exact solution
.1	168.600	232.556	292.297	302.143
.5	128.840	217.251	232.775	236.268
1.	98.532	183.027	193.863	196.297
2.	74.290	135.709	144.243	146.328

Case g = +10

$-i\nu$	Ordinary Padé	Variational 2x2 Matrix Padé	Variational 3x3 Matrix Padé	Exact solution
.1	169.139	335.595	339.285	341.624
.5	130.947	264.812	273.828	279.900
1.	101.311	216.637	226.235	231.011
2.	77.660	161.248	171.353	174.073

References

1 Alabiso, C., P. Butera and G.M. Prosperi, Resolvent operator, Padé approximation and bound states in potential scattering, Lett. Nuo. Cim. 3, (1970), 831.

2 Alabiso, C., P. Butera and G.M. Prosperi, Variational principles and Padé approximants: bound states in potential theory, Nucl. Phys, 31B (1971), 141.

3 Alabiso, C., P. Butera and G.M. Prosperi, Variational principles and Padé approximants: resonances and phase shifts in potential scattering, Nucl. Phys., 42B (1972), 493.

4 Alabiso, C., P. Butera and G.M. Prosperi, Variational Padé solution of the Bethe-Salpeter equation, Nucl. Phys., 46B (1972), 593.

5 Baker, G., The theory and application of the Padé approximant method, Adv. Theor. Phys., 1 (1965), 1.

6 Basdevant, J.L., and B.W. Lee, Padé approximation and bound states exponential potential, Nucl. Phys., B13 (1969), 182.

7 Benofy, L., J. Gammel and P. Mery, The off stell momentum as a variational parameter in calçulations of matrix Padé approximants in potential scattering, Phys. Rev., D13 (1976), 3116.

8 Benofy, L., and J. Gammel, Variational principles and matrix Padé approximants, these Proceedings.

9 Bessis, D., and M. Pusterla, Unitary Padé approximant in strong coupling field theory and application to the calculation of the ρ and δ Meson Regge trajectories, Nuo. Cim. 54A (1968), 243.

10 Bessis, D., Padé approximants in quantum field theory, Padé approximants and their applications, P.R. Graves-Morris, ed., Academic Press, New York (1973), 275.

11 Bessis, D., and J. Talman, Variational approach to the theory of operator Padé approximants, Rocky Mountain Journal of Math., 4 (1974), 151.

12 Bessis, D., P. Mery, and G. Turchetti, Variational bounds from matrix Padé approximants in potential scattering, submitted to Phys. Rev. D.

13 Bessis, D., P. Mery, and G. Truchetti, Angular momentum analysis of the four-nucleon Green's function, Phys. Rev. D10, (1974), 1992.

14 Bessis, D., Topics in the theory of Padé approximants, Padé approximants, P.R. Graves-Morris, ed., Institute of Physics, London (1973), 19.

15 Bessis, D., L. Epele, and M. Villani, Summation of Regularized perturbative expansions for singular interactions, J. Math. Phys., 15, (1975), 2071.

16 Caneschi, L., and R. Jengo, On the problem of the bound states in the framework of the Padé approximants, Nuo. Cim., 60A (1969), 25.

17 Cini, M., and Fubini, Non perturbation treatment of scattering in quantum field theory, Nuo. Cim. 11 (1954), 142.

18 Garibotti, C.R. and M. Villani, Continuation in the coupling constant for the total K and T matrices, Nuo. Cim., 54A (1969), 107; and Padé approximant and the jost function, Nuo. Cim., 61A (1969), 747.

19 Lippman, B.A., and J. Schwinger, Variational principles for scattering processes, Phys. Rev., 79 (1950), 469.

20 Masson, D., Analyticity in the potential strength, J. Math. Phys., 8 (1967), 2308.

21 Mery, P., Calcul de la diffusion nucléon-nucléon à basse energie par la methode des approximations de Padé en théorie Lagrangie des champs, Ph. D. thesis - Université de Luminy-Marseille (1976).

22 Mery, P., and G. Turchetti, Relativistic Green's function approximation to the low-energy nucleon-nucleon interaction, Phys. Rev. D11, (1975), 2000.

23 Nuttall, J., The convergence of the Kohn variational method, Ann. Phys., 52, (1969), 428.

24 Ortholani, F., G. Fratamico, and G. Turchetti, Exact solutions from the variational [1/1] matrix Padé approximant in potential scattering, Lett. Nuo. Cim. 17, (1976), 553.

25 Sugar, R., and R. Blankenbecler, Variational upper and lower bounds for multichannel scattering, Phys. Rev. 136B, (1964), 472.

26 Tani, S., Complete continuity of Kernel in generalized potential scattering, Ann. Phys., 37 (1966), 451.

27 Turchetti, G., Padé approximants in quantum field theory, Padé approximants and their applications, P.R. Graves-Morris, ed., Academic Press, New York (1973), 313.

28 Turchetti, G., Variational principles and matrix approximants in potential theory, Lett. Nuo. Cim., 15 (1976), 129.

P. Mery
U.E.R. Scientifique de Luminy et Centre de Physique Théorique
70, route Leon-Lachamp
Marseille, France

OPERATOR PADÉ APPROXIMANTS AND THREE-BODY SCATTERING

J.A. Tjon

The numerical solution of the problem of elastic scattering of a free particle by two particles in a bound state is discussed within the integral equation approach. The two-body interactions are chosen to be of a simple separable type. The convergence rate of Padé Approximants (PA) applied to the multiple scattering series is studied. Results are presented also for an operator PA in which off-shell momenta are used as variational parameters.

1 Introduction

Although the method of scalar Padé approximants (SPA) has been shown to be a useful technique for summing the multiple scattering series in the three-nucleon problem [1], calculations with realistic two-nucleon interactions would be very time consuming on present-day computers. For this reason it is interesting to explore alternative methods. In this paper we discuss the applicability of operator Padé approximants [2] (OPA) as a possible tool for studying three-body equations. For ordinary potential scattering the method of OPA has recently been demonstrated to be an attractive alternative [3].

In the next section we describe the equations basic to the problem. As a simple model we have chosen two-nucleon interactions of a separable type. It is expected, in view of the experience with SPA, that the convergence rate will not depend very sensitively on the type of interactions used. The situation in which we will be interested is the description of elastic scattering of nucleons by deuterons. Results are presented in the last section for the doublet and quartet channels.

2 Three-Body Equations

Let us assume that the two-nucleon interactions are represented by separable potentials of a Yamaguchi type acting only in the spin-singlet and triplet channels with angular momentum $\ell = 0$. The off-shell two-nucleon t-matrix is then given by

$$(1) \qquad t_\alpha(p,p';z) = g_\alpha(p)\, g_\alpha(p')\, \tau_\alpha(z) ,$$

where $g_\alpha(p) = [p^2 + \mu_\alpha^2]^{-1}$

$$\tau_\alpha(z) = \lambda_\alpha [1 + \pi^2 \lambda_\alpha / (\mu_\alpha(\sqrt{-z} + \mu_\alpha)^2)]^{-1}$$

with $\alpha = 1,2$ for $s = 1,0$ respectively. The parameters $\lambda_\alpha, \mu_\alpha$ are found from low energy nucleon-nucleon scattering and are given in ref.[4]. For separable interactions the three-body equations reduce to one-dimensional integral equations of the form

$$(2) \qquad \phi_\alpha(q) = \phi_\alpha^{(0)}(q) - \frac{16\pi}{\sqrt{3}} \sum_{\beta=1}^{n} C_{\alpha\beta} \int_0^\infty dq'\, V_{\alpha\beta}(q,q')\, \tau_\beta(E-q^2+i\varepsilon)\, \phi_\beta(q'),$$

where E is the total c.m. energy of the three particle system and $n = 1,2$ for $S = 3/2$, $1/2$ respectively. The $C_{\alpha\beta}$ are spin-isospin recoupling coefficients and are given by $C_{11} = -1/2$ for $S = 3/2$ and $C = \begin{pmatrix} \frac{1}{4} & -\frac{3}{4} \\ -\frac{3}{4} & \frac{1}{4} \end{pmatrix}$ for $S = 1/2$. The kernel of eq. (2) has the same structure as that of a Lippmann-Schwinger equation. For the potential matrix we have for total angular momentum zero

$$(3) \qquad V_{\alpha\beta}(q,q') = \int_{A(q,q')}^{B(q,q')} p'dp' \, \frac{g_\alpha(\bar{p}(p',q',q))\, g_\beta(p')}{p'^2 + q'^2 - E - i\varepsilon}$$

with $A = |2q - q'|/\sqrt{3}$, $B = |2q + q'|/\sqrt{3}$,

and $\bar{p}(p',q',q) = [p'^2 + q'^2 - q^2]^{1/2}$.

From eq. (3) we see that above the breakup threshold the potentials are complex. The inhomogeneous term of eq. (2) can be written as

$$(4) \qquad \phi_\alpha^{(0)}(q) = \frac{32\pi}{\sqrt{3}\, q_2} C_{\alpha 1} \int_{A(q,q_2)}^{B(q,q_2)} p'dp' \; g_\alpha(\bar{p}(p',q_2,q))\, \phi_D(p')\, \tau_\alpha(E-q^2+i\varepsilon),$$

where ϕ_D is the deuteron wave function normalized to 1 and q_2 is taken to be the on-shell momentum $\hat{q}$, which is given by $\hat{q} = \sqrt{E-E_D}$ with $E_D = -2.226$ MeV. Knowing the solution to eq. (2), the T-matrix describing the elastic scattering of nucleons on deuterons can be calculated from

(5) $$T = B_o + \bar{T} ,$$

where B_o is the Born term given by

(6) $$B_o = \frac{4}{\sqrt{3}\,\hat{q}^2} C_{11} \int_{A(\hat{q},\hat{q})}^{B(\hat{q},\hat{q})} p'dp' \; \phi_D(p')^2 \; (E_D - p'^2)$$

and $\bar{T}$ is expressed in terms of the solutions ϕ_α:

(7) $$\bar{T} = \sum_{\beta=1}^{n} C_{1\beta} \int_0^\infty dq' \; \phi_\beta(q') \; A_\beta(q')$$

with

(8) $$A_\alpha(q) = \frac{2}{\sqrt{3}\,q_1} \int_{A(q_1,q)}^{B(q_1,q)} p'dp' \; g_\alpha(p') \; \phi_D(\bar{p}(p',q,q_1))$$

and q_1 is taken to be $\hat{q}$. The relation between the T-matrix and the phase parameters is given by

(9) $$T = \frac{i}{\pi\hat{q}} \eta \, [e^{2i\delta} - 1] .$$

3 Calculations and Discussion

Our main concern is the solution of the integral equations (2). One method consists of iterating eq. (2) and forming the SPA from the series so obtained. The integrals which occur were done in the same way as described in ref.[5] by using Gaussian quadrature and removing the singularities in the integrand by subtractions. Typically 24 gaussian meshpoints were used. Numerical stability was checked by increasing this number of points and varying the distribution of points. The error was estimated to be of the order of 1/2 %. In table I and II the results are presented for the phase parameters for L = 0 at E = 3.25 MeV and 14.4 MeV lab energy. The diagonal SPA using the K-matrix method [5] are also given. As is seen the convergence rate is found to be good and confirms the results found previously [5,6].

We now turn to the OPA-calculations. In particular, we have studied the usefulness of the [1/1]-OPA. For these we need to know

	Doublet		Quartet	
	T-matrix	K-matrix	T-matrix	K-matrix
[1,1]	-92.86	-89.66	113.22	122.46
[2,2]	-50.29	2.85	109.10	108.46
[3,3]	-28.67	13.39	109.09	109.09
[4,4]	-22.02	-22.76	109.09	109.09
[5,5]	-21.28	-21.29		
[6,6]	-21.24	-21.24		
[7,7]	-21.24	-21.24		

Table I

	Doublet				Quartet			
	T-matrix		K-matrix		T-matrix		K-matrix	
	δ	η	δ	η	δ	η	δ	η
[1,1]	86.68	3.383	169.57	0.733	94.11	0.469	38.37	0.673
[2,2]	142.21	1.740	97.11	1.102	73.37	0.959	80.24	1.030
[3,3]	131.12	0.503	157.65	0.173	73.79	0.971	73.78	0.974
[4,4]	126.81	0.465	126.44	0.456	73.79	0.971	73.78	0.974
[5,5]	126.25	0.457	126.27	0.454				
[6,6]	126.34	0.458	126.32	0.454				
[7,7]	126.34	0.458	126.32	0.454				

Table II

the inhomogeneous term and the first iteration of eq. (2) for a set of off-shell momenta q_1 and q_2 introduced in eqs. (4) and (8). The results are shown in fig. 1 for E = 3.25 MeV for both the T-matrix and K-matrix method, with one off-shell momentum q taken into account. The [1/1] OPA consists simply of taking for $\bar{T}$ [3]

$$\bar{T} = \bar{T}_1 [\bar{T}_1 - \bar{T}_2]^{-1} \bar{T}_1 , \tag{10}$$

where $\bar{T}_1$ and $\bar{T}_2$ are matrices in the space of on-shell and off-shell points. They are determined from eq. (7) with ϕ_β taken

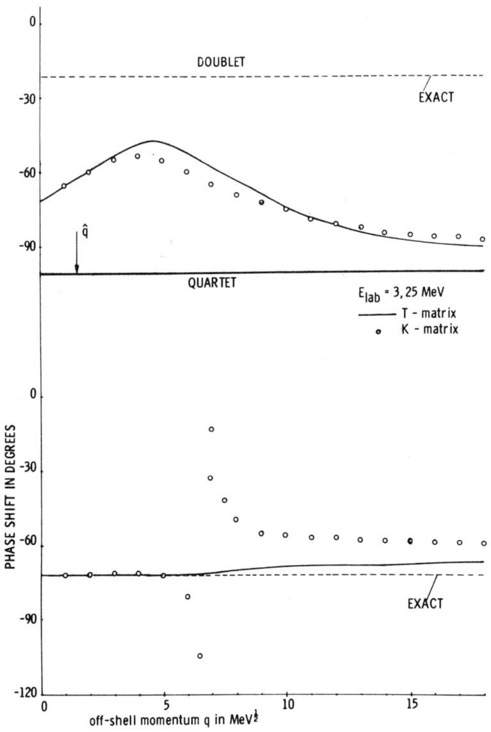

fig. 1

to be the inhomogeneous term and first iterate of eq. (2) respectively. Comparing table I and fig. 1 we see that for the quartet channel the introduction of q improves the result for low q-values. We also see that the K-matrix method behaves more erratically and in view of this the T-matrix method should perhaps be preferred in OPA. Moreover, using q as a variational parameter, we see that for the doublet channel OPA gives $\delta = -47^{\circ}$ to be compared with $\delta = -90^{\circ}$ for SPA and the exact result $\delta = -21.24^{\circ}$. An even more substantial improvement in the S = 1/2 channel is obtained by introducing in the initial and final states the additional channel with spin singlet in the two-nucleon states. This is simply done by replacing $C_{\alpha 1}$ and $C_{1\beta}$ in eqs. (4) and (7) by $C_{\alpha\beta}$. Fig. 2 shows the phase shifts at 3.25 MeV obtained by applying the OPA in the space of the spin channels α and off-shell momentum q.

As an example of how OPA behaves above breakup threshold we present some results at 14.4 MeV in figs. 3 and 4. For S = 3/2 we see from fig. 3 that there is virtually no improvement for the phase shift δ using OPA with one off-shell point q. However if we take the stationary value of δ with respect to q, the inelastici-

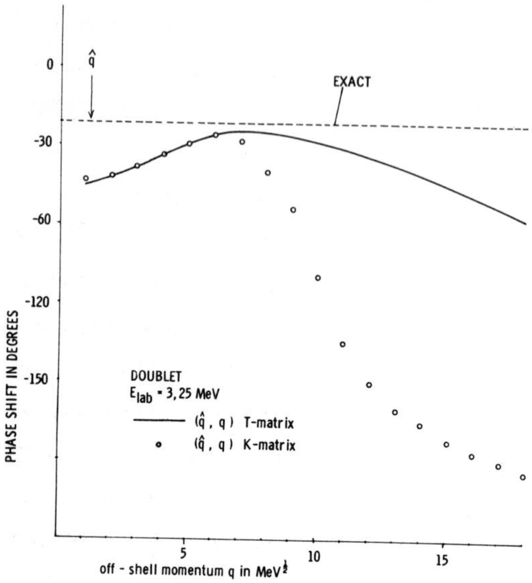

fig. 2

ty parameter η is considerably better. Furthermore including more off-shell points in OPA give rise to rapid convergence towards the exact answer. These results were obtained in the T-matrix formalism. Essentially the same features also hold when we use the K-matrix method. From fig. 4 we see that the inelasticity parameter η is reproduced poorly for

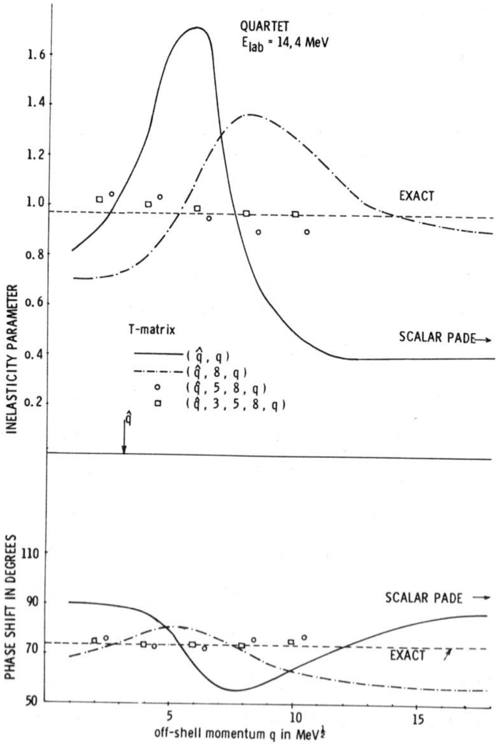

fig. 3

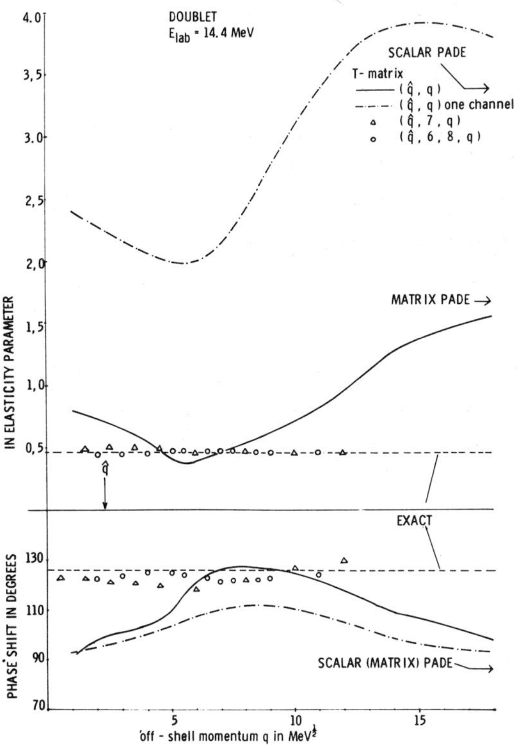

fig. 4

S = 1/2 if we only take one discrete channel into account. Including more off-shell momentum points in the calculation did not yield better results. If we include the second spin channel, the outcome drastically changes, the agreement with the exact solution becoming reasonable. In particular, if we use q as a variational parameter we see that at the point where δ becomes stationary, η is very close to the exact result. For comparison we exhibit also in the figure the values obtained from the ordinary [1/1] SPA and matrix PA in the space of the spin channels.

From these calculations we may conclude that OPA is superior to the [1/1] SPA, if we take an additional discrete channel in the doublet case into account. Furthermore the use of the off-shell momentum as a variational parameter gives remarkably good results for the inelasticity parameter. However the results for the phase shifts are in general less impressive than for the case of potential scattering [3]. In order to get more accurate results we have to include a larger set of off-shell momenta. Over the range between 0 and 50 MeV lab energy in general four off-shell points are sufficient to get reasonable results. It should however be mentioned that the convergence of the sequence of SPA is much more smooth than for OPA.

References

1 Tjon J.A. in Padé Approximants and their Applications edited by P.R. Graves-Morris, Academic Press, New York, 1973, p. 241.

2 Bessis D. in Padé Approximants edited by P.R. Graves-Morris, The Institute of Physics, London, 1973, p. 19.

3 Benofy L.P., J.L. Gammel and P. Mery, Phys. Rev. D13 (1976), 3111.

4 Aaron R., R.D. Amado and Y.Y. Yam, Phys. Rev. 140B (1965) 1291.

5 Kloet W.M. and J.A. Tjon, Ann. of Phys. 79 (1973), 407.

6 Brady T. and I.H. Sloan, Phys. Lett. 40B (1972), 55.

J.A. Tjon
Institute for Theoretical Physics
University of Utrecht
Utrecht, The Netherlands.

COMPUTATIONS

SOFTWARE FOR APPROXIMATIONS
OR
APPROXIMATION THEORY AS AN EXPERIMENTAL SCIENCE

L. Wayne Fullerton

Numerical analysis and approximation theory, in particular, can be an experimental science. This experimental nature is illustrated with several more-or-less new results. In the first half of this paper techniques for estimating the accuracy and significance of approximations are given. In the second half several generalizations of Chebyshev series that lead to nearly best approximations with respect to almost arbitrary weight functions and basis sets are presented.

1. Introduction

Conversational references to the experimental nature of numerical analysis usually emphasize the trial-and-error aspects of research. Certainly I do not dispute the trial-and-error nature of numerical analysis research, but I am most anxious to avoid illustrating the errors I have made. I call numerical analysis an experimental science in the same way that we all call physics or chemistry experimental sciences. There are two essential facets to an experimental science. First, theory or hypothesis suggests experiments that should be carried out. And second, experiments (conducted perhaps with computer programs) suggest new theoretical results. I wish primarily to emphasize this latter facet. In the next section, it is shown how computational experience can dictate the kind of numerical analysis that should be done. And in the third section, it is shown how experiments conducted with computer software can lead to new theoretical results.

2. Numerical Analysis for Software

Anyone who has used an approximation program probably has been annoyed by its inability to detect user errors. In order to compute an approximation, the user must supply function values that are somewhat more accurate than the approximation he desires. The more accurate values are often computed with a convenient ascending series for some argument values and an asymptotic series for other argument values. It is not uncommon to estimate incorrectly the number of terms needed in one of the series, so that the two series fail to match to the required accuracy. Alternatively, the user may incorrectly estimate the stability against roundoff of one of the series, so that it is inaccurate even though enough terms are used.

Now when a user requests a very accurate approximation with inaccurate function values, some approximation programs will do a great deal of work and possibly fail to derive any approximation. Even if the user supplies accurate function values, his approximation form may be so unstable that the approximation (if it can be derived) is not useful. These common experiences with approximation software dictate that the troublesome situations be detected so that perplexed users can be warned.

2.1 Input Function Accuracy

We wish to assess the error of a user-supplied function. The general methods in this section may be used to derive, for example, the relative error but in this case Generalized Chebyshev Series discussed in Section 3.2 must also be used. Let us, therefore, restrict consideration to the estimation of absolute errors and simply note that extension of the results here to arbitrarily weighted errors is straightforward.

Suppose we compute a high-order Chebyshev series approximation to the user-supplied function. Even though the series may contain 50 terms, only 10 terms may be significant. In such a case the error of the 10-term series would be nearly the same as the full 50-term series, and the magnitude of the last 40 terms would all be nearly the same. We can determine how many terms are significant by observing that an Nth order series

$$F(x) \simeq \sum_{i=0}^{N} f_i \, T_i(x)$$

is not only a near minimax approximation but also a discrete least squares approximation over the Chebyshev points $x_j = \cos \frac{j\pi}{N}$. Our strategy, then, is to estimate the number of terms to keep in the Chebyshev series in the same way that we estimate the number of terms to keep in any least squares approximation (cf. Ralston [5]).

The sum of the squares of the errors for an ℓ-th order series is

$$V_\ell = \sum_{j=0}^{N} \left[F(x_j) - \sum_{i=0}^{\ell} f_i \, T_i(x_j) \right]^2 .$$

If we estimate the value of $F(x_j)$ by the Nth order series and if we make use of orthogonality relations to eliminate cross products, we obtain

$$V_\ell = \sum_{j=0}^{N} \sum_{i=\ell+1}^{N} f_i^2 \, T_i^2 \, (x_j) = \frac{N+1}{2} \sum_{i=\ell+1}^{N} f_i^2 .$$

The standard error of one function value for an ℓ-th order series is given by

$$\sigma_\ell^2 = \frac{V_\ell}{N-\ell} = \frac{N+1}{2} \, \frac{1}{N-\ell} \sum_{i=\ell+1}^{N} f_i^2 .$$

We now compute these values for all ℓ. In order to evaluate the sum accurately, we start at $\ell=N$ for which the sum is zero and progressively decrease ℓ. Next we check in a forward direction for some $\sigma_{k+1} \geq \sigma_k$. We then have an estimate of the number of terms, k, to keep and also an estimate of the error, σ_k , of the user-supplied function.

The scheme we have described can be used to detect both random errors and discontinuities. The scheme works because we know the true function being approximated must have only very low-amplitude high "frequencies" and that it must have no discontinuities. Otherwise, a low-order polynomial approximation would be inappropriate. We have found an efficient method for assessing the accuracy of input functions as well as output Chebyshev series approximations. The requirement for such an accuracy estimate was dictated by computational experience, and well known numerical techniques fortunately provided the solution.

2.2 Stability of Approximation Form

Knowing only the accuracy of an approximation is insufficient, because the approximation may be unstable against roundoff. A ten-digit approximation is of little use if 100-digit accuracy is needed to evaluate the approximation. A significance loss of 90 digits is, of course, uncommon; however, even a loss of one digit of significance may be unacceptable. Anyone who derives an approximation for use in a full machine-precision special function routine will be most distraught to learn the approximation is unstable against roundoff error while he is testing the special function routine. He should be warned about the instability of the approximation when the approximation is derived. Once again, experience (or experiment) dictates the need for some numerical analysis research. The results are just as easily obtained as in the previous subsection.

The significance loss incurred during the evaluation of an approximation can be easily estimated when the approximation itself is derived, provided we do not try to do too much. A simple way of measuring the stability of an approximation is to calculate the number of significant digits that should be kept in each of the coefficients of the approximation so that the extra error introduced by rounding the coefficients is no larger than the weighted error of the approximation. Because every major computer represents floating point numbers with a nearly constant relative error, we need to calculate only one number, namely the number of significant digits to keep in each coefficient.

Suppose now we are given an approximation

$$A_n = \sum_{i=0}^{n} f_i \phi_i(x)$$

whose weighted error

$$\varepsilon = \max |\varepsilon(x)| = \max |w(x) [F(x) - A_n(x)]|$$

is nearly minimax. We require the orthogonal functions ϕ_i to be normalized so that $w^2(x) \phi_i^2(x) \leq 1.0$ as in Section 3.2. In the special case $w(x) \equiv 1$, the ϕ_i are just Chebyshev polynomials. We have chosen to analyze orthogonal series, because they presumably are the most stable form and, moreover, the easiest form to derive.

Assume the errors introduced by arithmetic operations and by evaluating the ϕ_i are negligible. Further assume the absolute error of the rounded coefficient f_i is Gaussian distributed with standard deviation σ_i. Of course, the errors are not really Gaussian distributed, but we need only an estimate of the required significance. An error of 50 percent in our estimate corresponds to only 0.3 significant figures and is perfectly acceptable. The standard deviation of the absolute error of the approximation evaluated with rounded coefficients is given by

$$\sigma_A^2(x) = \sum_{i=0}^{n} \left(\frac{\partial A_n}{\partial f_i} \right)^2 \sigma_i^2 .$$

Now let δ be the standard deviation of the relative error of each rounded coefficient so that $\sigma_i^2 = f_i^2 \delta^2$. Furthermore, recall that we want the weighted error introduced by the rounded coefficients to be less than the weighted error of the approximation, ε . Then we find

$$\varepsilon^2 = \max w^2(x)\, \sigma_A^2(x) = \max \left\{ \sum_{i=0}^{n} f_i^2\, w^2\, \phi_i^2 \right\} \delta^2 .$$

But the $\phi_i(x)$ are normalized so that $w^2(x)\, \phi_i^2(x) \leq 1$, and so

$$\delta^2 \geq \frac{\varepsilon^2}{\sum_{i=0}^{n} f_i^2} .$$

Finally, the number of significant figures, S, required to insure the effect of the rounding errors does not exceed the error of the approximation is

$$S = -\log_{10} \delta .$$

Stable approximations are those for which δ is a large number compared with ε, that is, the required number of significant figures should be small. Thus, stable approximations will have small leading coefficients -- the higher order coefficients are unimportant if the series converge reasonably quickly.

The extension of the analysis in this subsection to rational orthogonal series is straightforward, but the resulting expression for δ is not as elegantly simple as the result above.

3. Software for Numerical Analysis

In the previous section, the importance of numerical analysis applications to approximation programs used in a production mode was emphasized. Naturally, these programs become at the same time more useful and reliable as research tools. In this section, we emphasize the use of carefully designed programs to conduct numerical experiments that may lead to new theoretical results. Like a true experimental science, these theoretical results may immediately suggest new numerical experiments. Two (almost) new theoretical results are used to illustrate the utility of computer programs as research tools in the next two subsections.

3.1. Leveled Truncated Chebyshev Series

Truncated Chebyshev series are well known to be nearly best absolute error approximations in the uniform norm. Because Chebyshev series are near minimax approximations and because they are quite stable against roundoff errors, it is natural to express true minimax approximations in terms of Chebyshev polynomials. It is also natural to wonder what the error of a minimax approximation looks like in terms of Chebyshev polynomials. The Chebyshev series of the error is almost trivially calculated, especially if one is already expressing minimax approximations in terms of Chebyshev polynomials.

Consider, therefore, the dominant error terms of a second order polynomial minimax approximation to the exponential function on the interval $[-1, +1]$:

$$\varepsilon_2(x) = \ldots + .00013\ T_1 - .00553\ T_2 + .04434\ T_3 + .00547\ T_4 + .00054\ T_5 + \ldots$$

The main error term is, as expected, T_3. Note, though, that the neighboring error terms are of the same magnitude but opposite sign. If this happens only once or twice, it must be an accident. But it happens over and over. It even occurs for rational minimax approximations. Consider the Chebyshev series for the absolute error of a second order divided by a second order rational minimax approximation to the exponential:

$$\varepsilon_{2,2}(x) = \ldots + .000009\ T_3 - .000038\ T_4 + .000067\ T_5 + .000037\ T_6 + .000011\ T_7 + \ldots$$

As anticipated, T_5 is the dominant error term. And again neighboring error terms are of the same magnitude but opposite sign. Because the behavior we observe for these two cases occurs very frequently, we should consider an explanation.

A truncated Chebyshev series is ironically guaranteed to have a nonuniform error curve. If, for example, we truncate a Chebyshev series at fourth order, then the dominant error term will ordinarily be T_5. The next error term will be T_6, and this error term (if nonzero) will constructively interfere with T_5 in some places and destructively interfere in other places. We truncate a Chebyshev series to obtain a nearly best approximation, but at the same time we insure the error curve is nonuniform.

From the above numerical results we know what to do about the interference of higher order error terms with the dominant error term: we modify the truncated Chebyshev series so that lower order error terms of the same magnitude but opposite sign are introduced in the error expansion. This procedure works because the sum of the high and low order terms have zeroes exactly where the dominant error term has extremae. To see this effect, make the transformation $x = \cos\theta$. The dominant error term is then $T_m(x) = \cos m\theta$, and furthermore

$$T_{m-\ell}(x) - T_{m+\ell}(x) = \cos(m-\ell)\theta - \cos(m+\ell)\theta = 2\sin\ell\theta\ \sin m\theta \quad .$$

The nonzero low order error term aliases the high order error term and, therefore, reduces interference effects.

We have, then, derived a technique for leveling truncated Chebyshev series -- a technique suggested solely by Chebyshev series expansions of true minimax approximation errors. The leveled Chebyshev series should be regarded only as first order modifications to truncated Chebyshev series, because the introduction of the lower order error terms simply avoids the addition of more error at the extremae of the main error term. Nonetheless, the improvement is obtained at essentially no cost, and while a truncated Chebyshev series may deviate from a minimax approximation by perhaps 20 or 30 percent, the deviation of a leveled Chebyshev series is more likely to be only a few percent.

Economization of a power series [2] is a commonly employed method of obtaining a good approximation from a power series. In effect, the power series is converted to a Chebyshev series, then the small amplitude high order terms are dropped. One then obtains an economical approximation with fewer terms, but with little additional error. The results in this subsection could, however, be used to obtain a still better approximation with the same number of terms. Rather than truncating the Chebyshev series, the Chebyshev series should be leveled.

3.2. Generalized Chebyshev Series

Truncated Chebyshev series are nearly best approximations in the uniform norm. Unfortunately, they are only nearly best polynomial approximations and only in the sense of absolute error. It is natural to wonder about generalizations that would be good for arbitrary weight functions and non-polynomial bases. Originally this problem was motivated by the need for good starting values for the rational Remez iteration. However, before the rational problem is studied, we should solve the polynomial case.

Consider first the problem of finding an approximation $A_n(x)$ to the function $F(x)$ on $[-1, +1]$ with weight function $W(x) = 1$, such that the error

$$\varepsilon(x) = W(x)\ [F(x) - A_n(x)]$$

is near minimax. We know, of course, the solution is first to define some polynomials -- Chebyshev polynomials -- from the orthogonality condition

$$\int_{-1}^{1} \frac{T_m(x)\ T_n(x)}{\sqrt{1-x^2}}\ dx = 0\ ,\ m \neq n,$$

with the $T_n(x)$ normalized so that their extreme value is unity. Next we expand $F(x)$ in a series

$$F(x) = \sum f_i\ T_i(x)$$

with

$$f_i = \frac{1}{h_n} \int_{-1}^{1} \frac{F(x)\ T_i(x)}{\sqrt{1-x^2}}\ dx\ ,$$

where

$$h_n = \int_{-1}^{1} \frac{T_i^{\,2}(x)}{\sqrt{1-x^2}}\ dx\ .$$

When this series is truncated at n-th order, we obtain the desired approximation A_n.

In generalizing to arbitrary weights, it is reasonable to suppose a simple function of the weight must be included in the orthogonality condition. I incorrectly conjectured that the weight in the orthogonality condition might be $W(x)/\sqrt{1-x^2}$ or perhaps $\sqrt{W(x)}\ /\ \sqrt{1-x^2}$. The problem of finding the appropriate orthogonal polynomials and expansion coefficients can be posed essentially as a Gauss-Christoffel quadrature problem. Because a good Gauss-Christoffel quadrature program was available to me, I quickly learned that these conjectures did not lead to nearly best approximations. I did observe, however, that the quadrature weight containing $\sqrt{W(x)}$ was the worse, so I tried

$W^2(x) / \sqrt{1-x^2}$. That choice I found to be the correct one. Now that the correct generalization is known, it is easy -- embarrassingly easy -- to explain why.

We will be expanding $F(x)$ in a series of some orthogonal polynomials

$$F(x) = \sum f_i \phi_i(x) \quad ,$$

and when we truncate the series at n-th order, the weighted error will be roughly $W(x) \phi_{n+1}(x)$. We want this error to be an equal ripple curve, just like $T_{n+1}(x)$ would be. Thus, the analogue of the Chebyshev polynomial T_i is $W \phi_i$. And when we substitute this result in the orthogonality condition, we find the ϕ_i are given by

$$\int_{-1}^{1} \frac{W^2(x) \quad \phi_m(x) \; \phi_n(x)}{\sqrt{1-x^2}} \quad dx = 0 \quad , m \neq n.$$

See Gautschi [4] for a discussion of the derivation of orthogonal polynomials. We choose to normalize these polynomials so that the extremum of $W(x) \phi_i(x)$ is unity. Such a normalization allows one to assess readily the accuracy of a truncated series in these polynomials. The weighted error bound is simply the sum of the absolute values of all the coefficients dropped from the series, and this bound is usually close to the true weighted error.

Truncated generalized Chebyshev series often are within 20 or 30 percent of the corresponding true weighted minimax approximations. Because each approximation will usually have a unique weight function, the use of a general Gauss-Christoffel quadrature routine is not the best way to obtain the orthogonal polynomials and expansion coefficients. The integrals needed can be done efficiently by an automated Gauss-Chebyshev scheme, and the expansion coefficients can be derived at the same time as the recurrence coefficients for the orthogonal polynomials.

Generalization to non-polynomial bases is now straightforward, in principle. Instead of constructing the orthogonal functions ϕ_i from the basis x^i, it is necessary to orthogonalize the desired basis functions. In practice, the integrals are not so easily evaluated, and one must be certain that the basis functions form a Chebyshev set. For a brief discussion of the applications of these approximations see Fullerton [3].

Further generalization to orthogonal-Padé approximations, where we require

$$\frac{\sum_{i=0}^{m} p_i \phi_i}{\sum_{i=0}^{n} q_i \phi_i} = \sum_{i=0}^{m+n} f_i \phi_i(x) ,$$

with $q_0 \equiv 1$, are now easily obtained. We first discretize at the zeroes of $\phi_{m+n+1}(x)$, and solve the resulting linear equations for the p_i and q_i. As with Chebyshev-Padé approximations [1], we obtain either a degenerate approximation or a good one. If it is degenerate, we are not interested in the approximation because the next lowest order rational approximation will be nearly as good as the one we are trying to derive. If the discretized approximation is good, we can then use it as a starting value for finding differential corrections to the rational coefficients in the above nonlinear equation. The discretized solution usually is so close to the orthogonal-Padé solution, that only one or two iterations of the linearized version of the above equation are necessary. The orthogonal-Padé approximations always seem to be more accurate than the corresponding discretized approximations. Although the solution for orthogonal-Padé approximations outlined here is not nearly as elegant as the solution for Chebyshev-Padé approximations given by Clenshaw and Lord [1], at least we have obtained a solution.

4. Conclusions and Acknowledgements

The need for quality software dictates to some extent the type of numerical analysis that should be done, and several examples were given in Section 2. The results there led to programs that were not only much more useful production tools, but also much more useful research tools. Employing computer programs to conduct numerical experiments was shown in Section 3 to be an effective means of testing conjectures and deriving new theoretical results. I am confident that I would never have obtained the results in Section 3 unless I had had quality software tools available. Approximation theory and numerical analysis can be an experimental science.

Dr. D. D. Warner suggested the solution to the input function accuracy problem given in Section 2.1, and I am grateful to him for the suggestion. I am also grateful to Warner for numerous, long conversations that certainly led to some of the other results presented in this paper.

References

1 Clenshaw, C. W. and K. Lord, Rational approximations from Chebyshev series, Studies in Numerical Analysis (B. K. P. Scaife, editor), Academic Press, London, 1974, pp. 95-113.

2 Dahlquist, G. and A. Björck, Numerical Methods, Prentice - Hall, Englewood Cliffs, N. J., 1974, pp 125-126, (Translated by N. Anderson).

3 Fullerton, L. W., Portable special function routines, Proc. Workshop on Portability of Numerical Software, (W. Cowell, editor), Springer - Verlag, New York, (in Press).

4 Gautschi, W., Construction of Gauss-Christoffel quadrature formulae, Math. Comp., 22 (1968), pp. 251-270.

5 Ralston, A., A First Course in Numerical Analysis, McGraw - Hill, New York, 1965, pp. 234-235.

L. W. Fullerton
Group C3, MS 265
Los Alamos Scientific Laboratory
Los Alamos, New Mexico 87545

BEST RATIONAL APPROXIMATIONS WITH NEGATIVE POLES TO e^{-x} ON $[0,\infty)$

E.H. Kaufman, Jr. and G.D. Taylor

In this paper a theory for approximating e^{-x} on $[0,\infty)$ with rational functions having negative poles is developed. Numerical results suggest that the best uniform approximation to e^{-x} on $[0,\infty)$ from this class has only one pole and this is shown to be the case when using rational functions of this form which are linear polynomials divided by quadratic polynomials. Numerical results are given and compared to recent results of Saff, Schönhage and Varga.

1 Introduction

Let π_m denote the space of all real algebraic polynomials of degree less than or equal to m. For each m = 1,2,..., define $\mathcal{R}_m$ by

$$\mathcal{R}_m = \{R=P/Q: P \in \pi_{m-1}, Q(x)= \prod_{i=1}^{m} (q_i x+1), q_i \geq 0 \text{ for all } i\}.$$

Thus, $\mathcal{R}_m$ is the collection of all rational functions with denominator 1 or negative poles from $\mathcal{R}_m^{m-1}[0,\infty)$. Define λ_m by

$$(1.1) \quad \lambda_m = \inf\{||e^{-x} - R||_{L^\infty[0,\infty)}: R \in \mathcal{R}_m\}.$$

It is known that λ_m converges geometrically to zero (i.e. $\overline{\lim}_{m\to\infty} \lambda_m^{1/m} < 1$) since Saff, Schönhage and Varga [5] have proved that there exists a sequence $\{R_m\}_{m=1}^{\infty}$, with $R_m(x) = P_{m-1}(x)/(1+ \frac{x}{m})^m$, $P_{m-1} \in \pi_{m-1}$ such that

$$3 - 2\sqrt{2} \leq \overline{\lim}_{m\to\infty} ||e^{-x} - R_m||_{L^\infty[0,\infty)}^{1/m} \leq \frac{1}{2} .$$

In addition, since the poles of $R_m(x)$ are all real it follows that $R_m(z)$ must converge geometrically to e^{-z} in an infinite sector symmetric about the positive x-axis [6]. In what follows, we define

$$(1.2)\quad \mu_m = \inf\{\| e^{-x} - \frac{P(x)}{(1+\frac{x}{m})^m} \|_{L^\infty[0,\infty)} : P \in \pi_{m-1}\}.$$

An application of this theory is in the construction of numerical algorithms for solving linear systems of ordinary differential equations which arise from semi-discretization of linear parabolic partial differential equations (see [1], [5]). Numerically, this reduces to an iteration of the form

$$(1.3)\quad \underset{\sim}{w}^{(r)} = A^{-1}\underset{\sim}{k} + R_m(\Delta tA)\{\underset{\sim}{w}^{(r-1)} - A^{-1}\underset{\sim}{k}\}$$

where A is a nxn matrix (band), $\underset{\sim}{k}$ and $\underset{\sim}{w}^{(j)}$ are n dimensional vectors (n is related to the stepsize of the discretization), Δt is a scalar and R_m is the rational function defined above. Due to the special form of the denominator of R_m, $\underset{\sim}{w}^{(r)}$ can be obtained from the repeated inversion of $(I + \frac{\Delta t}{m} A)\underset{\sim}{g}_{\ell+1} = \underset{\sim}{g}_\ell$, $0 \leq \ell \leq m-1$ using an appropriately defined $\underset{\sim}{g}_0$. This is an attractive method numerically, since an LU factorization can be done for $I + \frac{\Delta t}{m} A$ only once and this factorization will preserve any band structure that is present.

One can construct a similar numerical method using a solution $R_m^*(x) = P_{m-1}^*(x)/\prod_{i=1}^{m}(q_i x+1)$ to (1.1). The apparent disadvantage of such a method compared to that of [5] is that $\underset{\sim}{w}^{(r)}$ is found from

$$\{\prod_{i=1}^{m}(I+q_i\Delta tA)\}\underset{\sim}{w}^{(r)} = \{\prod_{i=1}^{m}(I+q_i\Delta tA)\}A^{-1}\underset{\sim}{k} + P_{m-1}^*(\Delta tA)\{\underset{\sim}{w}^{(r-1)} - A^{-1}\underset{\sim}{k}\}$$

which will involve a greater number of operations (though, at most m LU factorizations and 2m substitutions). The advantage of this method is that λ_m might be smaller than μ_m giving increased accuracy. However, it appears (numerically) that R_m^* actually has $q_1 = q_2 = \dots = q_m$ and λ_m is approximately one half of μ_m for $m > 2$. Thus, using R_m^* gives a method that has the same desirable properties as that of (1.3) and increased accuracy.

In the next section we shall state some general facts concerning uniform approximation from $\mathcal{R}_m$ (these results will appear in a future paper [3]), give a theoretical treatment of best uniform approximation of (1.1) for the special case that $m = 2$ and state some conjectures. In the last section we will discuss our algorithm and present some numerical results.

2 Theoretical Results

In this section we begin by giving an existence theorem for approximating from $\mathcal{R}_m$ on $[0,\infty)$. This result is valid for a large class of functions (containing e^{-x}). We shall outline a proof for the special case that $m = 2$.

THEOREM 2.1. <u>There exists</u> $R^* \in \mathcal{R}_m$ <u>for which</u> $||e^{-x}-R^*(x)||_{L^\infty[0,\infty)} = \lambda_m$.

THEOREM 2.2. <u>There exists</u> $R^* \in \mathcal{R}_2$ <u>for which</u> $||e^{-x}-R^*(x)||_{L^\infty[0,\infty)} = \lambda_2$.

Proof. The proof begins by first observing that $\lambda_2 < \frac{1}{2}$. Thus, let $\{a_n\}$, $\{b_n\}$, $\{q_{1n}\}$ and $\{q_{2n}\}$ $(n=1,2,\dots)$ be sequences such that $q_{1n} \geq 0$, $q_{2n} \geq 0$ for all n and $\frac{1}{2} \geq ||e^{-x} - \frac{a_n x+b_n}{(q_{1n}x+1)(q_{2n}x+1)}|| \downarrow \lambda_2$ as $n \to \infty$, where we will no longer write the subscript $L^\infty[0,\infty)$ on the norm bars. Next, the proof is divided into two cases. The first case is when the sequences $\{q_{1m}\}$ and $\{q_{2n}\}$ are bounded. In this case it follows that the sequences $\{a_n\}$ and $\{b_n\}$ are also bounded and the desired result follows as in the standard rational approximation theory.

Thus, let us assume that the sequences $\{q_{1n}\}$ and $\{q_{2n}\}$ are not both bounded. By relabelling and extracting subsequences we may assume that $q_{1n} \uparrow \infty$ and $q_{1n} \geq q_{2n}$ for all n. In addition, by looking at the error curve at $x = 0$ and $x = \frac{1}{2}$, respectively, we have that $\frac{1}{2} \leq b_n \leq \frac{3}{2}$ for all n, and for all n sufficiently large (say $n \geq n_0$) that

$$(2.1)\quad 0 < \eta \leq \frac{\frac{1}{2} a_n}{(\frac{1}{2} q_{1n}+1)(\frac{1}{2} q_{2n}+1)} \leq M$$

where η and M are positive constants independent of n.

Next, we claim that the sequence $\{q_{2n}\}$ must be bounded. Indeed, if not then by passing to subsequences (and relabelling) we may assume that $q_{2n} \uparrow \infty$ as $n \to \infty$. Let $x_n = [\eta(\frac{1}{2}q_{1n}+1)(\frac{1}{2}q_{2n}+1)]^{-1}$ ε $[0,\infty)$. Note that $x_n \downarrow 0$ and that

$$\frac{a_n x_n + b_n}{(q_{1n}x_n+1)(q_{2n}x_n+1)} \geq \frac{a_n x_n}{(q_{1n}x_n+1)(q_{2n}x_n+1)} \to 2$$

as $n \to \infty$. For n sufficiently large this contradicts our assumption that $||e^{-x} - \frac{a_n x + b_n}{(q_{1n}x+1)(q_{2n}x+1)}|| \leq \frac{1}{2}$. Thus, $\{q_{2n}\}$ must be bounded. Since $a_n \to \infty$ by (2.1), we have, reciprocating (2.1), that there exists positive constants c_1 and c_2 independent of n such that for $n \geq n_1 \geq n_0$,

$$(2.2)\quad \frac{c_1}{\frac{1}{2} q_{2n}+1} \leq \frac{q_{1n}}{a_n} \leq \frac{c_2}{\frac{1}{2} q_{2n}+1}.$$

By (2.2) we may extract a subsequence (and relabel) for which $q_{2n} \to q^* \geq 0$ and $q_{1n}/a_n \to c^* > 0$ as $n \to \infty$. Hence, for fixed $x \,\varepsilon\, (0,\infty)$ we have that

$$\frac{a_n x + b_n}{(q_{1n}x+1)(q_{2n}x+1)} \to \frac{1}{c^*(q^*x+1)} = \frac{b^*}{q^*x+1}.$$

By continuity, $||e^{-x} - \frac{b^*}{q^*x+1}|| \leq \lambda_2$ completing the argument. ∎

The above proof (suitably modified) also establishes the following corollary where $\tilde{R}_m = \{R=P/Q\colon P \in \pi_{m-1},\ Q(x)=(qx+1)^m,\ q \geq 0\}$.

COROLLARY 2.3. There exists $\hat{R} \in \tilde{R}_m$ such that $\|e^{-x}-\hat{R}\| = \inf\{\|e^{-x}-R\|\colon R \in \tilde{R}_m\}$.

Next, we wish to turn to proving that for the m=2 case the best approximation from R_2 is actually contained in $\tilde{R}_2$. To do this we shall first show that neither of the coefficients in the denominator is zero and that the numerator and denominator do not have a common non-constant factor.

THEOREM 2.4. The best approximation to e^{-x} from R_2 is not of the form $\frac{ax+b}{qx+1}$, $q \geq 0$.

Proof. To prove this we use some computed results. First of all, running the Remes-Difcor algorithm as described in [2], we found the "best" approximation of the form $(ax+b)/(q_1x+q_2)$ (with $|q_1| \leq 1$, $|q_2| \leq 1$) to e^{-x} on $X = \{\frac{i}{25}\}_{i=0}^{500}$. This routine returned the values $a = -.0934450154$, $b = .6698426328$, $q_1 = 1.0$ and $q_2 = .6330537047$. It also returned four extreme points $x_1 = 0.0$, $x_2 = .44$, $x_3 = 2.76$ and $x_4 = 20.0$ such that $e^{-x_i} - (ax_i+b)/(q_1x_i+q_2) = (-1)^i e_i$ with $e_i > .058$ for all i=1,2,3,4. Thus, by a de la Vallée Poussin type argument we have that $\inf\{\|e^{-x}-r\|_{L^\infty[0,\infty)}\colon r \in R_1^1[0,\infty)\} \geq .058$. Next, setting $r^*(x) = (a^*x+b^*)/(p^*x+1)^2$ with $a^* = -.1853243706$, $b^* = 1.022709327$ and $p^* = .524169575$ we calculated $\gamma = \max\{|e^{-x}-r^*(x)|\colon x = i/1000 \text{ for } 0 \leq i \leq 20{,}000\}$ and found that $\gamma < .023$. Next, by dividing [0,20] into [0,z] and [z,20] where z is the zero of $r^*(x)$ we are able to show that $|E'(x)| \leq 3.2$ on [0,20] where $E(x) = e^{-x}-r^*(x)$. Using this and the above value of γ with Taylor's theorem for linear polynomials we can show that $|E(x)| \leq .0246$ on [0,20]. Since $E(x) > 0$ and $E'(x) < 0$ for $x \geq 20$, we have that $|E(x)| \leq E(20) < .022$ for

$x \geq 20$. This completes the proof.

Note that this proof also shows that r^* is a better approximation than the one calculated in [5] for m=2. Next, we turn to proving that for any best approximation in the m=2 case, the coefficients in the denominator coalesce; that is, $q_1 = q_2$.

THEOREM 2.5. <u>Any best approximation to</u> e^{-x} <u>from</u> R_2 <u>on</u> $[0,\infty)$ <u>belongs to</u> $\tilde{R}_2$; <u>that is, it is of the form</u> $(ax+b)/(qx+1)^2$ <u>with</u> $q \geq 0$. <u>Furthermore</u>, $q > 0$, <u>and the numerator and denominator have no non-constant common factors</u>.

Proof. The facts that $q > 0$ and the numerator and denominator have no non-constant common factors follow from Theorem 2.4. Let $R(x) = (p_1+p_2x)/(q_1x+1)(q_2x+1)$ be a best approximation to e^{-x} on $[0,\infty)$ from R_2 with $0 < q_1 < q_2$. We first claim that $e^{-x} - R(x)$ has at least 5 alternating extreme points in $[0,\alpha]$ where α is chosen such that $x \geq \alpha$ implies $|e^{-x}-R(x)| \leq \frac{1}{2}||e^{-x}-R(x)||_{L^\infty[0,\infty)} = \frac{1}{2}\lambda_2$. This follows from the fact that $R \in R_2^1[0,\alpha]$ and has defect zero, since if $e^{-x}-R(x)$ had fewer than 5 alternating extreme points, then the standard argument to prove alternation in $R_2^1[0,\alpha]$ can be used to find $\bar{R}(x)=(a+bx)/(1+cx+dx^2) \in R_2^1[0,\alpha]$ such that $||e^{-x}-\bar{R}(x)||_{L^\infty[0,\alpha]} < ||e^{-x}-R(x)||_{L^\infty[0,\alpha]}$ and with $|p_1-a|$, $|p_2-b|$, $|q_1+q_2-c|$ and $|q_1q_2-d|$ as small as desired. Thus, we can guarantee that $||e^{-x}-\bar{R}(x)||_{L^\infty[0,\infty)} < \lambda_2$ holds and that $\bar{R}(x)$ also has unequal negative poles (the discriminant of $1+cx+dx^2$ can be made arbitrarily close to that of $1+(q_1+q_2)x+q_1q_2x^2$). This, of course, is a contradiction showing that $e^{-x}-R(x)$ must have 5 alternating extreme points on $[0,\alpha]$.

Thus, $R(x)$ is the best approximation to e^{-x} on $[0,\alpha]$ from R_2^1 by the classical alternation theorem and also, therefore on $[0,\infty)$. Thus, we shall complete this proof by showing that the best approximation to e^{-x} from $\mathcal{R}_2^1[0,\infty)$ does not have real poles. To do this, we computed the "best approximation", $R(x) = (a+bx)/(1+cx+dx^2)$ to e^{-x} from $\mathcal{R}_2^1[0,20]$ on a 200,001 point equally spaced grid imposed on $[0,20]$. The computed results (rounded to 10 decimal places) were a = .9911236330, b = -.1577830783, c = .6704780400, d = .6494291043; the extreme points were $y_1=0$, $y_2=.2483$, $y_3=1.0852$, $y_4=3.2271$ and $y_5=13.1518$. The absolute errors at the extreme points were .0088763670 (they actually differed by less than 5×10^{-18}) and the sign of $e^{-x}-R(x)$ was positive at y_1. The discriminant of the denominator was -2.1481756150. By direct calculation, it can be easily seen that $E(x) = e^{-x}-R(x) > 0$ and $E'(x) < 0$ for $x \geq 20$. Thus, $||E(x)||_{L^\infty[0,\infty)} = ||E(x)||_{L^\infty[0,20]}$. Now, let us assume that there exists $\bar{R} \in \mathcal{R}_2^1$ having negative poles and for which $||e^{-x}-\bar{R}(x)|| < ||E(x)||$ holds, where for the remainder of this proof $||\cdot|| \equiv ||\cdot||_{L^\infty[0,20]}$. This will lead to a contradiction and give our desired result. We begin by noting that $|E'(x)| \leq 1$ for all $x \in [0,20]$ since $-e^{-x}$ and $-R'(x)$ have opposite signs for $x \in [0, -a/b]$, and $|R'(x)| \leq 1$, for $x \in [0,20]$, since the denominator is increasing faster than the absolute value of the numerator $x\in[0,20]$. For $x\in[-a/b,20]$ simply look at the ratio of the maximum of the numerator on this interval and the value of the denominator at $-a/b$. Thus, by the mean value theorem we have that for each $x \in [0,20]$, $|E(x)-E(\bar{x})| \leq .00005$ where $\bar{x}$ denotes a closest grid point to x. Let $\delta = .000054$, then $||E|| - \min\{|E(y_i)|: i=1,\ldots,5\} < \delta$ since $|E(y_i)|-|E(y_j)| < .000002$ for i, $j = 1,\ldots,5$. Since we are assuming that $||e^{-x}-\bar{R}(x)|| < ||E(x)||$, we must have that $||e^{-x}-\bar{R}(x)|| - \min\{|E(y_i)|: i = 1,\ldots,5\} < \delta$.

Now, there must exist i_o, $1 \leq i_o \leq 5$ such that $(-1)^{i_o}(R(y_{i_o}) - \bar{R}(y_{i_o})) > 0$ since $R \not\equiv \bar{R}$. Let us assume that $\max\{(-1)^i(R(y_i) - \bar{R}(y_i)): i = 1,\ldots,5\} = (-1)^5(R(y_5)-\bar{R}(y_5))$. Next, find $R^*(x) =$

$(a+\Delta a+(b+\Delta b)x)/(1+(c+\Delta c)x+(d+\Delta d)x^2)$ such that $R^*(y_i) = R(y_i) + (-1)^i\delta$ for $i = 1,\ldots,4$. To do this we must solve the linear system $\Delta a+\Delta by_i-\Delta cy_i(R(y_i)+(-1)^i\delta)-\Delta dy_i^2(R(y_i)+(-1)^i\delta) = (1+cy_i+dy_i^2)(-1)^i\delta$, $i=1,\ldots,4$. Solving this with Cramer's rule, with the determinants computed by cofactor expansion to avoid error magnification by divisions, gives $\Delta a = -.0000540000$, $\Delta b = .0004710533$, $\Delta c = -.0005063974$, $\Delta d = .0019944507$. Using this R^*, we have that $R(y_5) - R^*(y_5) = -.0000746489$ and $||R(y_i)-R^*(y_i)| - \delta| < 2 \times 10^{-18}$ for $i = 1,\ldots,4$. The discrimant of R^* was -2.156832218. Now, by construction, $(-1)^i(\bar{R}(y_i) - R^*(y_i)) < 0$, $i=1,\ldots,4$, and also, for $i = 1,\ldots,5$ we must have $(-1)^i(\bar{R}(y_i)-R(y_i)) < \delta$, since $||e^{-x}-\bar{R}(x)|| < \min\{|E(y_i)|: i = 1,\ldots,5\} + \delta$. Now, suppose $(-1)^5(\bar{R}(y_5)-R^*(y_5)) > 0$ (for, if not, then $\bar{R} \equiv R^*$ and we have our desired contradiction as R^* has non real poles). Then, we have that $\delta^* + R(y_5) = R^*(y_5) > \bar{R}(y_5)$ and the $\delta^* > (-1)^5(R(y_5)-\bar{R}(y_5)) \geq (-1)^i(R(y_5) - \bar{R}(y_i))$, $i = 1,\ldots,5$ so that $|R(y_i)-\bar{R}(y_i)| \leq \delta^*$, $i = 1,\ldots,5$. Letting $\bar{R}(y_i) - R(y_i) = \delta_i$, $i=1,\ldots,4$, we can estimate the coefficients of $\bar{R}$ from the equations $\bar{R}(y_i) = R(y_i) + \delta_i$, $i=1,\ldots,4$, where we know that $|\delta_i| \leq \delta^*$. Writing $\bar{R}(x) = (a+\Delta a+(b+\Delta b)x)/(1+(c+\Delta c)x+(d+\Delta d)x^2)$, this system is equivalent to

$$\Delta a+\Delta by_i-\Delta cy_i(R(y_i)+\delta_i)-\Delta d_iy_i^2(R(y_i)+\delta_i)=(1+cy_i+dy_i^2)\delta_i,$$
$$i = 1,\ldots,4.$$

Once again we resort to Cramer's rule to estimate Δc and Δd. Writing the determinant of the coefficients of this system as the sum of four determinants, one of which had no δ_i's in it (say D), we then computed D (= .1194079538) and estimated upper bounds for the three remaining determinants; subtracting these values from D showed that the determinant of the coefficients $\geq .1192288500$. Calculating upper bounds for the numerator determinants in the formulas for Δc and Δd and then estimating gives $|\Delta c| < .0012430121$,

$|\Delta d| \leq .0027602264$. Thus, letting D_1 and $\bar{D}_1$ denote the discriminant of the denominators of R and $\bar{R}$, respectively, we have that $|D_1 - \bar{D}_1| \leq .0127092755$. Treating the other cases where the maximum of $(-1)^i(R(y_i)-\bar{R}(y_i))$ occurs for i = 1,2,3, or 4, similarly, we found that $|D_1-\bar{D}_1| \leq .3959944800$, .1553098052, .0673900942 and .0219551498, respectively.

We conjecture that this result is true for all $m \geq 2$. We close this section by stating a local characterization and local uniqueness result which will be proved in a forthcoming paper [3].

Definition 2.6. $R(x) = (p_1+\ldots+p_m x^{m-1})/(qx+1)^m \in \tilde{R}_m$ <u>is a local best approximation to</u> e^{-x} <u>on</u> $[0,\infty)$ <u>if there exists a</u> $\delta > 0$ <u>such that if</u> $\bar{R}(x) = (\bar{p}_1+\ldots+\bar{p}_m x^{m-1})/(\bar{q}x+1)^m \in \tilde{R}_m$, $|p_i-\bar{p}_i| < \delta$, $i=1,\ldots,m$ <u>and</u> $|q-\bar{q}| < \delta$ <u>then</u> $||e^{-x}-R(x)|| \leq ||e^{-x}-\bar{R}(x)||$. <u>If, in addition, strict inequality holds whenever</u> $\bar{R}(x) \neq R(x)$ <u>then</u> R <u>is said to be locally unique</u>.

THEOREM 2.7. <u>Let</u> $m > 1$. <u>Then a nondegenerate</u> $R(x) = P(x)/Q(x) = (p_1+\ldots+p_m x^{m-1})/(qx+1)^m \in \tilde{R}_m$ (i.e. $R \neq 0$, P(x) <u>and</u> Q(x) <u>have no common factors and</u> $q > 0$) <u>is a best local approximation to</u> e^{-x} <u>from</u> $\tilde{R}_m$ <u>on</u> $[0,\infty)$ <u>if and only if</u> $e^{-x} - R(x)$ <u>has at least</u> $m + 2$ <u>alternating extreme points</u>. <u>Whenever this occurs</u>, R <u>is locally unique</u>.

We remark that numerical examples seem to suggest that there exist distinct $R_1, R_2 \in \tilde{R}_3$ satisfying this theorem.

3 Numerical Results

Our initial algorithm for computing approximations to e^{-x} from R_m and $\tilde{R}_m$ involved linearizing the denominator by Taylor's Theorem and setting up an iterative procedure, using the differential correction algorithm to compute an approximation at each inner stage. Precisely, for R_m set $g(q_1,\ldots,q_m,x) = \prod_{i=1}^{m}(q_i x+1)$

and define $\psi_j(q_1,\ldots,q_m,x) = x\prod_{\substack{i=1\\ i\neq j}}^{m}(q_i x+1)$ for $j=1,\ldots,m$, $\psi_0(q_1,\ldots,q_m,x) = g(q_1,\ldots,q_m,x) - \sum_{\nu=1}^{m} q_\nu \psi_\nu(q_1,\ldots,q_m,x)$. Thus, if $\bar{R}(x) = \bar{P}(x)/\prod_{i=1}^{m}(\bar{q}_i x+1)$, $0 \le \bar{q}_1 \le \bar{q}_2 \le \cdots \le \bar{q}_m$ is an approximation to e^{-x} at some step in the algorithm, then a new approximation $R(x) = (p_0+p_1x+\ldots+p_{m-1}x^{m-1})/\prod_{i=1}^{m}(q_i x+1)$ is found by calculating $p_0,\ldots,p_{m-1},q_1,\ldots,q_m$ that minimize $||e^{-x}-(p_0+\ldots+p_{m-1}x^{m-1})/(q_1\psi_1(\bar{q}_1,\ldots,\bar{q}_m,x)+\ldots+q_m\psi_m(\bar{q}_1,\ldots,\bar{q}_m,x)+\psi_0(\bar{q}_1,\ldots,\bar{q}_m,x))||$ over T a finite subset of $[0,N]$. Observe that the denominator in this problem is precisely the linearization of $g(q_1,\ldots,q_m,x)$ via Taylor's Theorem applied to the first m independent variables. This minimum can be calculated by the differential correction algorithm. Since this is a linearization of the problem we wish to solve, if we force an ordering on the $q_1,\ldots,q_m$ to get a unique solution, it seems reasonable to expect that if the initial approximation is sufficiently close to a best approximation then this algorithm will converge to that best approximation.

This approximation must be calculated on a large interval (the length of the interval needed seems to increase as a function of m, but not monotonically) to give a candidate for a best approximation to e^{-x} on $[0,\infty)$; and since we wish to get an accurate approximation of the continuous solution, we must use a fairly fine mesh so that card (T) will be large. Since the differential correction algorithm tends to become unstable as card (T) grows large, we decided to use the Remes-Difcor algorithm [2] for calculating the linearized minimum. We did this because this algorithm applies the differential correction algorithm to certain (small) subsets of T chosen in such a manner (depending upon alternation) that convergence to the solution on the whole space occurs. Thus, we had no a priori guarantee that this would work since a standard alternation theory does not exist for the linearized minimization problem due to the addition of the constraints on $\{q_i\}_{i=1}^{m}$. However, in spite of this, the results of the algorithm are acceptable in that the algorithm returned (or tried to return) a solution in which the q_i's coalesced and for which the error curve $e^{-x} - P(x)/\prod_{i=1}^{m}(qx+1)^m$ (the final coalesced approximation) alternated on m+2 points of T. Thus, by an alternation theorem we have proved [3], we have a best local approximation from $\tilde{R}_m$ and, as we conjectured earlier, therefore also from R_m. We also ran an algorithm of this character for the class $\tilde{R}_m$; in all cases it has given the same results as the above algorithm applied to R_m. A precise study of these algorithms remains to be done and we conjecture that convergence results can be proved for both R_m and $\tilde{R}_m$, at least using the differential correction algorithm for the inner minimization.

We have run these algorithms for various values of m using a grid with spacing .002 imposed on an interval [0,N], where N is chosen by trial and error so that the computed results make it apparent that the error norm on $[N,\infty)$ is smaller than the error on [0,N]. The computations were done on a UNIVAC 1106, which has roughly 18 digits of accuracy in double precision. Initially, we started with $\bar{p}_0=1$, $\bar{p}_1=\ldots=\bar{p}_{m-1}=0$, $\bar{q}_j = \frac{j}{m}$, $j=1,\ldots,m$ and ran the program with additional constraints $q_i \leq q_{i+1}$ - DIFF where DIFF is a nonnegative parameter. If DIFF > 0 we found that the computed q_i's immediately differed by exactly DIFF, and if DIFF was set equal to 0 the algorithm ran and the computed q_i's coalesced. The algorithm for $\tilde{R}_m$ had a linearization in which the denominator of the approximation is $q\cdot m\,x(\bar{q}x+1)^{m-1}$ + $[(1-m)\bar{q}x+1](\bar{q}x+1)^{m-1} \equiv q\psi_1(\bar{q},x) + \psi_0(\bar{q},x)$ where $\bar{q}$ is the value from the previous approximation. Here the initialization was $\bar{p}_0 = 1$, $\bar{p}_1 = \ldots = \bar{p}_{m-1} = 0$, $\bar{q} = 1/n$. Although we allowed this program to run for seven outer iterations, the coefficients nearly always stopped changing after four or five outer iterations, and the computed absolute values of the errors at the m + 2 extreme points agreed to at least fourteen significant figures. The results are shown in the table below, with the error of [5] given in the last column for comparison purposes. The sign attached to the last extreme point is the sign of E(x) at that point. It should be noted that it is possible (although unlikely) that in some cases there is a local best approximation other than ours which gives a smaller error. In the m = 3 case we have found another local best approximation (with q = 1.05109 and ||error|| = 1.33720 (-02) and in the m = 6 case there appear to be at least three local best approximations other than the one in the table.

Finally, we would like to thank Professor R.S. Varga for bringing [4] to our attention where some of the results of this paper and of [3] have also been obtained independently.

Table of Numerical Results

m	last ext. pt.	q	||error||	||error|| [5]
2	12.932+	.52416	2.27093 (-02)	2.49038 (-02)
3	37.250-	.27127	8.04713 (-03)	1.50535 (-02)
4	83.814+	.17797	3.30771 (-03)	7.85325 (-03)
5	80.802+	.27866	1.16064 (-03)	3.05486 (-03)
6	152.352-	.19296	4.26252 (-04)	8.89243 (-04)

References

1. Cody, W.J., G. Meinardus and R.S. Varga, Chebyshev rational approximations to e^{-x} on $[0,\infty)$ and applications to heat-conduction problems, J. Approximation Theory, 2 (1969), 50-65.

2. Kaufman, E.H., Jr., D.J. Leeming and G.D. Taylor, A combined Remes-Differential correction algorithm for rational approximation, submitted.

3. Kaufman, E.H., Jr. and G.D. Taylor, Uniform approximation with rational functions having negative poles, submitted.

4. Lau, T. C-Y., Rational exponential approximation with real poles, preprint.

5. Saff, E.B., A. Schönhage and R.S. Varga, Geometric convergence to e^{-z} by rational functions with real poles, Numer. Math., Vol. 25 (1976), 307-322.

6. Saff, E.B. and R.S. Varga, Angular overconvergence for rational functions converging geometrically on $[0,+\infty)$, Theory of Approximation with Applications (edited by Law and Sahney), Academic Press, New York, 1976, 238-256.

E.H. Kaufman, Jr.
Department of Mathematics
Central Michigan University
Mount Pleasant, Michigan 48859

G.D. Taylor*
Department of Mathematics
Colorado State University
Fort Collins, Colorado 80523

*Research sponsored by the Air Force Office of Scientific Research, Air Force Systems Command, USAF, under Grant No. AFOSR-76-2878.

ALGORITHMS FOR RATIONAL APPROXIMATIONS FOR THE GAUSSIAN HYPERGEOMETRIC FUNCTION

Yudell L. Luke

In previous studies, rational approximations for ${}_2F_1(a,b;c;-z)$ were examined in some detail. If a=1, complete a priori error analyses for the main diagonal Padé approximations and much more were presented. For general parameters, the rational approximations are not of the Padé class. It was shown that they converge, but a complete error description was not available. This deficiency is now corrected. Further, FORTRAN programs are available to evaluate the rational approximations by using the appropriate recursion formulas to generate the numerator and denominator polynomials as a number, and to also evaluate the coefficients which define these polynomials.

1 Introduction

We consider rational approximations for

$$(1)\quad E(z) = {}_2F_1(a,b;c;-z)$$

in the form of a ratio of two polynomials each of degree n. The rational approximations are not of the Padé class. They follow from Luke [3,v.2,p.96] or Luke [4,p.224] with p=2, q=1, $\alpha_1=a$, $\alpha_2=b$, $\rho_1=c$, $\alpha=\beta=0$, $\lambda=1$, f=g=0, $\gamma=z$ and then z replaced by -z and a=0. The latter a has nothing to do with the a in (1). In the sources cited, if a=1 and all other parameters, γ and z are as specified above, we get rational approximations where the numerator and denominator polynomials are of degree n-1 and n respectively. If a=1 in (1), main diagonal and first subdiagonal Padé approximations can be deduced from the same parent rational approximations given in the cited sources from which we get the rational approximants treated in this paper. The Padé approximations are thoroughly detailed in [3,4] and will not further concern us here.

In [3] it was shown that the rational approximations converged, but a complete a priori error analysis was wanting. In Section 2, the rational approximations and related data are presented, and in Section 3 a complete error analysis with asymptotic estimates are developed. Some examples are treated in Section 4.

FORTRAN programs are available to evaluate the Padé and non-Padé rational approximations by using the appropriate recursion formulas to generate the numerator and denominator polynomials as a number, and to also evaluate the coefficients which define these polynomials. The programming of the routines was done for use by the IBM 370/168 operating under OS/VS Release 1.7 on the FORTRAN IV H-Extended Compiler, Release 2.1. All computer programs are written for quadruple precision and real arithmetic. By making a few simple changes, one can have double or single precision. Further, it is shown how to get complex arithmetic along with any of the precisions noted above. Unfortunately, because of space limitations, these programs are not given here. They are available from the author on request.

2 Rational Approximations

We have

$$(2)\quad E(z) = \{A_n(z)/B_n(z)\} + R_n(z),$$

$$(3)\quad B_n(z) = L_n z^n {}_3F_2(-n,n+1,c;a+1,b+1;-1/z),$$

$$(4)\quad A_n(z) = L_n z^n \sum_{k=0}^{n} \frac{(-n)_k(n+1)_k(a)_k(b)_k}{(a+1)_k(b+1)_k(k!)^2} \times {}_4F_3\left(\begin{matrix}-n+k,n+1+k,c+k,1\\ 1+k,a+1+k,b+1+k\end{matrix}\middle| -\frac{1}{z}\right),$$

$$(5)\quad L_n = \frac{(a+1)_n(b+1)_n}{(n+1)_n(c)_n}.$$

Here $R_n(z)$ is the remainder which we discuss later.

For the polynomials $B_n(z)$ and $A_n(z)$, we have

$$B_0(z) = 1, \quad B_1(z) = 1 + \frac{(a+1)(b+1)z}{2c},$$

$$B_2(z) = 1 + \frac{(a+2)(b+2)z}{2(c+1)} + \frac{(a+1)_2(b+1)_2z^2}{12(c)_2},$$

$$B_3(z) = 1 + \frac{(a+3)(b+3)z}{2(c+2)} + \frac{(a+2)_2(b+2)_2z^2}{10(c+1)_2} + \frac{(a+1)_3(b+1)_3z^3}{120(c)_3},$$

(6) $$A_0(z) = 1, \quad A_1(z) = B_1(z) - \frac{abz}{c},$$

$$A_2(z) = B_2(z) - \frac{abz}{c}\left[1 + \frac{(a+2)(b+2)z}{2(c+1)}\right] + \frac{(a)_2(b)_2z^2}{2(c)_2},$$

$$A_3(z) = B_3(z) - \frac{abz}{c} \times \left[1 + \frac{(a+3)(b+3)z}{2(c+2)} + \frac{(a+2)_2(b+2)_2z^2}{10(c+1)_2}\right] + \frac{(a)_2(b)_2z^2}{2(c)_2}\left[1 + \frac{(a+3)(b+3)z}{2(c+2)}\right] - \frac{(a)_3(b)_3z^3}{6(c)_3}.$$

Both $A_n(z)$ and $B_n(z)$ satisfy the same recurrence formula

(7) $$B_n(z) = (1 + F_1z)B_{n-1}(z) + (E + F_2z)zB_{n-2}(z) + F_3z^3B_{n-3}(z), \quad n \geq 3,$$

$$F_1 = \frac{[3n^2 + (a+b-6)n + 2 - ab - 2(a+b)]}{2(2n-3)(n+c-1)},$$

$$F_2 = \frac{-[3n^2 - (a+b+6)n + 2 - ab](n+a-1)(n+b-1)}{4(2n-1)(2n-3)(n+c-2)_2},$$

$$F_3 = \frac{(n+a-2)_2(n+b-2)_2(n-a-2)(n-b-2)}{8(2n-3)^2(2n-5)(n+c-3)_3},$$

$$E = \frac{-(n+a-1)(n+b-1)(n-c-1)}{2(2n-3)(n+c-2)_2}.$$

The recurrence formula is stable in the forward direction.

3 The Error

We now turn to the remainder. Several forms are presented according to the nature of the parameters and variable. Although the representations are general, it is convenient to make certain assumptions to simplify the discussion. If a numerator parameter takes on a specialized form, then the numerator parameter is a . We assume that a is not a negative integer for otherwise the ${}_2F_1$ is a polynomial. Similarly, neither c−a nor c−b is allowed to be a negative integer for otherwise the ${}_2F_1$ is a polynomial in the variable z or z/(1+z), except for a binomial multiplier, in view of the Kummer relations. The general theory in [3] for rational approximations for ${}_pF_q(\alpha_p;\rho_q;z)$ supposes that if a numerator parameter α_h and a denominator parameter ρ_h coalesce, then these parameters are cancelled before writing the rational approximations. This is for convenience to assure that all hypergeometric forms are the lowest order possible. Thus in the present context, we forbid c=a. If c=a, $E(z) = (1+z)^{-b}$, and in this event one should use the Padé approximation as it is more accurate than the approximation in (2)-(5). Finally, if c and a are positive integers, then $c > a$; and if b also is a positive integer, then $c > b \geq a$.

The remainder is written in the form

$$(8)\quad R_n(z) = S_n(z)/B_n^*(z), \quad B_n^*(z) = z^{-n} L_n^{-1} B_n(z).$$

We first consider $B_n^*(z)$. From [3,4] we have

$$(9)\quad B_n^*(z) = \frac{\Gamma(a+1)\Gamma(b+1)z^{\frac{1}{2}} N^{2\sigma} e^{N\zeta}}{2\Gamma(c)\Gamma(\frac{1}{2})(1+z)^{\sigma+\frac{1}{2}}} [1+O(N^{-1})]$$

where

$$(10)\quad N^2 = n(n+1), \quad 2\sigma = c - a - b - 3/2,$$

$$e^{-\zeta} = [2+z\mp 2(1+z)^{\frac{1}{2}}]/z,$$

and the sign is chosen so that $|e^{-\zeta}| < 1$. This is possible for all z, $z \neq 1$, $|\arg(1+z)| < \pi$. (9) is used only for $|\arg z| < \pi$. We also have

$$(11)\quad B_n^*(z) = \frac{(-)^n \Gamma(a+1)\Gamma(b+1) x^{\frac{1}{2}} N^{2\sigma} (-e^{-\zeta})^N}{2\Gamma(c)\Gamma(\frac{1}{2})(1-x)^{\sigma+\frac{1}{2}}} [1+O(N^{-1})],$$

$$z = -x, \quad 0 \leq x < 1.$$

The expression

$$(12)\quad B_n^*(z) = \frac{\Gamma(a+1)\Gamma(b+1) z^{n+1} (n+\frac{1}{2})^{2\sigma}}{2\Gamma(c)\Gamma(\frac{1}{2})(1+z)^{\sigma+\frac{1}{2}} (ze^{-\zeta})^{n+\frac{1}{2}}} [1+O(N^{-1})],$$

$$|\arg(1+z)| < \pi,$$

is often most convenient. Frequently, it is desirable to have the form for $e^{-\zeta}$ with z replaced by $1/z$. In this case, we put

$$(13)\quad e^{-\xi} = 2z + 1 \mp 2(z^2+z)^{\frac{1}{2}},$$

where the sign is chosen so that $|e^{-\xi}| < 1$ which is possible for all z, $z \neq -1$, $|\arg(1+1/z)| < \pi$.

We now take up the structure of $S_n(z)$ which follows from the work of Fields [1,2] and Luke [3,4]. We write

$$(14)\quad S_n(z) = F(z)M_n(z) + H(z)G_n(z),$$

$$(15)\quad F(z) = \frac{-abz\Gamma(c)}{\Gamma(2-c)}\; {}_2F_1\left(\begin{matrix} a+1-c,\ b+1-c \\ 2-c \end{matrix}\middle| -z\right),$$

or

$$(16)\quad F(z) = \frac{-abz\Gamma(c)(1+z)^{c-a-b}}{\Gamma(2-c)}\; {}_2F_1\left(\begin{matrix} 1-a,\ 1-b \\ 2-c \end{matrix}\middle| -z\right),$$

$$(17)\quad M_n(z) = \frac{\Gamma(n+1-c)}{\Gamma(n+1+c)}\; {}_3F_2\left(\begin{matrix} c, c-a, c-b \\ c-n, n+1+c \end{matrix}\middle| -z\right),$$

$$(18)\quad H(z) = \frac{\Gamma(1-c)}{\Gamma(-a)\Gamma(-b)}\; {}_2F_1\left(\begin{matrix} a, b \\ c \end{matrix}\middle| -z\right),$$

$$(19)\quad G_n(z) = \frac{n!\Gamma(n+1-a)\Gamma(n+1-b)z^{n+1}}{\Gamma(n+2-c)(2n+1)!}$$

$$\times\; {}_3F_2\left(\begin{matrix} n+1, n+1-a, n+1-b \\ 2n+2, n+2-c \end{matrix}\middle| -z\right).$$

We also have need for asymptotic expressions for $M_n(z)$ and $G_n(z)$. For the latter, with all parameters and z fixed and $|\arg(1+z)|<\pi$,

$$(20)\quad G_n(z) = \frac{(\pi z)^{1/2}N^{2\sigma}e^{-N\zeta}}{(1+z)^{\sigma+1/2}}\;[1+O(N^{-1})]$$

where N,σ and $e^{-\zeta}$ are as in (10). A form equivalent to (20) is

$$(21)\quad G_n(z) = \frac{(\pi z)^{1/2}(n+\tfrac{1}{2})^{2\sigma}e^{-(n+\frac{1}{2})\zeta}}{(1+z)^{\sigma+1/2}}\;[1+O(n^{-1})].$$

Finally, if a,b,c and z are fixed and $|\arg(1+z)| < \pi$, then

$$(22)\quad M_n(z) = \frac{\Gamma(n+1-c)}{\Gamma(n+1+c)}\left[{}_3F_2^r\left(\begin{matrix} c, c-a, c-b \\ c-n, n+c+1 \end{matrix}\middle| -z\right) + O(n^{-2r-2})\right]$$
$$+\; O(e^{-n\zeta}).$$

If none of the numbers a,b or c is an integer, then from (14), (21) and (22), since $G_n(z)$ is subdominant to $M_n(z)$, we have with $|\arg(1+z)| < \pi$,

$$(23)\quad S_n(z) = \frac{F(z)\Gamma(n+1-c)}{\Gamma(n+1+c)} \times \left[{}_3F_2^r\left(\begin{matrix} c,c-a,c-b \\ c-n,n+c+1 \end{matrix}\middle| -z\right) + O(n^{-2r-2}) \right].$$

If either a or b is a positive integer, and c is not a positive integer, then $H(z) = 0$ because of the factor $[\Gamma(-a)\Gamma(-b)]^{-1}$. In this event (23) is valid.

If c is a positive integer, but neither a nor b is a positive integer, it can be shown that

$$(24)\quad S_n(z) = \frac{(-z)^c\Gamma(a+1)\Gamma(b+1)\Gamma(n+1-c)E(z)}{\Gamma(a+1-c)\Gamma(b+1-c)\Gamma(n+1+c)} \times \left[{}_3F_2^r\left(\begin{matrix} c,c-a,c-b \\ c-n,n+c+1 \end{matrix}\middle| -z\right) + O(n^{-2r-2}) \right],$$

where $E(z)$ is defined by (1) and again $|\arg(1+z)| < \pi$.

Now suppose first that either a or b is a positive integer, but c is not a positive integer. If both a and b are positive integers, let $a \le b$. Then $H(z) = 0$ and it is convenient to put (14) in the form

$$(25)\quad S_n(z) = F^*(z)M_n^*(z),$$

$$(26)\quad F^*(z) = -abz\Gamma(c)(1+z)^{c-a-b}\ {}_2F_1^{a-1}\left(\begin{matrix} 1-a,1-b \\ 2-c \end{matrix}\middle| -z\right),$$

$$(27)\quad M_n^*(z) = \frac{\Gamma(n+1-c)\Gamma(c-n)}{\Gamma(n+1+c)\Gamma(2-c)} \times \left[\frac{1}{\Gamma(c-n)}\ {}_3F_2\left(\begin{matrix} c,c-a,c-b \\ c-n,n+1+c \end{matrix}\middle| -z\right) \right].$$

Now let c approach a positive integer such that $c > b \ge a$. Then

$$(28)\quad \frac{\Gamma(c-n)}{\Gamma(2-c)} \to \frac{(-)^{n+1}\Gamma(c-1)}{\Gamma(n+1-c)},$$

$$(29)\quad \frac{1}{\Gamma(c-n)}\ {}_3F_2\left(\begin{matrix} c, c-a, c-b \\ c-n, n+1+c \end{matrix}\middle| -z\right)$$

$$\to \frac{(-)^{n+1-c} z^{-c}\Gamma(n+1+c)G_n(z)}{\Gamma(c)\Gamma(c-a)\Gamma(c-b)}.$$

It follows that

$$(30)\quad S_n(z) = \frac{(-z)^{1-c} ab\Gamma(c-1)(1+z)^{c-a-b}}{\Gamma(c-a)\Gamma(c-b)}$$

$$\times\ {}_2F_1^{a-1}\left(\begin{matrix} 1-a, 1-b \\ 2-c \end{matrix}\middle| -z\right) G_n(z).$$

Under these same conditions, an alternative but less attractive form for $S_n(z)$ is

$$(31)\quad S_n(z) = \frac{(-z)^{1-c} ab\Gamma(c-1)}{\Gamma(c-a)\Gamma(c-b)}$$

$$\times\left\{ {}_2F_1^{c-1-a}\left(\begin{matrix} a+1-c, b+1-c \\ 2-c \end{matrix}\middle| -z\right) + \frac{\Gamma(a)\Gamma(b)\Gamma(c-a)(-)^{c-a} z^{c-1} E(z)}{\Gamma(c-1)\Gamma(c)\Gamma(b+1-c)} \right\} G_n(z).$$

Asymptotic forms for $S_n(z)$ readily follow from the forms for $G_n(z)$. Asymptotic forms for $R_n(z)$ follow from the appropriate forms for $S_n(z)$ and use of (8) and (12). Thus, if neither c nor a (or nor both a and b) are simultaneously positive integers, then

$$(32)\quad R_n(z) = W(n^{2\sigma+2c} e^{n\zeta})^{-1}\,[1+\mathcal{O}(n^{-1})],$$

$$|\arg(1+z)| < \pi,$$

where W is free of n. Clearly,

$$(33)\quad \lim_{n\to\infty} R_n(z) = 0, \quad |\arg(1+z)| < \pi .$$

For a measure of the rate of convergence, we have

$$(34)\quad R_{n+1}(z)/R_n(z) = e^{-\zeta}[1+\mathcal{O}(n^{-1})].$$

If c and a are both positive integers, c > a, and in the event that b is also a positive integer, $c > b \geq a$, then from (8), (12) and (21), we find

$$(35)\quad R_n(z) = \frac{2\pi\Gamma(c)\Gamma(c-1)(-z)^{1-c}(1+z)^{c-a-b}}{\Gamma(c-a)\Gamma(c-b)\Gamma(a)\Gamma(b)}$$

$$\times \; {}_2F_1^{a-1}\left(\begin{matrix}1-a,1-b\\ 2-c\end{matrix}\middle| -z\right) e^{-(2n+1)\zeta}$$

$$\times \; [1+\mathcal{O}(n^{-1})], \quad |\arg(1+z)| < \pi ,$$

$$(36)\quad \lim_{n\to\infty} R_n(z) = 0, \quad |\arg(1+z)| < \pi ,$$

and

$$(37)\quad R_{n+1}(z)/R_n(z) = e^{-2\zeta}[1+\mathcal{O}(n^{-1})].$$

4 Numerical Examples

i) Let

a=0.8, b=0.6, c=1.5, z=0.8.

We compare values of $B_n^*(z)$, $B_n(z)$, $S_n(z)$ and $R_n(z)$ based on equations (8) and (14) and the asymptotic representations (9) and (23) with r=1 and both without order terms with values determined by the machine using the FORTRAN programs. The latter values are called true. To simplify the notation, unless a quantity is labeled true, then it is derived from the asymptotic representation(s) noted.

n	$B_n^*(z)$	$B_n(z)$	True $B_n(z)$
6	5.093(3)	8.776	8.733
7	2.868(4)	12.02	11.97
8	1.655(5)	16.46	16.41
9	9.733(5)	22.56	22.50
10	5.811(6)	30.92	30.85

n	$-S_n(z)$	$-R_n(z)$	True $-R_n(z)$
6	7.046(-2)	1.383(-5)	1.393(-5)
7	4.535(-2)	1.581(-6)	1.588(-6)
8	3.092(-2)	1.868(-7)	1.874(-7)
9	2.204(-2)	2.264(-8)	2.270(-8)
10	1.626(-2)	2.798(-9)	2.804(-9)

ii) Let

$a=0.8$, $b=0.6$, $c=2.0$, $z=0.8$.

Again we compare asymptotic values with true values as in example i) above, where in the present instance we use (24) instead of (23) with $r=1$ and without the order term. If $n=10$, the asymptotic values for $B_n^*(z)$, $B_n(z)$, $S_n(z)$ and $R_n(z)$ are 1.440(7), 25.77, 1.822(-4) and 1.265(-13), respectively. while the true values of $B_n(z)$ and $R_n(z)$ are 26.71 and 1.226(-13), respectively.

iii) Let $a=2.0$, $b=3.0$, $c=5.0$, $z=2.0$.

Again we compare asymptotic values with true values as in the previous examples, but in the present instance we estimate $R_n(z)$ using (35) without the order term. Thus, for $n=10$ and 20, we get the respective values, 3.22(-11) and 1.17(-22). The true respective values as determined by the machine program are 1.656(-11) and 0.862(-22). If we neglect the order term in (37), then for any n, $R_{n+1}(z)/R_n(z) = 0.718(-1)$. For $n=10$ and 20, the values determined from the machine run are 0.772(-1) and 0.729(-1), respectively.

References

1 Fields, J.L., A linear scheme for rational approximations, J. Approximation Theory 6 (1972), 161-175

2 Fields, J.L., Written communication, 1976.

3 Luke, Y.L., The Special Functions and Their Approximations, Vols. 1,2, Academic Press, New York, 1969.

4 Luke, Y.L., Mathematical Functions and Their Approximations, Academic Press, New York, 1975.

Y.L. Luke*
Department of Mathematics
University of Missouri
Kansas City, Missouri 64110

*Research supported by the Air Force Office of Scientific Research under Grant AFOSR 73-2520.

ON THE LIMITATION AND APPLICATION OF PADÉ APPROXIMATION TO THE MATRIX EXPONENTIAL

Charles Van Loan

Non-normality in the matrix A and its effect on Padé approximation of the matrix exponential is discussed. Against this background we remark upon the selection and efficient evaluation of the appropriate approximant. An application from control theory serves to illustrate some of the principles discussed.

1 The Limitations of Padé Approximation

We shall consider the problem of computing the exponential e^A of an n-by-n matrix A . Here n is small enough such that the explicit formation of e^A is feasible. One approach to this problem is to compute an eigenvalue decomposition $A = XBX^{-1}$ and then invoke the formula $e^A = Xe^BX^{-1}$. Parlett [5] has detailed a technique of this type based upon the Schur decomposition

$$(1.1) \quad Q^*A\,Q = \operatorname{diag}(\lambda_1,\ldots,\lambda_n) + N$$

Here, Q is unitary, $N = (n_{ij})$ is strictly upper triangular ($n_{ij} = 0,\ i \geq j$) , and $\lambda(A) = \{\lambda_1,\ldots,\lambda_n\}$ is the spectrum of A . If A is normal ($A^*A = AA^*$) , then it is easy to verify that $N = 0$. In this case the algorithm is particularly easy to implement and the computed e^A extremely accurate. However, as with all eigenvalue methods for computing the exponential, serious numerical difficulties can arise when A is a non-normal matrix having confluent or nearly confluent eigenvalues [4] .

A desire to avoid these difficulties accounts, in part, for the attractiveness of using Padé approximants $R_{pq}(A) = [D_{pq}(A)]^{-1}N_{pq}(A)$ where

$$N_{pq}(z) = \sum_{j=0}^{p} \frac{(p+q-j)!p!}{(p+q)!j!(p-j)!} z^j$$

$$D_{pq}(z) = \sum_{j=0}^{q} \frac{(p+q-j)!q!}{(p+q)!j!(q-j)!} (-z)^j$$

However, even though no eigenvalue computations are required when these approximants are used, the effects of non-normality and confluence can be present. If

$$A = \begin{bmatrix} 0 & 6 & 0 & 0 \\ 0 & 0 & 6 & 0 \\ 0 & 0 & 0 & 6 \\ 0 & 0 & 0 & 0 \end{bmatrix}$$

then $\| R_{11}(A) - e^A \| = 18$ even though $R_{11}(z)$ approximates the exponential exactly on the spectrum of A. (Here, as elsewhere, $\|\cdot\|$ denotes the 2-norm.)

In general, the bounds on the accuracy of an approximation $f(A)$ to e^A deteriorate with loss of normality and not just eigenvalue confluence. We refer the reader to the analysis of Wragg and Davies [11]. which makes this point clear. Their results are derived through exploitation of the Jordan canonical form. In [7] the author derived bounds on $\|f(A) - e^A\|$ which involved powers of $\|N\|$, where N is the matrix of (1.1). The size of N is related to Henrici's "departure from normality" [3] . Of course, if A is normal and $f(z)$ is defined on $\lambda(A)$, then it is easy to verify from (1.1) that

$$\| f(A) - e^A \| = \max_{z \in \lambda(A)} | f(z) - e^z |$$

This illustrates that for normal A, the accuracy of $f(A)$ depends solely upon the behavior of $f(z)$ on the spectrum of A.

All these remarks raise the possibility that the problem of computing e^A is inherently difficult for certain non-normal matrices. In [8] we examined the sensitivity of the map $A \to e^{At}$ in hopes of answering this question. Various upper bounds to the relative perturbation

$$\phi(t) = \frac{\| e^{(A+E)t} - e^{At} \|}{\| e^{At} \|}$$

were derived. For example, it was shown that

$$\phi(t) \leqslant t \|E\| M_S(t)^2 e^{M_S(t) \|E\| t}$$

where

$$M_S(t) = \sum_{k=0}^{n-1} \|Nt\|^k / k!$$

and N is the matrix of (1.1). This bound is typical in that it "deteriorates" as A departs from normality.

More light is shed on the sensitivity problem through the formulation of the exponential "condition number"

$$\nu(A,t) = \max_{\|E\|=1} \left\| \int_0^t e^{A(t-s)} E e^{As} ds \right\| \frac{\| A \|}{\|e^{At}\|} .$$

This quantity amounts to a normalized Frechet derivative of the map $A \to e^{At}$. One can show that for a a given $t \geq 0$ there exists a perturbation $\hat{E}$ such that

$$\phi(t) \cong \frac{\|\hat{E}\|}{\|A\|} \nu(A,t) .$$

This indicates that if $\nu(A,t)$ is large, as it tends to be when A is non-normal, then small relative changes in A can induce relatively large changes in e^{At}. We refer the interested reader to [8].

These observations must be borne in mind when assessing algorithms for computing e^{A}. In general, rounding errors of the order $\varepsilon\, \nu(A,1)$ can be expected in the computed version of e^{A} no matter what algorithm is used and, hence, no technique should be faulted for producing errors of this magnitude. Here, ε is the machine precision. The situation is analogous to that of solving linear systems $Ax = b$. In this setting even a "stable" algorithm such as Gaussian elimination with pivoting need not produce an accurate solution when A is ill-conditioned with respect to inversion [2].

Are methods involving Padé approximation to e^{A} stable in the sense of Gaussian elimination or do they introduce errors greater than the inherent sensitivity of the problem warrants? We know of no rigorous analysis which answers this question. However, it is clear that the success of a particular Padé technique depends a great deal upon the details of the implementation. To illustrate this point let us consider the following family of approximations to e^{A}:

$$F(A,p,q,j) = \left[R_{pq}(A/2^{j})\right]^{2^{j}}$$

Usually, $\|A\|/2^{j} \approx 1$. If A is non-normal and $\lambda(A)$

is in the open left-half-plane, then it is possible for $\|e^A\| \approx \|F(A,p,q,j)\|$ to very small while

$$\left\| e^{A/2^{j-k}} \right\| \approx \left\| \left[R_{pq}\left(\frac{A}{2^j} \right) \right]^{2^k} \right\|$$

is very big for some intermediate value of k, $1 \leq k < j$. This is known as the "hump" phenomena [4]. Since the rounding errors in the computed $F(A,p,q,j)$ may be of the order

$$\varepsilon \max_{1 \leq k \leq j} \left\| \left[R_{pq}\left(\frac{A}{2^j} \right) \right]^{2^k} \right\|$$

where ε is the machine precision, the resulting approximate to e^A may have large relative error.

If instead we estimate e^A with an approximation of the form $e^{-\lambda}F(A+\lambda I,p,q,j)$ where $\lambda \geq \|A\|$, then the effect of the "hump" is somewhat dissipated. This is because the smallness in the approximation to e^A is achieved through the "stable" scalar-matrix multiplication $e^{-\lambda}F(A+\lambda I,p,q,j)$ rather than through the subtractive cancellation which may transpire during the repeated squaring of $R_{pq}(A/2^j)$. (Subtractive cancellation can cause a severe loss of numerical accuracy.)

Ward [10] advocates translation by λ, not to minimize cancellation, but to draw the eigenvalues of $A + \lambda I$ closer to the origin. For this reason he sets $\lambda = -\text{trace}(A)/n$. This choice of λ, or any other, undoubtedly effects the accuracy of the computed approximant to e^A. Yet, it is interesting to note that Ward's rigorous and very useful error analysis does not in any way explicitly exploit the fact that a translation has been done. This reminds us once again that there is no rigorous analysis relating the accuracy of the computed version of $e^{-\lambda}F(A+\lambda I,p,q,j)$ to $\nu(A,1)$, non-normality, the "hump", etc. Until such an analysis is done, the precise limitations which these factors impose on Padé approximation will remain unclear.

2 The Selection and Evaluation of F(A,p,q,j)

Let us now turn our attention to some of the practical problems which arise in connection with $F(A,p,q,j)$. (Assume A is translated if necessary.) In [4] it was shown that if

$$\frac{\|A\|}{2^j} \leq \frac{1}{2}$$

then

$$F(A,p,q,j) = e^{A+E}$$

where

$$(2.1) \quad \frac{\|E\|}{\|A\|} \leq \frac{8\ \|A\|^{p+q}}{2^{j(p+q)}} \ \frac{p!\ q!}{(p+q)!(p+q+1)!} \equiv \varepsilon(p,q,j)\ .$$

$F(A,p,q,j)$ costs approximately $w(p,q,j)$ multiplicative operations where

$$w(p,q,j) = n^3\ [j + \max\{p,q\} + \tfrac{1}{3} - \tfrac{4}{3}\,\delta_{oq}]$$

This follows since one must form $A^2,\ldots,A^{\max\{p,q\}}$; solve the linear system $[D_{pq}(A/2^j)]X = N_{pq}(A/2^j)$; and compute X^{2^j} . (We assume that the linear system is solved using Gaussian elimination. Of course, there is no linear system if $q = 0$.)

Now consider the problem of choosing p,q, and j such that for a given $\varepsilon > 0$,

$$\frac{\|E\|}{\|A\|} \leq \varepsilon(p,q,j) \leq \varepsilon$$

Clearly, there are many p, q, and j for which this inequality holds, but it makes economic sense to select these parameters in such a way that $w(p,q,j)$ is minimized. It turns out that for practical values of $\|A\|$ and ε we need only consider diagonal approximants $(p=q)$. The simplified form of $\varepsilon(q,q,j)$ and $w(q,q,j)$ makes it easy to determine an "optimum" q and j [4]. For example, if $\|A\| = 100$ and $\varepsilon = 10^{-6}$, then $F(A,3,3,8)$ is the most efficient approximant under this method of assessing work.

We briefly mention that the operation count $w(q,q,j) = n^3 [j + q + \frac{1}{3}]$ can be somewhat reduced if the special structure of $R_{qq}(B)$ is exploited. ($B = A/2^j$.) To illustrate, suppose $q = 2r+1$ where r is a positive integer and note

$$R_{qq}(B) = \frac{\sum_{k=0}^{r} c_{2k}(B^2)^k + B \sum_{k=0}^{r} c_{2k+1}(B^2)^k}{\sum_{k=0}^{r} c_{2k}(B^2)^k - B \sum_{k=0}^{r} c_{2k+1}(B^2)^k}$$

where $c_k = (2q-k)!k!/(2q)!k!(q-k)!$. It is obvious that to evaluate $R_{qq}(B)$, one need only compute B^2, $(B^2)^2, \ldots, (B^2)^r$; do a matrix multiplication by B ; and solve a linear system. Thus, $F(A,q,q,j)$ can be evaluated in approximately $n^3[j + \frac{q}{2} + \frac{4}{3}]$ operations rather than $n^3[j + q + \frac{1}{3}]$ operations as predicted above. We refer the reader to [6] for apt remarks concerning the efficient evaluation of matrix polynomials. Of course, the modified work functions which arise from "fast" schemes for evaluating the numerator and denominator can be used to determine "optimum" q and j just as in [4].

3 Approximating Integrals Involving e^A

In the course of solving the optimal linear regulator problem with step input, one is led to the following integrals:

$$(3.1) \quad H(\Delta) = \int_0^\Delta e^{As} B \, ds$$

$$(3.2) \quad Q(\Delta) = \int_0^\Delta e^{A^T s} Q_c e^{As} \, ds$$

$$(3.3) \quad M(\Delta) = \int_0^\Delta e^{A^T s} Q_c H(s) \, ds$$

$$(3.4) \quad W(\Delta) = \int_0^\Delta H(s)^T Q_c H(s) \, ds$$

Here, $\Delta > 0$, Q_c is a symmetric, positive definite nxn matrix, and B is an nxp matrix with $n \geq p$.

These integrals can be expressed as power series in Δ with computable matrix coefficients. One method for approximating (3.1)-(3.4) is to evaluate truncated versions of these series. See [1] and the references therein . In [9] we noted that if

$$C = \begin{bmatrix} -A^T & I & 0 & 0 \\ 0 & -A^T & Q_c & 0 \\ 0 & 0 & A & B \\ 0 & 0 & 0 & 0 \end{bmatrix}$$

and

$$e^{C\Delta} = \begin{bmatrix} F_1(\Delta) & G_1(\Delta) & H_1(\Delta) & K_1(\Delta) \\ 0 & F_2(\Delta) & G_2(\Delta) & H_2(\Delta) \\ 0 & 0 & F_3(\Delta) & G_3(\Delta) \\ 0 & 0 & 0 & F_4(\Delta) \end{bmatrix}$$

then

$$H(\Delta) = G_3(\Delta)$$
$$Q(\Delta) = F_3(\Delta)^T G_2(\Delta)$$
$$M(\Delta) = F_3(\Delta)^T H_2(\Delta)$$
$$W(\Delta) = [B^T F_3(\Delta)^T K_1(\Delta)] + [B^T F_3(\Delta)^T K_1(\Delta)]^T$$

These results follow by noting that $F_3(\Delta) = e^{A\Delta}$ and that the submatrices which "make up" the upper triangle of $e^{C\Delta}$ are various convolutions. For example,

$$G_3(\Delta) = \int_0^{\Delta} e^{-A^T(\Delta-s)} Q_c e^{As}\, ds$$

These observations suggest that approximations H, Q, M, and W to $H(\Delta)$, $Q(\Delta)$, $M(\Delta)$, and $W(\Delta)$ can be obtained by approximating $e^{C\Delta}$ by $F(C\Delta,q,q,j)$. The computational process involves selecting q and j (say by the methods of §2), forming $F(C\Delta,q,q,j)$, and then combining its various submatrices. The errors in the resulting approximations are easy to bound by using (2.1).

If the problem is of low enough dimension, then an easy course of action is to just input $C\Delta$ to any matrix exponential program such as Ward's [10]. If efficiency is of interest then the special structure of $C\Delta$ had better be exploited. Some of the ways this can be done are detailed in [9] .

References

1 Armstrong, E.S. and A.K. Caglayan, An algorithm for the weighting matrices in the sampled-data optimal linear regulator problem, NASA Technical Note, TN D-8372, 1976.

2 Forsythe, G.E. and C.B. Moler, Computer Solution of Linear Algebraic Systems, Prentice Hall, Englewood Cliffs, New Jersey, 1967.

3 Henrici, P., Bounds for iterates, inverses, spectral variation and fields of values of non-normal matrices, Numerische Math., 4(1962), 24-40.

4 Moler, C.B. and C. Van Loan, Nineteen ways to compute the exponential of a matrix, Cornell Computer Science Technical Report TR-76-283, 1976.(To appear SIAM Review.)

5 Parlett, B.N., Computation of functions of triangular matrices, Memo No. ERL-M481, Electronics Research Laboratory, University of California, Berkeley, 1974.

6 Paterson, M.S. and L.J. Stockmeyer, On the number of nonscalar multiplications necessary to evaluate polynomials, SIAM J.Comp., 2(1973), 60-66.

7 Van Loan, C, A Study of the matrix exponential, University of Manchester Numerical Analysis Report 7, 1974.

8 Van Loan, C, The sensitivity of the matrix exponential, Cornell Computer Science Technical Report TR 76-270, 1976.(To appear in SIAM J.Num. Analysis.)

9 Van Loan, C., Computing integrals involving the matrix exponential, Cornell Computer Science Technical Report TR 76-298, 1976.

10 Ward, R.C., Numerical Computation of the matrix exponential with accuracy estimate, Union Carbide Corp. Nuclear Division Technical Report UCCND CSD 24, Knoxville Tennessee, 1975.

11 Wragg, A., and C. Davies, Computation of the exponential of a matrix I: theoretical considerations, JIMA, 11(1973), 369-375

C.Van Loan*
Dept.Computer Science
Cornell, Ithaca, NY, 14853

*Supported by NSF grant MCS76-08686.

STATISTICAL ROUNDOFF ERROR ANALYSIS OF A PADÉ ALGORITHM FOR COMPUTING THE MATRIX EXPONENTIAL

Robert C. Ward

In this paper we present a statistical roundoff error analysis of an algorithm to compute the matrix exponential based on diagonal Padé approximations with appropriate scaling and squaring. An à posteriori estimate for the expectation and variance of the final error is produced. The results of this analysis has been incorporated into the algorithm and test results are presented.

1 Introduction

In a survey paper by Moler and Van Loan [5], several methods for computing the matrix exponential are discussed and analyzed. For each of these methods, there exist classes of matrices for which inaccurate approximations may result. It is not always known whether such failures result from the inherent sensitivity of the problem or from the instability of the algorithm. In either case, it is desirable for the algorithm to indicate this failure to the user. Ward [9] describes and analyses an algorithm to compute the matrix exponential based on diagonal Padé approximations. He computes an à posteriori bound on the size of the final error, including the effects of both truncation and roundoff, and returns to the user the minimum number of digits accurate in the norm of the computed exponential matrix. Thus, users can frequently determine that the algorithm has approximated the matrix exponential to their desired accuracy. However, users may be falsely notified of a failure because the error bound may be, and usually is, a severe overestimate of the actual error.

In this paper we present a statistical error analysis of Ward's algorithm and produce an à posteriori estimate for the expectation and variance of the final error. From the central limit theorem in probability theory (Parzen [6]), we expect the 95 per cent confidence bound to be a considerable reduction from the strict error bound, thus partially eliminating overly pessimistic results.

After the notation and assumptions are discussed in Section 2, the statistical error analysis is presented in Section 3. The results of some test cases are then presented in Section 4.

2 Notation and Assumptions

The base of the computer used for the calculations is denoted by b and t denotes the number of digits (base b) used to represent the mantissa of a floating-point number. A computed number is denoted by a symbol with a bar over it and the corresponding exact number by the symbol without the bar.

For n a positive integer, $R^{n,n}$ denotes the set of all real n x n matrices $A = [a_{ij}]$. Then, $||A||$ for A in $R^{n,n}$ denotes the 1-norm of A and is defined as the nonnegative quantity $\max_j \sum_{i=1}^{n} |A_{ij}|$. If R^n denotes the set of all real column n-vectors $\underline{x} = [x_1, x_2, \ldots, x_n]^T$, then $U^n(c)$ denotes the subset of R^n defined by $\{\underline{x} \in R^n : \sum_{i=1}^{n} |x_i| \leq c\}$.

We denote the expectation of a random variable ε by $E[\varepsilon]$, the variance by $Var[\varepsilon]$, and the probability density function by $pdf[\varepsilon]$. The conditional expectation and conditional variance of a random variable ε, given (the value of) a random variable η, are denoted by $E[\varepsilon|\eta]$ and $Var\ [\varepsilon|\eta]$, respectively. From Parzen [6], we have the following formulae:

(2.1) $E[\varepsilon] = E[E[\varepsilon|\eta]]$

(2.2) $Var[\varepsilon] = E[Var[\varepsilon|\eta]] + Var[E[\varepsilon|\eta]]$

where $E[\varepsilon]$ and $E[\varepsilon^2]$ must be bounded for (2.1) and (2.2) to be valid, respectively.

A basic assumption which we use throughout this paper is that roundoff errors are independent random variables. Henrici [3] offers a very eloquent argument in support of this assumption and the excellent agreement between his theoretical results and empirical results provides additional support. The importance of this assumption is illustrated by two observations. First, there would not be a basis for statistical analysis if roundoff errors were not random variables, and second, statistical error analysis would be untractable if roundoff errors were not assumed to be independent.

We also assume that the mantissas are logarithmically distributed; that is, their probability density function is

$$\text{pdf}[\mu] = \frac{1}{\mu \ln b}$$

where $b^{-1} \leq \mu \leq 1$. Benford [1], Pinkham [7], and Knuth [4] derive this distribution by observing empirical data, studying its invariance under scale change, and investigating the distribution of the leading digits of the positive integers, respectively. There appears to be general agreement to Knuth's conclusion that the logarithmic distribution is to be regarded as a very close approximation to the true distribution.

The preceding assumption also provides information about the distribution function for the roundoff errors. Feldstein and Goodman [2] show that the kth digit of logarithmically distributed mantissas for $k \geq 2$ is approximately uniformly distributed with the approximation becoming more accurate as k increases. Thus, the customary assumption that roundoff errors are uniformly distributed between -1/2 and 1/2 units in the last place of the result is a consequence of the logarithmic distribution assumed for the mantissas.

Using the above assumptions, Tsao [8] computes the probability density function, the expected value, and the variance for the relative error in a single operation or in the computer storage of a number not representable by t digits (base b).

Letting ε be the random variable representing this relative error, Tsao obtains

$$(2.3)\quad \mathrm{pdf}[\varepsilon] = \begin{cases} \dfrac{b^{t-1}}{\ln b}(b-1) & -1/2\ b^{-t} \le \varepsilon \le 1/2\ b^{-t} \\ \dfrac{1}{\ln b}\left(\dfrac{1}{2|\varepsilon|} - b^{t-1}\right) & 1/2\ b^{-t} \le |\varepsilon| \le 1/2\ b^{-t+1} \end{cases}$$

$$(2.4)\quad E[\varepsilon] = 0$$

$$(2.5)\quad \mathrm{Var}[\varepsilon] = \frac{b^2-1}{24 \ln b}\, b^{-2t}.$$

As a means of comparison, the strict error bounds are $-\frac{1}{2}b^{-t+1} \le \varepsilon \le \frac{1}{2}b^{-t+1}$. Random variables representing the relative errors in arithmetic operations and in computer storage are denoted by ε with super- and subscripts and ε with superscripts, respectively.

Finally, we assume that the columns of the matrix to which the diagonal Padé approximation is applied (cf. matrix B of Section 3) are independent and uniformly distributed over $U^n(1)$. For example, each point in $U^2(1)$ (shown below) is equally likely to form the jth column of B for n = 2.

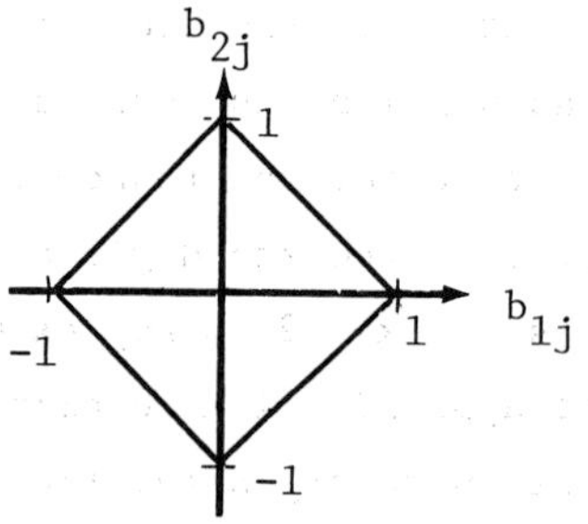

3 Statistical Error Analysis

The basic features of Ward's algorithm [9] are scaling the matrix by a power of 2, approximating the exponential of the scaled matrix by a diagonal Padé approximation, and removing the scaling by repeated squaring. Origin shifts and diagonal balancing are employed prior to scaling to reduce the size of the scale factor. Eliminating some of the minor steps and denoting the numerator polynomial of the pth diagonal Padé approximation by $Q_p(X)$, Ward's algorithm for computing e^A can be briefly stated as follows:

a) Balance A; i.e. $A' = D^{-1}AD$

b) Compute $B = A'2^{-m}$ such that $||B|| \leq 1$

c) Compute $Q_p(B)$ and $Q_p(-B)$

d) Compute $e^B = Q_p^{-1}(-B)\ Q_p(B)$

e) Compute $e^{A'} = (e^B)^{2^m}$

f) Compute $e^A = De^{A'}D^{-1}$

Steps a) and b) produce little error. Thus, only the error in steps c) through f) are analyzed by applying the error analysis techniques illustrated in Wilkinson [10].

We begin by establishing the following lemma which will be used to obtain the expectation and variance of a random variable from the conditional expectation and conditional variance.

LEMMA 3.1. <u>Let</u> $\underline{x} = \{x_1, x_2, \ldots, x_n\} \in U^n(c)$ <u>be a uniformly distributed random variable</u>, $\{i_1, i_2, \ldots, i_k\}$ <u>be a sequence of distinct positive integers such that</u> $1 \leq i_j \leq n$ for $1 \leq j \leq k \leq n$, $\{e_1, e_2, \ldots, e_k\}$ <u>be a sequence of even positive integers</u>, <u>and</u> o <u>be an odd positive integer</u>. <u>Then</u>,

$$(3.1)\quad E[x_{i_1}^{o}] = 0$$

$$(3.2)\quad E[x_{i_1}^{o}, x_{i_2}^{e_2}, x_{i_3}^{e_3}, \ldots, x_{i_k}^{e_k}] = 0$$

$$(3.3)\quad E[x_{i_1}^{e_1}, x_{i_2}^{e_2}, \ldots, x_{i_k}^{e_k}] = \frac{e_1!\ e_2!\ \ldots\ e_k!\ n!}{(n+e)!}\ c^e$$

<u>where</u> $e = \sum_{i=1}^{k} e_i$.

Proof. Since $U^n(c)$ is symmetric about the origin, (3.1) and (3.2) follow from the integration of odd functions. The uniform probability law is independent of x_i; therefore,

$$E[x_{i_1}^{e_1}, x_{i_2}^{e_2}, \ldots, x_{i_k}^{e_k}] = E[x_1^{e_1}, x_2^{e_2}, \ldots, x_k^{e_k}]$$

and

$$E[x_1^{e_1}, x_2^{e_2}, \ldots, x_k^{e_k}] = \frac{2^n}{R} \int_0^c x_1^{e_1} \int_0^{c-x_1} x_2^{e_2} \cdots \int_0^{c-\sum_{i=1}^{k-1} x_i} x_k^{e_k} \int_0^{c-\sum_{i=1}^{k} x_i} \cdots \int_0^{c-\sum_{i=1}^{n-1} x_i} dx_n\, dx_{n-1} \cdots dx_1$$

where $R = 2^n \int_0^c \int_0^{c-x_1} \cdots \int_0^{c-\sum_{i=1}^{n-1} x_i} dx_n\, dx_{n-1} \cdots dx_1$.

Repeated application of $\frac{1}{(n-\ell)!} \int_0^{c-\sum_{i=1}^{j-1} x_i} x_j^{e_j} (c-\sum_{i=1}^{j} x_i)^{n-\ell} dx_j =$

$\frac{e_j!}{(n-\ell+e_j+1)!} (c-\sum_{i=1}^{j-1} x_i)^{n-\ell+e_j+1}$ to the above integration yields

(3.3). ●

Step c) is the computation of $Q_p(B)$ and $Q_p(-B)$ where

$$(3.4) \quad Q_p(X) = \sum_{r=0}^{p} c_r(X)$$

$$(3.5) \quad c_r = \frac{(2p-r)!\ p!}{(2p)!\ r!\ (p-r)!} .$$

The basic operation is then to compute $\overline{B^r}$.

THEOREM 3.1. Let ε be a random variable with probability density function given by (2.3) and let η_{ij}^r be the random variable $\overline{B_{ij}^r} - B_{ij}^r$. Then, for $r \geq 2$,

$$E[\eta_{ij}^r] = 0 + O(b^{-2t})$$

$$\mathrm{Var}[\eta_{ij}^r] = \{2^{r-1}n^{-r} + O(n^{-r-1})\}\ \mathrm{Var}[\varepsilon] + O(b^{-4t}).$$

Proof. From Wilkinson [10], we have

$$\eta_{ij}^r = \sum_{k=1}^{n}[B_{ik}\eta_{kj}^{r-1}] + \sum_{k=1}^{n}[B_{ik}B_{kj}^{r-1}(\varepsilon^{(k)} + \sum_{\ell=1}^{n+2-k}\varepsilon_\ell^{(k)})] + O(b^{-2t}).$$

Thus,

$$E[\eta_{ij}^r|B] = \sum_{k=1}^{n}B_{ik}E[\eta_{kj}^{r-1}] + [\sum_{k=1}^{n}(n+3-k)B_{ik}B_{kj}^{r-1}]E[\varepsilon] + O(b^{-2t})$$

$$\mathrm{Var}[\eta_{ij}^r|B] = \sum_{k=1}^{n}(B_{ik})^2\mathrm{Var}[\eta_{kj}^{r-1}]$$

$$+ \{\sum_{k=1}^{n}(n+3-k)(B_{ik})^2(B_{kj}^{r-1})^2\}\mathrm{Var}[\varepsilon] + O(b^{-4t}).$$

Observing that $\eta_{kj}^1 = B_{kj}\ \varepsilon^{(j)}$, the results follow from (2.1), 2.2) (2.4), and Lemma 3.1. ●

The following theorem gives the result of the statistical error analysis of Step c):

THEOREM 3.2 Let ε be a random variable with probability density function given by (2.3) and let $\gamma_{ij}^{\pm}$ be the random variable $\overline{Q_p(\pm B)_{ij}} - Q_p(\pm B)_{ij}$. Then,

$$E[\gamma_{ij}^{\pm}] = 0 + O(b^{-2t})$$

$$\mathrm{Var}[\gamma_{ij}^{\pm}] = \{\delta_{ij} + [2\tfrac{3}{4} - \frac{1}{8p} + O(p^{-2})]n^{-2} + O(n^{-3})\}\mathrm{Var}[\varepsilon]+O(b^{-4t}).$$

where δ_{ij} is the kronecker delta function.

Proof. From (3.4) and Wilkinson [10], we have

$$\gamma_{ij}^{\pm} = \sum_{r=1}^{p}[c_r\eta_{ij}^r] + \sum_{r=1}^{p}[c_rB_{ij}^r(\varepsilon^{(r)}+\sum_{k=1}^{r+2}\varepsilon_k^{(r)})]+\delta_{ij}c_o\varepsilon_1^{(0)}+O(b^{-2t}).$$

Thus,

$$E[\gamma_{ij}^{\pm}|B] = \sum_{r=1}^{p}c_rE[\eta_{ij}^r] + [\sum_{r=1}^{p}(r+3)c_rB_{ij}^r+\delta_{ij}c_o]E[\varepsilon] + O(b^{-2t})$$

$$\mathrm{Var}[\gamma^{\pm}_{ij}|B] = \sum_{r=1}^{p} c_r^2 \mathrm{Var}[\eta^r_{ij}] + [\sum_{r=1}^{p} (r+3)c_r^2 (B^r_{ij})^2 + \delta_{ij} c_o^2]\mathrm{Var}[\varepsilon] + O(b^{-4t}).$$

The results follow from (2.1), (2.2), (2.4), (3.5), Lemma 3.1, and Theorem 3.1. ●

We now examine Step d) which computes the approximation to e^B by Gaussian elimination; that is,

$$(3.6) \quad e^B = Q_p^{-1}(-B)Q_p(B) + T_e(p)$$

where $T_e(p)$ is the truncation error matrix. Using similar analysis techniques to those found in the theorems above and observing that $Q_p(-B)$ is diagonally dominant, we have

$$(3.7) \quad (\overline{Q_p(-B)} + F)\ \overline{e^B} = \overline{Q_p(B)}$$

$$(3.8) \quad E[F_{ij}] = 0 + O(b^{-2t})$$

$$(3.9) \quad \mathrm{Var}[F_{ij}] = [2(i-j)+2|i-j|+\delta_{ij}12i+(i^2+j^2)n^{-2}+O(in^{-2})]\mathrm{Var}[\varepsilon] + O(b^{-4t})$$

with ε obeying (2.3). From (3.6) and (3.7), we have

$$\overline{e^B} - e^B = (\overline{Q_p(-B)}+F)^{-1}\{(Q_p(B)-\overline{Q_p(B)})+[(Q_p(-B)-\overline{Q_p(-B)})-F]e^B - T_e(p)\}.$$

Using Theorem 3.2 and (3.8),

$$(3.10) \quad E[(\overline{e^B} - e^B)_{ij}] = \sum_{k=1}^{n} E[(\overline{Q_p(-B)}+F)^{-1}_{ik}]T_e(p)_{kj} + O(b^{-2t})$$

$$(3.11) \quad \mathrm{Var}[(\overline{e^B}-e^B)_{ij}|e^B] = \sum_{k=1}^{n} \{\mathrm{Var}[(\overline{Q_p(-B)}+F)^{-1}_{ik}]+E^2[(Q_p(-B)+F)^{-1}_{ik}]\} \{\mathrm{Var}[(\overline{Q_p(B)}-Q_p(B))_{kj}] + \sum_{\ell=1}^{n} (\mathrm{Var}[(Q_p(-B)-\overline{Q_p(-B)})_{k\ell}] + \mathrm{Var}[F_{k\ell}])(e^B)^2_{\ell j}\} + \mathrm{Var}[(\overline{Q_p(-B)}+F)^{-1}_{ik}](T_e(p))^2_{kj} + O(b^{-4t}).$$

LEMMA 3.2. Let F be given by (3.7), (3.8), and (3.9). Then,

$$(3.12) \quad E[(\overline{Q_p(-B)}+F)^{-1}_{ij}] = \{1 + \tfrac{1}{4}n^{-2} + O(n^{-3})\}\delta_{ij} + O(b^{-2t})$$

$$(3.13)\quad \mathrm{Var}[(\overline{Q_p(-B)}+F)^{-1}_{ij}] = \{2(i-j) + 2|i-j| + \delta_{ij}12i + (i^2+j^2)n^{-2} + O(in^{-2})\}\mathrm{Var}[\varepsilon] + O(b^{-4t})$$

where ε obeys (2.3).

Proof. We have

$$(3.14)\quad (\overline{Q_p(-B)} + F)^{-1} = \overline{Q_p^{-1}(-B)} \sum_{k=0}^{\infty} [-F\ \overline{Q_p(-B)}^{-1}]^k$$

Using (3.8),

$$E[(\overline{Q_p(-B)} + F)^{-1}_{ij} \mid \overline{Q_p(-B)}] = \overline{Q_p^{-1}(-B)}_{ij} + O(b^{-2t})$$

From (2.1) and Theorem 3.2, we have

$$E[(\overline{Q_p(-B)} + F)^{-1}_{ij} \mid Q_p(-B)] = Q_p^{-1}(-B)_{ij} + O(b^{-2t})$$

Recalling (3.4) and using Lemma 3.1, we have result (3.12).

Expanding the sum in (3.14), the expression for the variance becomes

$$\mathrm{Var}[(\overline{Q_p(-B)}+F)^{-1}_{ij}] + \sum_{q=1}^{n} E[\overline{Q_p^{-1}(-B)}^2_{iq}]\{\sum_{\ell=1}^{n} E[\overline{Q_p^{-1}(-B)}^2_{\ell j}]\mathrm{Var}[F_{q\ell}]+\ldots\}.$$

The result (3.13) is obtained from (3.9) using Theorem 3.2, Lemma 3.1, and (3.4) to compute the expectation of $\overline{Q_p^{-1}(-B)}^2_{ij}$. ●

Applying Lemma 3.2 to (3.10) and (3.11), we have

$$(3.15)\quad E[(\overline{e^B} - e^B)_{ij}] = T_e(p)_{ij} + O(b^{-2t})$$

$$\mathrm{Var}[(\overline{e^B}-e^B)_{ij}|e^B] = \{2[(e^B_{ij})^2+\delta_{ij}] + \sum_{\ell=1}^{n}[2(i-\ell)+2|i-\ell|+\delta_{i\ell}12i +(i^2+\ell^2)n^{-2}+O(in^{-2})](e^B_{\ell j})^2\}\mathrm{Var}[\varepsilon] + O(b^{-4t})$$

where $T_e(p)^2_{kj}$ is of order b^{-2t}. The variance may be obtained by considering (3.6). As we did with $Q_p^{-1}(-B)$ and $Q_p(B)$, we can compute the expectation of e^B_{ij} using Lemma 3.1. Thus,

$$(3.16)\quad \mathrm{Var}[(\overline{e^B}-e^B)_{ij}] = \{2(i-j)+2|i-j|+\delta_{ij}12i + 4\tfrac{i}{n}\}\mathrm{Var}\ [\varepsilon].$$

Step e) is the squaring process to remove the scaling

introduced in Step b). Let the matrix θ^r be defined by

$$\theta^r = \overline{(e^B)^{2^r}} - (e^B)^{2^r}.$$

From Wilkinson [10], we have

$$\theta^r_{ij} = \sum_{k=1}^{n} [((e^B)^{2^{r-1}})_{ik}\theta^{r-1}_{kj} + \theta^{r-1}_{ik}((e^B)^{2^{r-1}})_{kj}]$$

$$+ \sum_{k=1}^{n} [((e^B)^{2^{r-1}})_{ik}((e^B)^{2^{r-1}})_{kj} \sum_{\ell=1}^{n+2-k} \varepsilon^{(k)}_{\ell}] + O(b^{-2t})$$

Making the assumption that the columns of $(e^B)^{2^r}$ for $1 \le r \le m$ are independent and uniformly distributed over $U^n(||(e^B)^{2^r}||)$, we have from Lemma 3.1

$$E[\theta^r_{ij}] = \delta_{1r} 2T_e(p)_{ij} + O(b^{-2t)}$$

$$\mathrm{Var}[\theta^r_{ij}] = \{\sum_{k=1} \mathrm{Var}[\theta^{r-1}_{kj}] + \mathrm{Var}[\theta^{r-1}_{ik}]\}||(e^B)^{2^{r-1}}||^2[2n^{-2}+O(n^{-3})]$$

$$+ (2n^{-2} + O(n^{-3}))||(e^B)^{2^{r-1}}||^4 \mathrm{Var}[\varepsilon] + O(b^{-4t}).$$

Step f) can be similarly analyzed with the following results:

$$E[(\overline{e^A} - e^A)_{ij}] = \frac{d_i}{d_j} E[\theta^m_{ij}] + O(b^{-2t})$$

$$\mathrm{Var}[(\overline{e^A} - e^A)_{ij}] = (\frac{d_i}{d_j})^2 \mathrm{Var}[\theta^m_{ij}] + 4(\frac{d_i}{d_j})^2||(e^B)^{2m}||^2 \mathrm{Var}[\varepsilon] + O(b^{-4t})$$

where $D = \mathrm{diag}\{d_1, d_2, \ldots, d_n\}$.

We can now give an à posteriori bound for the expectation and variance of the roundoff error for the complete algorithm by monitoring the size of $||(e^B)^{2^r}||$. Note that this computation is $O(n^2)$ while the squaring process is $O(n^3)$.

4 Test Results

Using the results of the previous section, Ward's algorithm has been modified to output the number of digits accurate in the norm of the computed exponential matrix at the 95% confidence level. Thus, two integers are now returned to the user with one integer based on strict error bounds (see Ward [9]) and one based on statistical analysis.

Two test cases have been run on the IBM 360/91 computer at Oak Ridge National Laboratory. These test cases are presented below.

Test Case 1:

$$A = \begin{bmatrix} 4 & 2 & 0 \\ 1 & 4 & 1 \\ 1 & 1 & 4 \end{bmatrix}$$

The matrix is defective and nonderogatory. The computed exponential matrix is given in Ward [9] and is accurate to 15 digits in its norm. The strict error bound indicates 14 accurate digits and the 95% confidence bound indicates 15 accurate digits.

Test Case 2:

$$A = \begin{bmatrix} -131 & 19 & 18 \\ -390 & 56 & 54 \\ -387 & 57 & 52 \end{bmatrix}$$

The matrix has distinct eigenvalues and the resulting shifted, balanced matrix (cf. matrix A' of Section 3) has a large norm. The computed exponential matrix is given in Ward [9] and is accurate to 12 digits in its norm. The strict error bound indicates 6 accurate digits and the 95% confidence bound indicates 11 accurate digits.

References

1 Benford, F., The law of anomalous numbers, Proc. Amer. Philos. Soc., 78 (1938), 551-572.

2 Feldstein, A. and R. Goodman, Convergence estimates for the distribution of trailing digits, J. ACM, 23 (1976), 287-297.

3 Henrici, P., Discrete Variable Methods in Ordinary Differential Equations, John Wiley, New York, 1962.

4 Knuth, D. E., The Art of Computer Programming, Vol. 2: Seminumerical Algorithms, Addison - Wesley, Reading, Mass., 1969.

5 Moler, C. B. and C. F. Van Loan, Nineteen ways to compute the exponential of a matrix, SIAM Rev., to appear.

6 Parzen, E., Stochastic Processes, Holden-Day, San Francisco, 1962.

7 Pinkham, R. S., On the distribution of first significant digits, Ann. Math. Statist., 32 (1961), 1223-1230.

8 Tsao, N., On the distributions of significant digits and roundoff errors, Comm. ACM, 17 (1974), 269-271.

9 Ward, R. C., Numerical computation of the matrix exponential with accuracy estimate, SIAM J. Numer. Anal., to appear.

10 Wilkinson, J. H., Rounding Errors in Algebraic Processes, Prentics-Hall, Englewood Cliffs, N.J., 1963.

Robert C. Ward
Mathematics and Statistics Research Department
Computer Sciences Division
Union Carbide Corporation, Nuclear Division
Oak Ridge, Tennessee 37830

*Prime contractor for the U.S. Energy Research and Development Administration

OPEN PROBLEMS

Summary of Panel Discussion

"DIRECTIONS FOR RESEARCH"

Panel:

L.W. Fullerton
J.L. Gammel
P. Henrici
W.L. Shepherd
R.C. Ward

Chairman:

R.S. Varga

A significant portion of the panel discussion emphasized the importance of communicating results with engineers and workers in the field. R.S. Varga stated that, "We as mathematicians tend to write our papers for one another. We have a responsibility to communicate with practitioners, even our theoretical results." In the same spirit, L.W. Fullerton claimed, "It's important to codify our own knowledge and to make it simple so that it's easily documented." More conferences of the present type which give members of the academic community a chance to communicate results to laboratory workers was called for by R. Ward.

The significance and meaning of mathematical work in the larger context of science was addressed by P. Henrici. He stated that, "We are attracted to proving theorems which please the intellect. However, we are not here to prove theorems but to do something with all of this - to influence the world in some modest way." He encouraged theoreticians to spend more time on applications, and to do more numerical experimentation utilizing our powerful present day computational tools. "We should look out of the window and see how we can do something outside our chosen fields."

Henrici's remarks sparked a discussion of the role and

importance of theorem proving. W. Shepherd stressed the point that we need to know how to interpret our measurements, and one way to know is to prove theorems, i.e., to investigate the validity of the mathematical models involved and to logically work out the consequences. R.S. Varga remarked that many of us "wear two hats". "At times we need and like to prove theorems," he said. "But at other times we must solve problems; we must make things work. The latter activity is perhaps more of an experimental science. There is not enough time to rigorously prove theorems when one is, for example, designing a nuclear reactor." Similar comments were echoed by Y.L. Luke.

Concerning specific recommendations for future investigative efforts, the following were some of the suggestions made:

1) Develop software for the computation of matrix functions comparable in quality to packages which are available for eigenvalue and eigenvector problems.

2) Develop methods for approximating data in real time situations as for example the tracking of a space vehicle. In this setting massive amounts of data are being collected and the analysis must be almost instantaneous.

3) In the free boundary problem of hydrodynamics (e.g., the rise of a bubble in a fluid), study the approximations of the double power series which arises. While Chisholm approximants have been useful in this regard, there are certain conditions under which this approximation method is not successful.

4) In the study of critical phenomena, develop algorithms which combine the analyses of several series simultaneously so as to extract more information.

5) Develop computer packages for obtaining rational approximations to special functions. (Y.L. Luke is presently preparing one such package which will apply to incomplete Gamma functions, Bessel functions, and others).

MINIMAL REALIZATION FROM DATA SETS - A PROBLEM OF RATIONAL APPROXIMATION

David R. Audley

The minimal realization problem of mathematical system theory is introduced. Some of the key ideas of this theory are presented with an eye towards providing references to sources of more complete exposition. One particular minimal realization problem, data set realization, is presented in some detail with a sketch of an associated solution. References for further reading on this problem are also given. Where appropriate, comparison of minimal realization problems with problems in rational approximation are made. Finally, a question of long standing annoyance is presented. This is the problem of constructing stable minimal realization.

1 Introduction

Fundamental to scientific inquiry is the desire to understand and control some part of the universe. However, little of the universe is so simple that its comprehension and manipulation can be achieved without abstraction; that is, the replacement of some essential characteristics by a model of similar but simpler structure. Models are thus central to the procedures of scientific inquiry [8].

Of particular interest are mathematical (as opposed to physical) models; for these the abstraction is to a useful formulation of the essential characteristics of that phenomena or system being investigated. There are a number of mathematical model structures (or formulations) that engineers find useful in representing physical phenomena and processes. For situations where the physics allow satisfactory representation by a dynamical, linear, stationary model, the matrix state variable is a useful formulation:

$$(1.1)\quad \begin{aligned} \dot{x}(t) &= Ax(t) + Bu(t) \\ y(t) &= Cx(t) \end{aligned} \qquad t \geq 0,$$

the vectors x, u and y are n, p and r dimensional, respectively, and Eq. (1.1) is referred to by the triple $(A,B,C)_n$ and $(A,B,C)_n$ is said to be of dimension n. It is assumed that $x(0) = 0$ so that by taking the Laplace transform of both sides of Eq. (1.1)

$$(1.2)\quad Y(s) = C(sI-A)^{-1}BU(s) = H(s)U(s),$$

where

$$(1.3)\quad H(s) = C(sI-A)^{-1}B$$

is the r×p matrix transfer function, U(s), X(s) and Y(s) are the Laplace transforms of u(t), x(t) and y(t), respectively; the definition of the Laplace transform of x(t) is $\mathcal{L}[x(t)] = X(s) \triangleq \int_0^\infty x(\tau)e^{-s\tau}d\tau$, and u(t), x(t) and y(t) are sufficiently well behaved for the Laplace integral to converge.

Often the engineer starts with the matrix transfer function model: an "external" description of the input/output behavior that does not explicitly display the system state behavior. However, since much of modern control theory utilizes the matrix formulation, Eq. (1.1), for design and computation schemes, it is of considerable interest to transform the external description, Eq. (1.3), to the triple of matrices (A,B,C) in Eq. (1.1). This brings us to the following problem statement.

MINIMAL REALIZATION PROBLEM. Given some description of an r×p (strictly) proper rational function (matrix), G(s), determine a triple $(A,B,C)_n$ where the square matrix A is such that dim A = n = minimum and $C(sI-A)^{-1}B$ satisfies the description.

EXAMPLE 1.1. This is one well known example of the minimal realization problem. Writing H(s) as a formal power series

(1.4) $H(s) = \sum_{i=1}^{\infty} Y_i s^{-i}; \quad Y_i = CA^{i-1}B \quad i = 1,2,\ldots$

gives the infinite sequence $\{Y_1, Y_2, \ldots\}$ of Markov parameters associated with H(s). From this data (what has been called) minimal complete realizations of H(s) can be obtained [5],[9]. This is a complete realization of lowest dimension and it is unique up to a change of variables.

If we take only the first M terms, $\{Y_1, Y_2, \ldots, Y_M\}$, then (what has been called) order M minimal partial realizations of H(s) can be obtained [5],[10]. This partial realization may not be unique up to a change of variable. The order M should not be confused with the dimension of the partial realization.

It should be noted that it is straightforward to show the following. If H(s) admits a minimal complete realization of dimension n, then it is sufficient that $M \geq 2n$ for a minimal partial realization of order M of H(s) to be a minimal complete realization of H(s). A stronger result may be stated using the degree of the minimal polynomial of the minimal complete realizations of H(s) (see [9]).

It is clear that a solution to this minimal realization problem provides a solution to a modified Padé approximation problem. Indeed, the order M minimal partial realization, $(A,B,C)_n$, provides a strictly proper rational function, i.e. a rational vanishing at $s = \infty$,

(1.5) $C(sI-A)^{-1}B = \dfrac{P(s)}{q(s)}$,

where for the $r \times p$ matrix polynomial P(s) and the scalar polynomial q(s), deg P(s) < deg q(s) = minimum, and Eq. (1.5) matches the first M terms in the expansion Eq. (1.4).

In the next section, minimal realization of a different description of the matrix transfer function is discussed. Finally, in Section 3 the open problem concerning stability is introduced.

2 Minimal Realization of Data Sets

As proposed previously in [1] and [3], an order M data set may be obtained from H(s), Eq. (1.3), in the following way. Corresponding to a set of finite complex numbers $\lambda_1,\ldots,\lambda_m$, is the set of r×p matrices

$$(2.1)\quad Y_{ij} = \lim_{s\to\lambda_i} \frac{d^j}{ds^j} H(s) = H^{(j)}(\lambda_i), \qquad \begin{array}{l} j=0,1,\ldots,k_i-1 \\ i=1,\ldots,m \end{array}$$

where $\sum_{i=1}^{m} k_i = M$. It is assumed that each Y_{ij} is finite and that the data set is symmetric. That is, if $H^{(j)}(\lambda_i) = Y_{ij}$ is in the set, then $H^{(j)}(\overline{\lambda}_i) = \overline{Y}_{ij}$ is in the set, where $\overline{\lambda}_i$ is the complex conjugate of λ_i. For a discussion of the application of data sets see [1].

A data set of order M clearly specifies an equivalence class of state variable models: all those triples (A,B,C) such that $C(sI-A)^{-1}B$ satisfies the values given in Eq. (2.1). In keeping with the earlier terminology, an order M minimal partial realization of H(s) is defined as a triple of least dimension such that $C(sI-A)^{-1}B = G(s)$ satisfies a given data set of order M. The similarity of definition is intentional; however, note that now "order M" refers to the number of data set points to be satisfied by G(s) rather than the number of common Markov parameters (that G(s) has with H(s)).

Similar to the minimal realization problem using Markov parameters, an order M minimal partial realization of a data set (Eq. (2.1)) provides a solution to a problem of rational approximation. In particular, it provides a solution to a modified matrix Cauchy interpolation problem. Indeed, a realization provides a strictly proper rational function, Eq. (1.5), where for the r×p matrix polynomial P(s) and the scalar polynomial q(s), deg P(s) < deg q(s) = minimum, and Eq. (1.5) matches, or interpolates the prescribed order M data set, i.e.,

$$\lim_{s\to\lambda_i} \frac{d^j}{ds^j} G(s) = Y_{ij}$$

for all i,j of Eq. (2.1). The realization algorithm is presented next. For proof, see [1].

REALIZATION ALGORITHM. Given an order M data set obtained from H(s) (of the form Eq. (2.1)), the following provides an order M minimal partial realization of H(s).

1. PROCESSING THE ORDER M DATA SET.

From the prescribed order M data set; viz., Eq. (2.1), compute the $r\times p$ matrix polynomial B(s) (with real coefficients) such that degree B(s) < M and B(s) satisfies the data set

$$\text{(2.2)} \quad B^{(j)}(\lambda_i) = Y_{ij}, \qquad \begin{array}{l} j=0,\ldots,k_i-1 \\ i=1,\ldots,m \end{array}$$

(B(s) is an interpolating polynomial of the Full Hermite type, [4].) Also, compute the scalar polynomial a(s) (with real coefficients) defined as

$$\text{(2.3)} \quad a(s) = \prod_{i=1}^{m} (s-\lambda_i)^{k_i},$$

where degree a(s) = M. Forming the proper rational function

$$\text{(2.4)} \quad \frac{B(s)}{a(s)} = \frac{B_{M-1}s^{M-1} + B_{M-2}s^{M-2} + \ldots + B_1 s + B_0}{s^M + a_{M-1}s^{M-1} + \ldots + a_1 s + a_0} = \sum_{i=1}^{\infty} V_i s^{-i}$$

provides the infinite sequence $\{V_1, V_2, \ldots\}$ of which the finite subsequence $\{V_1, V_2, \ldots, V_M\}$ is needed.

2. DETERMINE THE MINIMAL EXTENSIONS OF $\{V_1, V_2, \ldots, V_M\}$ [5],[10].

The generally nonunique minimal extension sequences,

$\{\tilde{V}_{M+1},\ldots,\tilde{V}_{M'}\}$, of the sequence $\{V_1,V_2,\ldots,V_M\}$ are found by the method of Kalman [5] or Tether [10]. This provides for the ($N' \times N$ block) Hankel matrix, $\underline{\underline{H}}_{N',N}$, (associated with the sequence $\{\tilde{V}_1,\ldots,\tilde{V}_M,\tilde{V}_{M+1},\ldots,\tilde{V}_{M'}\}$ of $r\times p$ matrices, where $\tilde{V}_k = V_k$, $k=1,\ldots,M$) that satisfies the rank condition

$$(2.6)\quad \operatorname{rank} \underline{\underline{H}}_{N',N} = \operatorname{rank} \underline{\underline{H}}_{N'+1,N} = \operatorname{rank} \underline{\underline{H}}_{N',N+1},$$

where N',N are the smallest integers such that Eq. (2.6) holds; $M' = N+N'$; and once the extension sequence is fixed through M', it is uniquely determined thereafter (i.e., for $\{\tilde{V}_{M'+1},\tilde{V}_{M'+2},\ldots\}$).

Among all minimal extensions of $\{V_1,V_2,\ldots,V_M\}$ only those that do not provide "unattainable points" are allowed [3],[7]. If all minimal extensions provide a realization with unattainable points, a new data set must be selected. A procedure for determining if a given minimal extension sequence provides unattainable points is discussed in [1].

3. AN ORDER M MINIMAL PARTIAL REALIZATION.

The following provides an order M minimal partial realization:

$$(2.7)\quad A = E^{n}_{rN'}\, P\left[\sigma \underline{\underline{H}}_{N',N}\right] M E^{pN}_{n}$$

$$(2.8)\quad B = E^{n}_{rN'}\, P\left[\underline{\underline{H}}_{N',N}\right] E^{pN}_{p}$$

$$(2.9)\quad C = E^{r}_{rN'}\left[-a_0 \underline{\underline{H}}_{N',N} - a_1 \sigma \underline{\underline{H}}_{N',N} - \cdots - a_{M-1}\sigma^{M-1}\underline{\underline{H}}_{N',N} - \sigma^{M}\underline{\underline{H}}_{N',N}\right] M E^{pN}_{n}$$

where P and M are nonsingular $rN' \times rN'$ and $pN \times pN$ matrices, respectively, such that

$$(2.10)\quad P\left[\underline{\underline{H}}_{N',N}\right] M = \begin{bmatrix} I_n & 0 \\ 0 & 0 \end{bmatrix};$$

n = dimension A; the shift operator is defined (as in [6])

$$\sigma^k \underline{\underline{H}} = \begin{bmatrix} \tilde{V}_{1+k} & \tilde{V}_{2+k} & \cdots \\ \tilde{V}_{2+k} & \tilde{V}_{3+k} & \cdots \\ \vdots & \vdots & \end{bmatrix};$$

and the editing matrix is defined (as in [6])

$$E_n^m = \begin{cases} m \times n \text{ matrix } \begin{bmatrix} I_m 0 \end{bmatrix} \text{ if } m < n \\ m \times n \text{ matrix } \begin{bmatrix} I_n \\ 0 \end{bmatrix} \text{ if } m > n \\ m \times m \text{ unit matrix } I_m \text{ if } m = n. \end{cases}$$

REMARK 2.1 (on obtaining minimal complete realizations).

The realization algorithm is (of course) applicable in determining minimal complete realizations when sufficient data exists. In fact, identical to the Markov parameter case: If H(s) admits a minimal complete realization of dimension at most n, then this realization can be uniquely determined (up to a change of variables) from any data set of order 2n obtained from H(s). For a proof of this remark, see [1].

EXAMPLE 2.1. Consider the data set:

$$\{\lambda_1 = 1,\ Y_1 = [1\ \ 1];\quad \lambda_2 = -1,\ Y_2 = [-1\ \ 1]\}.$$

A minimal partial realization of order 2 will be determined. From step 1,

$$\left.\begin{aligned} B(s) &= [1\ \ 0]s + [0\ \ 1] \\ a(s) &= s^2 - 1, \end{aligned}\right\} \longrightarrow \begin{aligned} V_1 &= [1\ \ 0] \\ V_2 &= [0\ \ 1]. \end{aligned}$$

Step 2 provides the Hankel matrix $H_{2,1} = \begin{bmatrix} 1 & 0 \\ 0 & 1 \end{bmatrix}$; rank $H_{2,1} = 2$

satisfying Eq. (2.6); $M' = 1+2 = 3$ and $\tilde{V}_3 = [x_1 \quad x_2]$ for arbitrary x_1, x_2. Once x_1 and x_2 are fixed,

$$\tilde{V}_k = x_1\tilde{V}_{k-2} + x_2\tilde{V}_{k-1} \qquad k = M' + 1, M' + 2, \ldots .$$

It is found that (see [1]) if $x_2 \neq \pm(1 - x_1)$ for the minimal extension sequences, no unattainable points occur. Step 3 provides the minimal partial realization of order 2 as follows. The matrices P and M are both the 2 χ 2 identity. So Eqs. (2.7)-(2.9) provide

$$A = I_2 \cdot I_2 \cdot \left[\sigma \underline{\underline{H}}_{2,1}\right] \cdot I_2 \cdot I_2 = \begin{bmatrix} 0 & 1 \\ x_1 & x_2 \end{bmatrix} \qquad x_2 \neq \pm(1 - x_1)$$

$$B = I_2 \cdot I_2 \cdot \left[\underline{\underline{H}}_{2,1}\right] \cdot I_2 = \begin{bmatrix} 1 & 0 \\ 0 & 1 \end{bmatrix}$$

$$C = [1 \quad 0] \cdot \left[\underline{\underline{H}}_{2,1} - \sigma \underline{\underline{H}}_{2,1}\right] \cdot I_2 \cdot I_2$$

$$= [1 \quad 0] \begin{bmatrix} 1-x_1 & -x_2 \\ -x_1x_2 & 1-x_1-x_2^{\,2} \end{bmatrix}$$

$$= \begin{bmatrix} 1-x_1 & -x_2 \end{bmatrix} \qquad x_2 \neq \pm(1 - x_1)$$

As a check

$$G(s) = C(sI-A)^{-1}B = \frac{\left[s-x_1s-x_2 \qquad -x_2s-x_1+1\right]}{s^2-x_2s-x_1},$$

where $x_2 \neq \pm(1 - x_1)$, and it follows that $G(\lambda_1) = Y_1$, $G(\lambda_2) = Y_2$. Note that if the restriction $x_2 \neq \pm(1 - x_1)$ is violated by fixing $x_1 = 0$ and $x_2 = 1$,

$$G(s) = \begin{bmatrix} \frac{1}{s} & -\frac{1}{s} \end{bmatrix}$$

and the point Y_1 is unattainable with G(s).

3 The Stability Problem

At this point a discussion of data set realization could take several directions. Here, however, it seems appropriate to introduce a major unsolved problem of this theory.

If a given order M data set is obtained from an asymptotically stable system, it is desired that the resulting order M minimal partial realization also be asymptotically stable. Also, an order M data set which corresponds to an unstable system should yield an unstable order M minimal partial realization. These conditions are not automatically satisfied. In fact, with respect to a particular data set, an asymptotically stable n dimensional system may have no lower dimensional asymptotically stable minimal partial realizations!

EXAMPLE 3.1. We shall compute minimal partial realizations of the asymptotically stable system with transfer function

$$H(s) = \frac{s+10}{(s+1)^2}$$

using three different data sets. Of course, the only nontrivial minimal partial realizations are of dimension one.

(i) The Markov parameter data set of order two is $\{1,8\}$. It is straightforward to show that the transfer function of the corresponding minimal partial realization is

$$G(s) = \frac{1}{s-8} .$$

Clearly this realization is unstable.

(ii) Consider the order 2 data set given by $\{H(1) = 11/4;$ $H(2) = 4/3\}$. An order 2 minimal partial realization may be found to be $(1/17,\ 1,\ 44/12)$. When this realization is used in Eqs. (1.1) or (1.3), it is clear that the realization is again unstable.

(iii) Finally, if the order 2 data set $\{H(1/2) = 14/3,$ $H(2) = 4/3\}$ is used in constructing an order 2 minimal partial realization, it is found to be (-1/10, 1, 56/20). It is clear that an asymptotically stable partial realization has now been obtained.

There is some characteristic of data sets that provide for stable partial realizations that other data sets do not possess. Some a priori test is needed to determine if a given order M data set provides for a stable, order M minimal partial realization. Such a test would be useful for eliminating troublesome points (in the data set) while providing a criteria for adding new, useful points that would allow stable realizations.

Note, a data set which provides for a stable realization need not provide only for stable realizations. Recall that in the example of section 2, suitable choice of parameters x_1 and x_2 provide a stable realization.

An alternate approach in generating stable, order M partial (data set) realizations is to determine nonminimal partial realizations which satisfy the desired stability property. This corresponds to selecting nonminimal extension sequences for the set $\{V_1,...,V_M\}$ and checking to see if the resulting partial realization will be asymptotically stable or unstable. A procedure for doing this is discussed further in [2].

References

1 Audley, D.R., A method of constructing minimal approximate realizations of linear input/output behavior, Automatica, 13 (1977) May.

2 Audley, D.R., On the H-matrix representation and system realization, Ph.D. Dissertation, The Johns Hopkins Univ., Baltimore, MD (1972). Tech Report ARL 73-0118, Aerospace Research Laboratories, Wright-Patterson AFB, OH (1973).

3 Audley, D.R., S.L. Baumgartner and W.J. Rugh, Linear system realization based on data set representations. IEEE Trans. Automat. Contr., AC-20 (1975). Tech Report JHU 74-6, Dept of Elect. Engr., The Johns Hopkins Univ., Baltimore, MD (1974).

4 Davis, P.J., Interpolation and Approximation, Blaisdell, NY (1963), 28.

5 Kalman, R.E., On minimal partial realization of a linear input/output map, Aspects of Network and System Theory, Holt, Rinehart and Winston, NY (1971), 385-407.

6 Kalman, R.E., P.L. Falb and M.A. Arbib, Topics in Mathematical System Theory, Chapters 2, 10, McGraw-Hill, NY (1969).

7 Meinguet, J., On the solubility of the Cauchy interpolation problem, Approximation Theory, Academic Press, NY (1970).

8 Rosenblueth, A. and N. Wiener, The role of models in science, Philosophy of Science, 12 (1945), 316-321.

9 Silverman, L.M., Realization of linear dynamical systems, IEEE Trans. Automat. Contr., AC-16 (1971), 554-567.

10 Tether, A.J., Construction of minimal linear stave-variable models from finite input/output data, IEEE Trans. Automat. Contr., AC-15 (1970), 427-436.

David R. Audley
Directorate of Aerospace-Mechanics Sciences
Frank J. Seiler Research Laboratory
U.S. Air Force Academy, Colorado 80840

SOME DIRECTIONS FOR GUN FIRE CONTROL PREDICTORS

V. Benokraitis and Č. Masaitis

This paper presents two applications of rational approximation that arise in gun fire control. No attempt is made to present mathematically well-posed problems nor to discuss methods for their solution. Our only intent is to indicate where rational approximation may possibly be applied. As an aid to the reader, we include a short annotated bibliography of recent U.S. Army research efforts at Aberdeen Proving Ground in fire control methodology. Further references are contained in the cited reports.

1 Introduction

The essential components of a gun fire control system consist of processes for tracking, estimation, and prediction of target motion. Tracking involves the observation of target position and motion by keeping the target within a sensor's view. Estimation, sometimes also referred to as filtering or smoothing, is the process which determines the present position of the target and possibly first derivatives of target motion from noisy observations of past target positions. Prediction is concerned with establishing the future position and possibly some velocity components of the target from noisy observations of past target motion.

The prediction process may be further classified as open or closed loop. An open-loop predictor determines the position of the target a bullet's time of flight into the future. A closed-loop predictor performs the open-loop function, senses bullet-target miss distance, and feeds back corrections to initiate new gun commands.

In the remainder of this paper we first concentrate on a particular open-loop predictor and then turn our attention to a possible closed-loop design.

2 Open-Loop Predictor

Assume that we are given a process x(t), denoting the smoothed position of the target, as well as its derivative x'(t). We wish to predict the position of the target a time λ into the future. Specifically, we wish to determine $x(t+\lambda)$ in terms of a linear combination of x(t) and x'(t):

$$x(t+\lambda) \approx a\,x(t) + b\,x'(t). \tag{2.1}$$

Our procedure will be to determine the constants a and b such that the mean square error e is minimized, where

$$e = E\left\{(x(t+\lambda) - [a\,x(t) + b\,x'(t)])^2\right\}.$$

Here E denotes the expected value operator.

It follows that the constants a and b are such that the error $x(t+\lambda) - [a\,x(t) + b\,x'(t)]$ is orthogonal to the data x(t) and x'(t). That is, we have

$$\begin{cases} E\left\{(x(t+\lambda) - [a\,x(t) + b\,x'(t)])\,x(t)\right\} = 0 \\ E\left\{(x(t+\lambda) - [a\,x(t) + b\,x'(t)])\,x'(t)\right\} = 0. \end{cases} \tag{2.2}$$

Defining the autocovariance $R(\lambda)$ of x(t) by

$$R(\lambda) = E\left\{x(t+\lambda)\,x(t)\right\} \tag{2.3}$$

we obtain the successive derivatives from (see [2], p. 396)

$$\begin{cases} R'(\lambda) = -E\left\{x(t+\lambda)\,x'(t)\right\} \\ R''(\lambda) = -E\left\{x'(t+\lambda)\,x'(t)\right\}. \end{cases} \tag{2.4}$$

Thus (2.2) reduces to

$$(2.5)\quad \begin{cases} R(\lambda) - a\,R(0) + b\,R'(0) = 0 \\ -R'(\lambda) + a\,R'(0) + b\,R''(0) = 0\,. \end{cases}$$

But since we assume $x'(t)$ exists, we have [2] $R'(0) = 0$, and the solution of (2.5) becomes

$$(2.6)\quad a = \frac{R(\lambda)}{R(0)}\,, \quad b = \frac{R'(\lambda)}{R''(0)}\,.$$

Since target data are samples of discrete points, we approximate the relations (2.3)-(2.4) by

$$R(\lambda) \approx \frac{1}{n} \sum_{i=1}^{n} x(t_i+\lambda)\, x(t_i)$$

$$R'(\lambda) \approx -\frac{1}{n} \sum_{i=1}^{n} x(t_i+\lambda)\, x'(t_i)$$

$$R''(\lambda) \approx -\frac{1}{n} \sum_{i=1}^{n} x'(t_i+\lambda)\, x'(t_i)\,.$$

Combining these approximations with (2.6) and (2.1), we note that the desired prediction for $x(t+\lambda)$ is a rational function of smoothed values $x(t_i)$ and $x'(t_i)$, $i = 1,2,\ldots, n$. Now, since these smoothed values are functions of observed values y_i, say, we look for a predictor which is a rational function of observed values:

$$x(t+\lambda) \approx \frac{P(y_1,y_2,\ldots, y_n)}{Q(y_1,y_2,\ldots, y_n)}$$

for some polynomials P and Q. We address the rational approximation community to directly construct such an "optimal" rational predictor.

3 Closed-Loop Predictor

Next, let us consider a closed-loop prediction scheme. The procedure is based on a paper by Grubbs [1] of the Ballistic Research Laboratory. The paper is concerned with a one-dimensional application for setting or adjusting machines.

Let x be the desired measurement for a certain machined item. Let the first measurement of this item be $y_1 = x + \varepsilon_0$ with error ε_0. After producing the first item, we adjust the machine so that the next measurement becomes $y_2 = y_1 - (y_1 - x) + \varepsilon_1$. Thus, we correct the next item by the difference $z_1 = y_1 - x$. The third item is then corrected by 1/2 the latest difference $z_2 = y_2 - x$, i.e., $y_3 = y_2 - 1/2\,(y_2 - x) + \varepsilon_2$. In general,

$$y_{n+1} = y_n - \frac{1}{n}(y_n - x) + \varepsilon_n.$$

Thus

$$y_{n+1} - x = (y_n - x) - \frac{1}{n}(y_n - x) + \varepsilon_n$$

or

$$(3.1)\quad \begin{cases} z_{n+1} = \left(\frac{n-1}{n}\right) z_n + \varepsilon_n, \; n = 1, 2, \ldots \\ z_1 = y_1 - x = \varepsilon_0 . \end{cases}$$

The fractional corrections in the first relation of (3.1) were determined by a minimum variance analysis (see [1]).

If we denote the target position vector by x, the bullet impact point vector by y, the difference vector by z = y - x, and the error vector by ε, we may formally "extend" Grubbs' result by writing

$$z_{n+1} = A_n z_n + \varepsilon_n$$

where A_n is a matrix function of n (time).

In the light of (3.1), we expect the elements of A_n to be rational functions of time n. Assuming the error covariance matrix is known, the problem becomes, then, a matter of choosing a family of functions A_n and determining an element of the family for which, say,

$$||E(z_n)|| = f(A_1, A_2, \ldots, A_N)$$

is minimized for $0 < n \leqslant N$, where N is the number of shots fired at the target. Again, we leave further analysis to the experts.

4 Conclusion

We offer the suggestions in Sections 2 and 3 fully aware of present fire control systems which are variously restricted by assumptions requiring, for example, stationary processes or uncorrelated noise. Since target maneuvers are rarely so cooperative, exploring new directions can easily be justified.

Acknowledgment

Comments and suggestions by Dr. Stephen S. Wolff are gratefully acknowledged.

References

1 Grubbs, F.E., An optimum procedure for setting machines or adjusting processes, Industrial Quality Control, 11 (1954), 1-4.

2 Papoulis, A., Probability, Random Variables, and Stochastic Processes, McGraw-Hill Book Company, New York, 1965.

Annotated Bibliography

1 Burke, H.H. and T.R. Perkins, Air defense gun fire control systems: status, simulation, and evaluation methodology, U.S. Army Materiel Systems Analysis Activity (AMSAA) Technical Report No. 161, February 1976, Aberdeen Proving Ground (APG), Maryland. (Evaluation and description of conventional and modern antiaircraft fire control systems.)

2 Leathrum, J.F., An approach to fire control system computations and simulations, AMSAA Technical Report No. 126, April 1965, APG, Md. (Summary of mathematical and computational foundations of modern fire control systems, target models, and Kalman estimators. New approach to fire control computations based on hybrid coordinate system.)

3 Reed, H.L., Jr., A simple model of gunfire predictors, U.S. Army Ballistic Research Laboratory (BRL) Report No. 1775, April 1975, APG, Md. (Derivation of general expression for optimum maneuver of a target bounded by Nth derivative of its path when attacked by weapon with optimum linear predictor.)

4 Reed, H.L., Jr., Limitations of the r.m.s. criterion for fire control, BRL Report No. 1805, July 1975, APG, Md. (Discussion of a class of non-Gaussian target maneuvers not amenable to usual r.m.s. analysis which is the basis of Wiener's and Kalman's prediction theories.)

5 Wolff, S.S. and P.R. Schlegel, Fire control predictors for non-Gaussian target maneuvers, BRL Report No. 1875, April 1976, APG, Md. (Derivation of equation for conditional distribution of future target position given present state of target in one-dimensional motion with "random telegraph wave" acceleration. Generalization of model to allow intervals of constant target acceleration to be an ordinary renewal process leading to principal result: there exists a distribution for those intervals that minimizes the maximum hit probability.)

Vitalius Benokraitis
USA Ballistic Research Lab
Aberdeen Proving Gd., MD 21005

Česlovas Masaitis
USA Ballistic Research Lab
Aberdeen Proving Gd., MD 21005

SOME OPEN PROBLEMS CONCERNING POLYNOMIALS AND RATIONAL FUNCTIONS

E.B. Saff and R.S. Varga

We gather below a list of some open problems and conjectures concerning polynomials and rational functions, along with related references and comments.

1. If $\pi_{n,n}$ denotes the set of all real rational functions of degree n, set

$$\lambda_{n,n} := \inf_{r_n \in \pi_{n,n}} \| e^{-x} - r_n(x) \|_{L_\infty[0,+\infty)} .$$

Based on numerical results (cf. Cody, Meinardus, and Varga, J. Approximation Theory 2 (1969), 50-65) it is conjectured that

$$\lim_{n\to\infty} \lambda_{n,n}^{1/n} = \frac{1}{9} .$$

For related upper and lower bounds for $\lambda_{n,n}^{1/n}$, see Q.I. Rahman and G. Schmeisser, these Proceedings, and the references given there.

2. Let $f(z) = \sum_{k=0}^{\infty} a_k z^k$ be an entire function of finite order $\rho > 0$, with $a_k > 0$ for all $k \geq 0$, and with f being of perfectly regular growth. If $s_n(z;f) := \sum_{k=0}^{n} a_k z^k$, it is known (cf. Meinardus and Varga, J. Approximation Theory 3 (1970), 300-309) that

$$\lim_{n\to\infty} \left\| \frac{1}{f(x)} - \frac{1}{s_n(x;f)} \right\|_{L_\infty[0,+\infty)}^{1/n} = \frac{1}{2^{1/\rho}} .$$

A particular function for which the above is valid is $f(z) = e^z$. If we denote generically

$$\lambda_{0,n}(f) := \inf_{p_n \in \pi_n} \left\| \frac{1}{f} - \frac{1}{p} \right\|_{L_\infty[0,+\infty)} ,$$

where π_n is the collection of real polynomials of degree $\leq n$, then Schönhage (J. Approximation Theory 7 (1973), 395-398) showed that

$$\lim_{n\to\infty} [\lambda_{0,n}(e^z)]^{1/n} = \frac{1}{3}.$$

It is conjectured that for any entire function f of finite positive order ρ, having positive Maclaurin coefficients and possessing perfectly regular growth that

$$\lim_{n\to\infty} [\lambda_{0,n}(f)]^{1/n} = \frac{1}{3^{1/\rho}}.$$

For related results, see A.R. Reddy and O. Shisha, J. Approximation Theory 12 (1974), 425-434.

3. Given any τ with $0 \leq \tau \leq 2$, and given any $K > 0$, define

$$S(\tau;K) := \{z=x+iy: |y| < Kx^{1-(\tau/2)}, x > 0\},$$

and for any θ with $0 \leq \theta < 2\pi$, define

$$S(\tau;K;\theta) := \{z: ze^{-i\theta} \in S(\tau;K)\}.$$

Given any entire function $f(z) = \sum_{k=0}^{\infty} a_k z^k$ with order $\rho>\tau$, it is conjectured that there is no $S(\tau;K;\theta)$ which is devoid of all zeros of all partial sums $s_n(z;f) := \sum_{k=0}^{n} a_k z^k$, $n \geq 1$.

For the case when $\tau = 0$, $S(0;K;\theta)$ is just a proper sector with vertex at $z = 0$, and the result of F. Carlson (Arkiv for Mathematik, Astronomi O. Fysik, Bd. 35A, No. 14 (1948), 1-18) precisely gives the truth of the conjecture for $\tau = 0$. Thus what is conjectured is a generalization of Carlson's result. The above conjecture is related to the width conjecture of E.B. Saff and R.S. Varga (cf. Pacific J. Math. 62 (1976), 523-549). Note also that for $f(z) = e^z$, for which $\rho = 1$, it is known (cf. Saff and Varga, SIAM J. Math. Anal. 7 (1976), 344-357) that no partial sum $s_n(z) = \sum_{k=0}^{n} z^k/k!$ for any $n \geq 1$ has zeros in the parabolic region

$$P_1 := \{z=x+iy:\ y^2 \le 4(x+1),\ x > -1\}.$$

Because the boundary points of this parabola satisfy $|y| \sim 2x^{1/2}$, as $x \to +\infty$, we see that the above conjecture cannot in general be valid with $\tau = \rho$.

4. Let

$$\tilde{\lambda}_{n-1,n} := \inf_{p_{n-1}\in\pi_{n-1}} \{\| e^{-x} - p_{n-1}(x)/\prod_{i=1}^{n}(1+b_i x)\|_{L_\infty[0,+\infty)}:$$

$$b_i > 0 \text{ for } i=1,2,\ldots,n\},$$

where π_{n-1} denotes the collection of all real polynomials of degree $\le n-1$. It is known (cf. Saff, Schönhage, and Varga, Numer. Math. 25 (1976), 307-322) that these restricted rational approximations possess geometric convergence; specifically

$$\overline{\lim_{n\to\infty}}\, (\tilde{\lambda}_{n-1,n})^{1/n} \le \frac{1}{2}.$$

An open problem is to determine the exact geometric convergence rate of these restricted approximations. Next, it is known (cf. Kaufman and Taylor, these Proceedings) that there is a rational function

$$\tilde{r}_{n-1,n}(x) := \tilde{p}_{n-1}(x)/\prod_{i=1}^{n}(1+\tilde{b}_i(n)x)$$

with $b_i(n) > 0$, $1 \le i \le n$, such that

$$\tilde{\lambda}_{n-1,n} = \|e^{-x} - \tilde{r}_{n-1,n}\|_{L_\infty[0,+\infty)}.$$

<u>We and others</u> (E. Kaufman, T.C.-Y. Lau, A. Schönhage and G. Taylor) <u>conjecture that</u>

(i) $\tilde{b}_1(n) = \tilde{b}_2(n) = \ldots = \tilde{b}_n(n)$, for all $n \ge 1$,

(ii) $\lim_{n\to\infty} \tilde{b}_n(n) \cdot n = 1.$

Numerical results of Kaufman and Taylor (these Proceedings) and T.C.-Y. Lau (preprint) confirm (i) above for small values of n.

5. In the statement of the open problems of this section the following notation is used: $C_r[-1,1]$ denotes the collection of all real continuous functions on $[-1,1]$; π_ν^r is the collection of all polynomials of degree at most ν having real coefficients; π_ν^c is the analogous collection of polynomials with complex coefficients; $\pi_{n,n}^r$ is the collection of rational functions of the form p/q with $p,q \in \pi_n^r$; $\pi_{n,n}^c$ is the analogous collection with $p,q \in \pi_n^c$. For any real or complex-valued function g defined on $[-1,1]$ we also set

$$||g|| := \sup_{x\in[-1,1]} |g(x)|.$$

Finally for any $f \in C_r[-1,1]$ we put

$$E_{n,n}^r(f) := \inf_{R\in\pi_{n,n}^r} ||f-R||; \quad E_{n,n}^c(f) := \inf_{R\in\pi_{n,n}^c} ||f-R||;$$

and denote the corresponding collections of best approximates by

$$B_{n,n}^r(f) = \{R \in \pi_{n,n}^r : ||f-R|| = E_{n,n}^r(f)\},$$

$$B_{n,n}^c(f) = \{R \in \pi_{n,n}^c : ||f-R|| = E_{n,n}^c(f)\}.$$

It is well-known that $|B_{n,n}^r(f)|$ (the cardinality of $B_{n,n}^r(f)$) is one, i.e. uniqueness holds. In contrast, it has been shown by Saff and Varga (Bull. Amer. Math Soc., 83 (1977) No. 3) that there are classes of functions $f \in C_r[-1,1]$ for which $E_{n,n}^c(f) < E_{n,n}^r(f)$ and $|B_{n,n}^c(f)| \geq 2$. See also Bennett, Rudnick, and Vaaler (these Proceedings), and Arden Ruttan (these Proceedings).

We now list some open questions concerning real versus complex rational approximation to real functions on the real interval $[-1,1]$.

(i) Does there exist an $f \in C_r[-1,1]$ for which $E_{n,n}^c(f) = E_{n,n}^r(f)$, but for which $|B_{n,n}^c(f)| \geq 2$?

(ii) Find $\gamma(n) := \inf\{E^c_{n,n}(f)/E^r_{n,n}(f) : f \in C_r[-1,1]$ with $f \notin \pi^r_{n,n}(f)\}$.

(iii) Does there exist an entire function $f \in C_r[-1,1]$ such that $E^c_{n,n}(f) < E^r_{n,n}(f)$ for <u>every</u> $n \geq 1$? It has been shown by Saff and Varga (Bull. Amer. Math. Soc., <u>83</u> (1977) No. 3) that there exists an entire function $f \in C_r[-1,1]$ such that $E^c_{n,n}(f) < E^r_{n,n}(f)$ for <u>infinitely many</u> n.

(iv) Is $|B^c_{n,n}(|x|)| \geq 2$ for all $n \geq 1$?

6. A well-known conclusion of the Kakeya-Eneström Theorem is that if $p_n(z) = \sum_{i=0}^{n} a_i z^i$ with $a_i > 0$ for all $0 \leq i \leq n$, then the zeros $\hat{z}$ of $p_n(z)$ satisfy

$$|\hat{z}| \leq \max_{1 \leq i \leq n} \left(\frac{a_{i-1}}{a_i}\right).$$

In this form, the above result is sharp for every $n \geq 1$. When applied, say, to the Padé numerators $P_{n,\nu}(z)$ of type (n,ν) to e^z, the Kakeya-Eneström inequality yields

$$|\hat{z}| \leq n(\nu+1) \quad \text{for all } n \geq 1, \text{ all } \nu \geq 0.$$

On the other hand, it is known (Saff and Varga, these Proceedings) that the n zeros $\hat{z}$ of $P_{n,\nu}(z)$ satisfy

$$|\hat{z}| < n + \nu + \frac{4}{3},$$

which is a better estimate for these zeros than the Kakeya-Eneström bounds, except for small values of n and ν. This suggests the following open problem which would generalize the Kakeya-Eneström Theorem. Given $p_n(z) := \sum_{i=0}^{n} a_i z^i$ with $a_i > 0$ for all $0 \leq i \leq n$, assume in addition that all zeros $\hat{z}$ of $p_n(z)$ lie in the symmetric sector

$$S(\psi) := \{z : |\arg z| \geq \psi > 0, \text{ where } -\pi \leq \arg z \leq \pi\}.$$

Then, derive a bound $M(\psi; a_i)$ such that $|\hat{z}| \leq M(\psi; a_i)$ for all zeros $\hat{z}$ of p_n and such that

$$M(\psi;a_i) \leq \max_{1\leq i\leq n} \left(\frac{a_{i-1}}{a_i}\right) .$$

7. Let K be a compact set in the complex plane without isolated points, and let $||\cdot||_K$ denote the uniform norm on K. For each $n \geq 1$ denote by $T_n(K;z) = z^n + \cdots$ the Chebyshev polynomial of degree n for K, i.e.,

$$||T_n(K;z)||_K = \inf\{||P_n||_K : P_n \text{ monic of degree } n\}.$$

It is well-known that for $K = [-1,1]$ there holds for <u>any</u> polynomial p of degree $\leq$ n

$$||p'||_{[-1,1]} \leq ||p||_{[-1,1]} \cdot \frac{||T_n'||_{[-1,1]}}{||T_n||_{[-1,1]}} = n^2||p||_{[-1,1]},$$

where $T_n(x) := T_n([-1,1];x)$ is the classical Chebyshev polynomial of degree n. This suggests the following open question: For what types of sets K is it true that for any polynomial p of degree $\leq$ n

$$||p'||_K \leq ||p||_K \cdot \frac{||T_n'(K;z)||_K}{||T_n(K;z)||_K} \quad ?$$

For example, does the above inequality hold when K is a finite union of disjoint compact intervals?

E.B. Saff*
Department of Mathematics
University of South Florida
Tampa, Florida 33620

R.S. Varga**
Department of Mathematics
Kent State University
Kent, Ohio 44242

*Research supported in part by the Air Force Office of Scientific Research under Grant AFOSR-74-2688.

**Research supported in part by the Air Force Office of Scientific Research under Grant AFOSR-74-2729, and by the Energy Research and Development Administration (ERDA) under Grant E(11-1)-2075.

RATIONAL APPROXIMATION AT WHITE SANDS MISSILE RANGE

W. L. Shepherd

1 Introduction

In the program library at the White Sands Missile Range (WSMR) Computer Facility, there are computer algorithms for computing values of rational function approximations to a number of special functions. Many of these computer programs are used frequently. It should be emphasized that, as they stand, they constitute a valuable WSMR resource, but I do not propose to discuss them further. Rather, I want to mention two specific problems where I believe further research could eventually result in something valuable to the Army. The following two sections describe them briefly.

2 System Function for a Linear Digital Filter

The filters considered are of the forms

$$(1) \quad y_n = \sum_{i=1}^{M} c_i y_{n-i} + \sum_{i=o}^{N} d_i x_{n-i} \quad , \qquad n = 0, 1, \ldots$$

with

$$x_i = y_i = 0 \qquad \text{for } i < 0 \quad ,$$

or of the form

$$(2) \quad y_n = \sum_{i=o}^{\infty} a_i x_{n-i} \quad , \qquad n = 0, \pm 1, \ldots$$

$$x_i = 0 \qquad \text{for } i < 0 \quad .$$

There are more general forms, but these suffice for illustration. In fact, (1) with $c_i = 0$, $i = 1, \ldots, M$ and $N = \infty$ is (2). (The sequence $\{x_i\}_{i=0}^{\infty}$, $\{y_i\}_{i=0}^{\infty}$ is called input, output, respectively, of the filter. The special sequence with $x_o = 1$, $x_i = 0$ for $0 < i$ is the impulse input and the corresponding output is the impulse response.) The system (transfer) function for (2) is

$$(3) \quad F(\omega) = H(e^{j\omega}) = \frac{\sum_{i=0}^{N} d_i \, e^{-j\omega i}}{1 - \sum_{k=1}^{M} c_k \, e^{-j\omega k}} ,$$

and H(z) is the z-transform. The z-transform of $\{a_k\}_{k=0}^{\infty}$ is simply $\sum_{i=0}^{\infty} a_i (z^{-1})^i$. The use of (3) in the design of digital filters has been quite extensive.

There are several modern engineering textbooks, e.g., [1], presenting the theory and applications of these filters, and many recent advances can be found in the last 15 years issues of [2]. Other papers, especially [3], at this conference address some problems associated with digital filters.

A particular problem I think might be worth further investigation by someone especially interested in Padé approximants is the application of the theory of Padé approximants to getting suitable and simpler rational approximations to the above mentioned z-transforms for an already specified filter.

3 Special Function Computation Via Interpolation and Rational Approximation

In this section is presented an example of advantage in using more than one approximation method in a general purpose computer program for computing values of a special function.

$$P(x) := \frac{1}{2} + \int_0^x \frac{1}{\sqrt{2\pi}} \, e^{-t^2/2} \, dt$$

$$G(x) := P^{-1}(x)$$

The problem is to devise a procedure for computing G(x) for

$\frac{1}{2} \le x < 1$. [5] gives much information on rational approximation to the inverse of the closely related error function and references to additional work. Since the graph $(x, G(x))$ has a vertical asymptote $x = 1$, interpolation for x near 1 is not feasible. For x not near 1, a simple interpolation suffices. [4] gives an interpolating quadratic spline which agrees at every other knot with G and G' over most of $[\frac{1}{2}, 1)$, with optional rational approximation for x near 1. Numerical experience indicates that the resulting computer program is very fast, and I am working on an improved version of this procedure. My experience indicates that this kind of combination of methods is likely to be worth considering in a variety of applications.

References

1. Oppenheim, A. V. and R. W. Shafer, Digital Signal Processing, Prentice-Hall, 1957.
2. IEEE Transactions on Acoustics, Speech and Signal Processing.
3. Chui, C. K., P. W. Smith and L. Y. Su, A minimization problem related to Padé synthesis of recursive digital filters, these Proceedings.
4. Shepherd, W.L. and J.N. Hynes, Table look-up and interpolation for a normal random number generator, to appear in Transactions of the Twenty-Second Conference on Design of Experiment in Army Research, Development and Testing.
5. Blair, J.M., C.A. Edwards and J.H. Johnson, Rational Chebyshev approximations for the inverse of the error function, Mathematics of Computation, 30, No. 136, (October 1976), 827-830.

W. L. Shepherd
Instrumentation Directorate
US Army White Sands Missile Range
White Sands Missile Range, New Mexico 88002

A
B 7
C 8
D 9
E 0
F 1
G 2
H 3
I 4
J 5